Functional Oxides

Inorganic Materials Series

Editors:

Professor Duncan W. Bruce
Department of Chemistry, University of York, UK

Professor Dermot O'Hare
Chemistry Research Laboratory, University of Oxford, UK

Dr Richard I. Walton
Department of Chemistry, University of Warwick, UK

Series Titles

Functional Oxides
Molecular Materials
Porous Materials
Low-Dimensional Solids
Energy Materials

Functional Oxides

Edited by

Duncan W. Bruce
University of York, UK

Dermot O'Hare
University of Oxford, UK

Richard I. Walton
University of Warwick, UK

A John Wiley and Sons, Ltd, Publication

This edition first published 2010
© 2010 John Wiley & Sons, Ltd

Registered office
John Wiley & Sons Ltd, The Atrium, Southern Gate, Chichester, West Sussex, PO19 8SQ, United Kingdom

For details of our global editorial offices, for customer services and for information about how to apply for permission to reuse the copyright material in this book please see our website at www.wiley.com.

The right of the author to be identified as the author of this work has been asserted in accordance with the Copyright, Designs and Patents Act 1988.

Wiley also publishes its books in a variety of electronic formats. Some content that appears in print may not be available in electronic books.

Designations used by companies to distinguish their products are often claimed as trademarks. All brand names and product names used in this book are trade names, service marks, trademarks or registered trademarks of their respective owners. The publisher is not associated with any product or vendor mentioned in this book. This publication is designed to provide accurate and authoritative information in regard to the subject matter covered. It is sold on the understanding that the publisher is not engaged in rendering professional services. If professional advice or other expert assistance is required, the services of a competent professional should be sought.

The publisher and the author make no representations or warranties with respect to the accuracy or completeness of the contents of this work and specifically disclaim all warranties, including without limitation any implied warranties of fitness for a particular purpose. This work is sold with the understanding that the publisher is not engaged in rendering professional services. The advice and strategies contained herein may not be suitable for every situation. In view of ongoing research, equipment modifications, changes in governmental regulations, and the constant flow of information relating to the use of experimental reagents, equipment, and devices, the reader is urged to review and evaluate the information provided in the package insert or instructions for each chemical, piece of equipment, reagent, or device for, among other things, any changes in the instructions or indication of usage and for added warnings and precautions. The fact that an organization or Website is referred to in this work as a citation and/or a potential source of further information does not mean that the author or the publisher endorses the information the organization or Website may provide or recommendations it may make. Further, readers should be aware that Internet Websites listed in this work may have changed or disappeared between when this work was written and when it is read. No warranty may be created or extended by any promotional statements for this work. Neither the publisher nor the author shall be liable for any damages arising herefrom.

Library of Congress Cataloging-in-Publication Data

Functional oxides / edited by Duncan W. Bruce, Dermot O'Hare, Richard I. Walton.
    p. cm.
Includes bibliographical references and index.
ISBN 978-0-470-99750-5
1. Organic oxides. 2. Inorganic compounds. I. Bruce, Duncan W. II. O'Hare, Dermot. III. Walton, Richard I.
QD181.O1F87 2010
546'.7212—dc22

                                   2009041495

A catalogue record for this book is available from the British Library.

ISBN: 978-0-470-99750-5 (Cloth)

Set in 10.5/13 Sabon by Integra Software Services Pvt. Ltd, Pondicherry, India.

Contents

Inorganic Materials Series Preface

Back in 1992, two of us (DWB and DO'H) edited the first edition of *Inorganic Materials* in response to the growing emphasis and interest in materials chemistry. The second edition, which contained updated chapters, appeared in 1996 and was reprinted in paperback. The aim had always been to provide the reader with chapters that, while not necessarily comprehensive, nonetheless gave a first-rate and well-referenced introduction to the subject for the first-time reader. As such, the target audience was from first-year postgraduate student upwards. Authors were carefully selected who were experts in their field and actively researching their topic, so were able to provide an up-to-date review of key aspects of a particular subject, whilst providing some historical perspective. In these two editions, we believe our authors achieved this admirably.

In the intervening years, materials chemistry has grown hugely and now finds itself central to many of the major challenges that face global society. We felt, therefore, that there was a need for more extensive coverage of the area and so Richard Walton joined the team and, with Wiley, we set about a new and larger project. The *Inorganic Materials Series* is the result and our aim is to provide chapters with a similar pedagogical flavour but now with much wider subject coverage. As such, the work will be contained in several themed volumes. Many of the early volumes concentrate on materials derived from continuous inorganic solids, but later volumes will also emphasise molecular and soft matter systems as we aim for a much more comprehensive coverage of the area than was possible with *Inorganic Materials*.

We approached a completely new set of authors for the new project with the same philosophy in choosing actively researching experts, but also with the aim of providing an international perspective, so to reflect the diversity and interdisciplinarity of the now very broad area of inorganic materials chemistry. We are delighted with the calibre of authors who have agreed to write for us and we thank them all for their efforts and cooperation. We believe they have done a splendid job and that their work will make these volumes a valuable reference and teaching resource.

DWB, York
DO'H, Oxford
RIW, Warwick
January 2010

Preface

Metal oxides, particularly those containing one or more transition elements, for many years have been the foundation of solid-state inorganic chemistry. Here, the synthetic skill to manipulate the reactivity of diverse chemical elements, often at extreme temperatures and pressures, went hand-in-hand with developments in structural characterisation, including both spectroscopic and diffraction methods. A very good, and indeed already well-documented example, is the case of the cuprate superconductors, discovered in the early 1980s, which led to increasing complex structural chemistry and which continues to push the frontiers of knowledge of electronic properties of the solid-state. The interplay between the synthetic and structural work of chemists and the property measurement and theory of physicists led to the rapid development in understanding of a unique group of materials. When one also considers the role of the materials scientist in device fabrication of such electronic materials, the area is seen to be truly interdisciplinary.

Oxides continue to be the focus of much attention, and increasingly the area is driven by target properties. In this volume we have been largely concerned with properties arising from electronic structure but other applications, in catalysis or in optical media, are equally as important and are researched equally actively. The role of the chemist in synthesis is still paramount, and indeed it is very apparent that the scope for novel compositions and structures is far from being exhausted. More than ever the goal of a particular desirable property and the need to understand structure–property relationships is always in mind in contemporary research.

A complete review of the field of oxides would probably be impossible in a single volume, so instead we have selected five topical areas of functional oxides that illustrate their importance in modern materials chemistry. These highlight structural chemistry, magnetic properties, electronic properties, ionic conduction but also other emerging areas of importance in energy, such as thermoelectricity.

We approached five leading groups at the cutting edge of research to review these representative areas of functional oxides. We are very pleased that they agreed to write chapters for us, and that they have done such a good job in clearly explaining complex topics in an accessible way. We hope you will agree that these chapters provide an excellent introduction to what is an international field of great breadth.

DWB, York
DO'H, Oxford
RIW, Warwick
January 2010

List of Contributors

Edmund Cussen Department of Pure and Applied Chemistry, University of Strathclyde, Glasgow, Scotland

John E. Greedan McMaster University, Department of Chemistry, Hamilton, Ontario, Canada

Martha Greenblatt Department of Chemistry and Chemical Biology, Rutgers University, New Jersey, USA

P. Shiv Halasyamani Department of Chemistry, University of Houston, Houston, Texas, USA

Sylvie Hébert Laboratoire CRISMAT, UMR 6508 CNRS et ENSICAEN, Caen Cedex, France

Antoine Maignan Laboratoire CRISMAT, UMR 6508 CNRS et ENSICAEN, Caen Cedex, France

Tapas Kumar Mandal Department of Chemistry and Chemical Biology, Rutgers University, New Jersey, USA

1

Noncentrosymmetric Inorganic Oxide Materials: Synthetic Strategies and Characterisation Techniques

P. Shiv Halasyamani

Department of Chemistry, University of Houston, Houston, Texas, USA

1.1 INTRODUCTION

Materials that are crystallographically noncentrosymmetric (NCS), or acentric, are of current interest attributable to their functional properties, including piezoelectricity, ferroelectricity, and second-harmonic generation. Numerous relationships occur between these properties and crystal classes.[1] These relationships are shown in Figure 1.1, along with several well-known materials. It is instructive if we examine this figure more closely. If we examine the left-side of Figure 1.1, the symmetry dependent property we encounter is enantiomorphism, and the chiral crystal classes. All chiral materials must crystallise in one of eleven crystal classes, 1 (C_1), 2 (C_2), 3 (C_3), 4 (C_4), 6 (C_6), 222 (D_2), 32 (D_3), 422 (D_4), 622 (D_6), 23 (T), or 432 (O). Materials found in any of these crystal classes have a 'handedness', and a nonsuperimposable mirror image. The well-known

Functional Oxides　Edited by Duncan W. Bruce, Dermot O'Hare and Richard I. Walton
© 2010 John Wiley & Sons, Ltd.

Noncentrosymmetric Crystal Classes

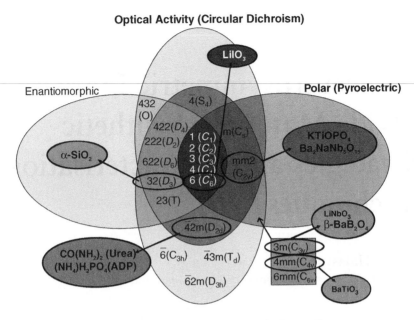

Figure 1.1 The relationships with respect to symmetry-dependent property between the noncentrosymmetric crystal classes are given along with representative compounds. Note that only five crystal classes, 1 (C_1), 2 (C_2), 3 (C_3), 4 (C_4), and 6 (C_6) have the proper symmetry for all of the symmetry dependent properties. Adapted from Halasyamani and Poeppelmeier, 1998 [71]. Copyright 1998 American Chemical Society

chiral material α-SiO_2[2, 3] crystallises in crystal class 32 (D_3). If we examine the right-side of Figure 1.1, we encounter the ten polar crystal classes, 1 (C_1), 2 (C_2), 3 (C_3), 4 (C_4), 6 (C_6), m (C_s), $mm2$ (C_{2v}), $3m$ (C_{3v}), $4mm$ (C_{4v}), and $6mm$ (C_{6v}). Materials found in these crystal classes have a permanent dipole moment. In fact $LiIO_3$,[4, 5] which crystallises in crystal class 6 (C_6) is both chiral and polar. The other materials shown: $KTiOPO_4$ (KTP)[6] and $Ba_2NaNb_5O_{15}$ ($mm2$ for both),[7] $LiNbO_3$ [8, 9] and β-BaB_2O_4 ($3m$ for both),[10, 11] and $BaTiO_3$ ($4mm$) are all 'purely' polar. They all have a dipole moment, but are not chiral. Examples are also given of materials, $CO(NH_2)_2$ (urea)[12] and $(NH_4)H_2PO_4$ (ammonium dihydrogen phosphate, ADP)[13] that crystallise in crystal class $\bar{4}2m$, that are neither chiral nor polar, but are still noncentrosymmetric. Other symmetry-dependent properties that are of importance are second-harmonic generation and piezoelectricity. Except for materials that are found in crystal class 432 (O), all NCS materials exhibit the correct symmetry for second-harmonic generation and piezoelectric behaviour.

Determining if a crystalline material is centrosymmetric or noncentrosymmetric is usually straightforward. From Friedel's law it is known that, during the diffraction process, if the incident wavelength is small compared with the absorption edge of any atom in the crystal, a centre of symmetry is introduced between oppositely related reflections. In other words $I(hkl) = I(-h-k-l)$. Friedel's law fails when the incident wavelength is similar to an atom's absorption edge. This anomalous scattering, when the imaginary part of the scattering factor becomes large, has been exploited to address a host of crystallographic problems.[14] Also, with the diffraction data the intensity distribution between a centric and acentric crystal differs. Statistical indicators of centricity have been developed by Wilson and Howell,[15, 16] but have been shown to be incorrect if the structure contains heavy atoms on special positions. Marsh has emphasised the importance of weak reflections if the centricity is in question.[17, 18] If weak reflections are removed, the statistical distribution tests can be strongly biased toward an acentric indication. Marsh also argues that when the diffraction data do not provide a clear choice between centrosymmetric and noncentrosymmetric space groups the centrosymmetric space group is preferred, even if disorder occurs.[17] The Platon suite of programs, specifically Addsym, can be used on refined structures to check for missing symmetry, *e.g.* inversion centres, as well as mistakes in crystal system or Laue class.[19]

1.2 STRATEGIES TOWARD SYNTHESISING NONCENTROSYMMETRIC INORGANIC MATERIALS

In the past decade or so a number of strategies have been described whose aim was to increase the incidence of acentricity in any new material. In one manner or another, each of these strategies involves crystal engineering.[20] One question that needs to be addressed is why there are so few (relatively) NCS materials? It is estimated that only ~15% of all inorganic materials are NCS. This would indicate that in the vast majority of inorganic materials, the 'building blocks' of the structure are centrosymmetric, *i.e.* made up of regular polyhedra. These regular polyhedra are usually related by inversion symmetry. Thus, in order to design inorganic NCS materials, two challenges must be overcome. First, the building blocks of the structure must necessarily be intrinsically acentric. In other words, there must be a distortion that requires or forces the metal cation not to be on an inversion centre. If local centricity occurs, macroscopic centricity is observed. Secondly, these building blocks must be connected or related in the structure by noninversion-type symmetry. In other words, it is not sufficient to have only acentric polyhedra; these polyhedra must be related by acentric symmetry elements. Numerous researchers have developed strategies to address both issues.

The purpose of this chapter is to discuss noncentrosymmetric materials, their synthetic strategies as well as their symmetry dependent properties. We will begin by discussing the various strategies employed in synthesising new NCS materials, and then move on to physical property characterisation. Although we will be unable to discuss in detail all of the proposed strategies for synthesising NCS materials, we will describe the major ideas in the field. Finally, we will discuss the outlook for this field with multifunctional materials in mind.

1.3   ELECTRONIC DISTORTIONS

One manner in which the incidence of acentricity may be increased in any oxide material is to use cations susceptible to second-order Jahn-Teller (SOJT) distortion.[21–27] These cations are octahedrally coordinated d^0 transition metals (Ti^{4+}, Nb^{5+}, W^{6+}, *etc.*), and cations with nonbonded electron-pairs (Sn^{2+}, Se^{4+}, Pb^{2+}, *etc.*). With the octahedrally coordinated d^0 transition metal cations, SOJT effects are observed when the empty d-orbitals of the metal mix with the filled p-orbitals of oxygen. In extended structures, this mixing results in a host of nearly degenerate electronic configurations that can be removed through the spontaneous distortion of the d^0 transition metal cation. This distortion can occur toward an edge (local C_2 direction), face (local C_3 direction), or corner (local C_4 direction), or between these 'special' directions (see Figure 1.2). The distortion results in unequal M-O bond distances, resulting in a MO_6 octahedron that is acentric. With the lone-pair cations, the original work of Sidgwick and Powell,[28] followed by the VSEPR theory of Gillespie and Nyholm,[29] attempted to rationalise the coordination geometry of the lone-pair cation. It was, however, Orgel[30] who explained the structural distortion and polarisation through the mixing of the metal s- and

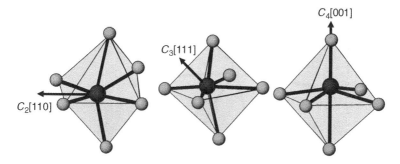

Figure 1.2 Out-of-centre distortion of the octahedrally coordinated d^0 cation along the local C_2 [110], C_3 [111], or C_4 [001] direction. Reprinted with permission from Halasyamani, 2004 [38]. Copyright (2004) American Chemical Society

p-orbitals. This traditional view of metal cation s-p orbital mixing has been shown to be incomplete. A number of researchers have shown that the interaction of the s- and p-orbitals of the metal cation *with* the oxide anion p-states is critical for lone-pair formation.[31–37] Regardless of how the lone-pair is created, its structural consequences are profound (see Figure 1.3). The lone-pair 'pushes' the oxide ligands toward one side of the metal cation resulting in a noncentrosymmetric coordination environment. The lone-pair cation coordination environment may be considered as pre-distorted,[38] as these cations are almost always found in local NCS environments. Halasyamani *et al.*,[38–50] Norquist *et al.*,[51–53] and others[54–63] have used SOJT-distorted cations in the design and synthesis of new NCS materials.

Halasyamani *et al.* have synthesised a variety of new NCS oxide materials that contain both an octahedrally coordinated d^0 transition metal and a lone-pair cation. These materials include $Na_2TeW_2O_9$,[41] $BaTeMo_2O_9$,[43] $K_2TeW_3O_{12}$,[64] $TlSeVO_5$,[49] and $(NH_4)_2Te_2WO_8$.[47] In doing so, they were able to increase the incidence of acentricity in any material to nearly 50%. They also demonstrated that when a d^0 transition metal oxide octahedron is linked to a lone-pair polyhedron, the d^0 cation is displaced away from the oxide ligand that links the two polyhedra. Thus, the lone-pair polyhedra serve to reinforce the SOJT distortion of the d^0 cation.[38] Additionally, with the octahedrally coordinated d^0 cations, a continuous symmetry measures approach[65–67] has been used to quantify the magnitude and direction of the distortion.[68] They were able to divide the d^0 transition metals into three categories: strong (Mo^{6+} and V^{5+}), moderate (W^{6+}, Ti^{4+}, Nb^{5+}, and Ta^{5+}), and weak (Zr^{4+} and Hf^{4+}) distorters (see Figure 1.4).[68] In addition, the preferred

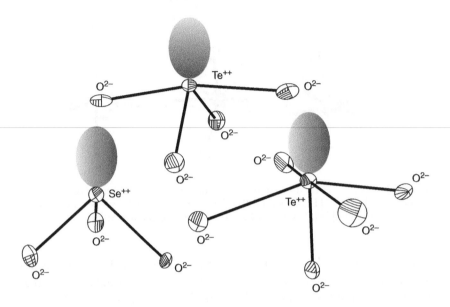

Figure 1.3 ORTEP (50% probability ellipsoids) diagram for lone-pair cation polyhedra

direction of the distortion for each d^0 cation was examined and discussed. With respect to directional preferences, for V^{5+}, distortions toward a vertex or edge are common. Interestingly, for V^{5+}, face-directed distortions are never observed. For Mo^{6+} and Hf^{4+} only edge- and face-directed distortions are observed, whereas with the other four cations, W^{6+}, Ti^{4+}, Nb^{5+}, and Ta^{5+}, the three directions, vertex, edge, and face are observed in similar proportions.

Norquist *et al.* have developed a novel strategy to design and synthesise new NCS compounds by using a SOJT-distorted cation, Mo^{6+}, in combination with chiral organic amines.[51–53] As stated previously, the first challenge in synthesising NCS materials is to use inherently asymmetric NCS polyhedra. In using the SOJT-distorted Mo^{6+}, Norquist *et al.* synthesised materials where the d^0 cation is substantially displaced from the centre of its oxide octahedra. Thus, each of the MoO_6 octahedra is inherently NCS. The second challenge, ensuring that the octahedra are not related by inversion centres, was successfully addressed by using single enantiomer templating agents. An example of this strategy involves the synthesis of $[(S)\text{-}C_5H_{14}N_2][(MoO_3)_3(SO_4)] \cdot H_2O$ and $[(R)\text{-}C_5H_{14}N_2][(MoO_3)_3(SO_4)] \cdot H_2O$.[52] These materials were synthesised as pure

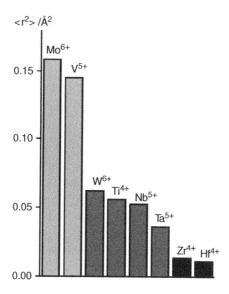

Figure 1.4 Average magnitude of the off-centre distortions for individual octahedrally coordinated d^0 transition metal cations. Reprinted with permission from Ok *et al.*, 2006 [68]. Copyright (2006) American Chemical Society

chiral compounds by using reaction gels in which a single enantiomer of 2-methylpiperazine was present, either as the $[(S)\text{-}C_5H_{14}N_2]^{2+}$ or $[(R)\text{-}C_5H_{14}N_2]^{2+}$ cation. By using single enantiomer species, the cancellation of any local d^0 cation distortions through extra-framework inversion centres is prohibited, since the structure would need to contain both S and R cations. Norquist *et al.* are able to chemically control the presence or absence of each enantiomer. If only the S-enantiomer is present, the chiral molecules can never be related to one another by inversion centres since the requisite R-enantiomer is absent. Thus, in the crystal structure, inversion centres are prohibited and the compound is constrained to be NCS.

Mao *et al.* have also developed a novel approach in utilising SOJT-distorted cations to design NCS materials.[54] They incorporate borate tetrahedra, BO_4 groups, in conjunction with asymmetric SeO_3 polyhedra. Other acentric borate materials with BO_3 polyhedra will be discussed later in the chapter. Mao *et al.* recently reported on the synthesis of $Se_2(B_2O_7)$.[54] The material exhibits a three-dimensional crystal structure consisting of corner-shared BO_4 tetrahedra that are linked to SeO_3 polyhedra. The material can be considered as an open-framework compound, with helical tunnels that propagate along the c-axis. The helices are oriented in a right-handed manner, and the tunnels themselves are based on B_6Se_4 10-member rings. The lone-pair on the Se^{4+} cations is directed toward the tunnels.

1.3.1   Metal Oxyfluoride Systems

Poeppelmeier *et al.* have developed a strategy for designing and synthesising materials using octahedrally coordinated d^0 transition metal oxide-fluoride anions. Specifically, these are anions of the type $[MO_xF_{6-x}]^{2-}$ ($x = 1$, $M = V^{5+}$, Nb^{5+}, Ta^{5+}; $x = 2$, $M = Mo^{6+}$, W^{6+}) (see Figure 1.5).[48, 69–78] Similar to the 'pure' d^0 oxides systems discussed earlier, the metal cation in the centre of the oxyfluoride octahedra spontaneously displaces toward a corner ($x = 1$) or edge ($x = 2$) to form short $M=O$ bonds.

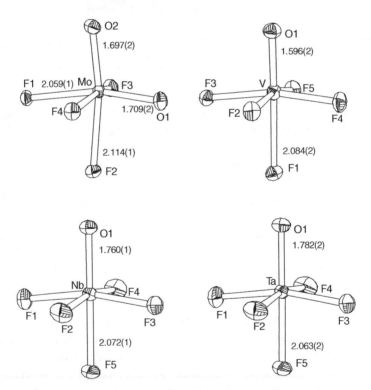

Figure 1.5   ORTEP (50% probability ellipsoids) for $[MoO_2F_4]^{2-}$, $[VOF_5]^{2-}$, $[NbOF_5]^{2-}$, and $[TaOF_5]^{2-}$ octahedra. Note that the cation is distorted toward the oxygen ligand(s). Reprinted with permission from Welk *et al.*, 2000 [75]. Copyright (2000) American Chemical Society

This spontaneous distortion is inherent to the oxyfluoride anion and is a result of metal-$d\pi$–oxygen-$p\pi$ orbital interactions. For example, in anions such as $[NbOF_5]^{2-}$, the Nb^{5+} cation distorts toward the oxide ligand, resulting in a short Nb-O bond and a long, *trans* Nb-F bond. This distortion is analogous to those observed in $KTiOPO_4$ (KTP)[6] and $BaTiO_3$.[79] There are also two challenges in synthesising NCS materials based on $[MOF_5]^{2-}$ ($M = V^{5+}$, Nb^{5+}, Ta^{5+}) or $[MO_2F_4]^{2-}$ ($M = Mo^{6+}$, W^{6+}) anionic octahedra. The first is to prevent crystallographic disorder between the oxide and fluoride ligands. Crystallographic disorder between the oxide and fluoride ligands can impose a centre of symmetry on the d^0 cation, rendering the structure centrosymmetric. The second challenge is to have these crystallographically ordered anions arranged in a NCS manner with respect to each other. By closely examining all of the intra-octahedral distortions, Poeppelmeier *et al.* were able to overcome the first challenge and successfully order the oxide and fluoride ligands.[72, 75–77] The second challenge was accomplished by using the $[NbOF_5]^{2-}$ anions in a hybrid inorganic-organic compound as well as more recently in a solid-state material.[48] The researchers were able to successfully meet both challenges by not only examining the primary distortion, the spontaneous displacement of the d^0 transition metal toward the oxide ligand, but also by focusing on the secondary distortion, the interaction of the ordered $[MO_xF_{6-x}]^{2-}$ anion with the extended crystal structure. With all of the ordered anions, an uneven amount of residual negative charge is observed on the ligands. They demonstrated that coordination within the structure is directed by the most anionic ligands. With the $[NbOF_5]^{2-}$ and $[TaOF_5]^{2-}$ anion, the most negative charges are found on the oxide and *trans*-fluoride ligands; thus coordination is directed in a *trans* fashion. Interestingly, this type of *trans*-directing is also observed in the $[WO_2F_4]^{2-}$ anion; however, both the $[MoO_2F_4]^{2-}$ and $[VOF_5]^{2-}$ anions are *cis*-directors.[75] By investigating and understanding these directional effects, both at the local, primary level and at the more macroscopic, secondary level, Poeppelmeier *et al.* have been able to design and synthesise NCS materials by aligning the ordered $[MO_xF_{6-x}]^{2-}$ anions in an acentric manner.

1.3.2   Salt-Inclusion Solids

Another novel strategy for synthesising and designing NCS materials has been described by Hwu *et al.*[80–83] He has focused on salt-inclusion solids whose framework consists of mixed ionic and covalent sub-lattices. Hwu *et al.* have utilised a combination of acentric polyanions, such as those found in silicates, along with first-order Jahn-Teller cations and chlorine-centred, acentric secondary building units (SBUs). We will describe each of these groups in more detail. With the acentric polyanions, moieties such as $[P_2O_7]^{4-}$, $[Si_2O_7]^{6-}$, and $[V_2O_7]^{4-}$ are used. These polyanions are not only inherently acentric, but are also polar. Acting in a cooperative manner with these anions are the first-order Jahn-Teller distorted cations, for example Mn^{3+} (d^4) and Cu^{2+} (d^9). Attributable to the first-order Jahn-Teller effect the coordination of these cations is inherently asymmetric. One of the most novel features of Hwu's strategy is the use of anion-based acentric SBUs. Specifically the acentric SBU $ClA_{6-n}M_n$ ($A = Cs$, Ba; $M = Mn$, Cu; $n = 1, 2$) is utilised (see Figure 1.6). This SBU has a templating effect on the framework, resulting in the observed

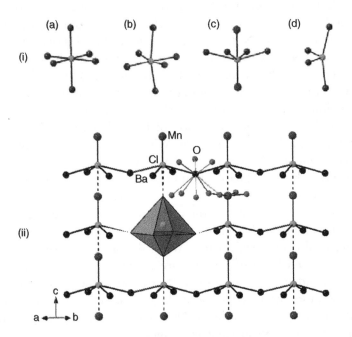

Figure 1.6 Figure depicting (i) Coordination surrounding chlorine in (a) NaCl, (b) $Cs_2Cu_7(P_2O_7)_4 \cdot 6CsCl$, (c) $Ba_2Mn(Si_2O_7)Cl$, and (d) $Ba_6Mn_4Si_{12}O_{34}Cl_3$; (ii) A slab of the $(Ba_2Mn)Cl$ lattice in $Ba_2MnSi_2O_7Cl$ showing the origin of the polar lattice – alternating short and long Mn-Cl linkages; (iii) A cage view of the Si_2O_7 unit residing in the centre of the anti-ReO_3 type $(Ba_2Mn)Cl$ lattice in $Ba_2MnSi_2O_7Cl$ (left) with the Si_2O_7 unit comprised of two corner-shared SiO_4 tetrahedra that are eclipsed (right); (iv) Coordination around chlorine in (a) NaCl, (b) $Cs_2Cu_7(P_2O_7)_4 \cdot 6CsCl$, and (c) $Ba_2MnSi_2O_7Cl$. Reprinted with permission from Mo and Hwu, 2003 [81] and Mo *et al.*, 2005 [82]. Copyright (2003) and (2005) American Chemical Society

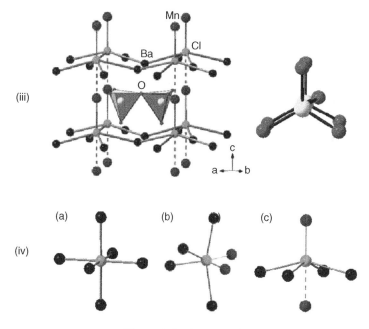

Figure 1.6 *(continued)*

NCS structure. Hwu *et al.* have successfully used these ideas to synthesise a variety of NCS salt-inclusion solids including $Ba_2Mn(Si_2O_7)Cl$, $Cs_2Cu_7(P_2O_7)_4 \cdot 6CsCl$, $Cs_2Cu_5(P_2O_7)_3 \cdot 3CsCl$, $Ba_6Mn_4Si_{12}O_{34}Cl$, and $Ba_6Fe_5Si_{11}O_{34}Cl_3$.[80–83]

1.3.3 Borates

In addition to the aforementioned materials, NCS borates have attracted a great deal of attention. Several excellent reviews on NCS borates have been written,[84–86] so only a brief outline will be given here. The first NCS borate to gain widespread use was β-BaB_2O_4 (BBO).[11] Although BBO can undergo unfavourable phase-transitions, the material has an exceptional optical transmission range, ~ 190–3500 nm, as well as a high damage threshold, 5 GW/cm^2. The fundamental idea behind synthesising NCS borates is the inclusion of the $[BO_3]^{3-}$ anion group in the structure. It has been shown that this group is most often observed with 1 (C_1) site symmetry.[87] In addition, delocalised π-type bonds are observed perpendicular to the BO_3 plane that when coupled with MO_6 ($M = d^0$ transition metal) octahedra often result in large nonlinear susceptibilities. As with the other systems discussed, the orientation of the $[BO_3]^{3-}$ anions and their density in the unit cell profoundly influences the nonlinear optical properties. Large nonlinear susceptibilities are thought to occur when a large number of these borate group are aligned in a crystal structure.[85] In BBO, the BO_3 groups from a B_3O_6 ring whose plane is perpendicular to the polar axis. The rings themselves are slightly mis-aligned reducing their maximum theoretically possible nonlinear optical susceptibility. This misalignment is also observed in $Sr_2Be_2B_2O_7$, where the BO_3 groups are linked to BeO_4 tetrahedra to form sheets. These sheets are stacked in a co-planar manner along the c-axis. Numerous other NCS borates have been synthesised, specifically those in the huntite family, $MM'_3(BO_3)_4$ (M = lanthanide; M' = Al, Ga, Sc, Cr, or Fe), orthoborates, ABe_2BO_3F (A = Na or K), $SrAl_2(BO_3)_2O$ (SBBO), and $BaCaBO_3F$ (BCBF), polyborates, MM' $(B_3O_5)_3$ (M = Ba or Sr; M' = Li or Na), $CsLi(B_3O_5)_2$ (CLBO), and $Na_4Li(B_3O_5)_5$, and pyroborates, $AMOB_2O_5$ (A = K, Rb, or Cs; M = Nb or Ta).[84–86] Recently a NCS borate, $Li_6CuB_4O_{10}$, has been reported.[88] The material melts congruently indi-cating large single crystals could be grown.

1.3.4 Noncentrosymmetric Coordination Networks

In addition to electronic distortions, salt inclusion materials, and borates, coordination networks that are acentric have also been designed. The design and synthesis of NCS coordination networks have been developed by Rosseinsky et al.,[89–95] Lin et al.,[96–102] and others.[103–107] All of these researchers use various crystal engineering strategies to design NCS chiral frameworks. Rosseinsky et al. created chiral frameworks based on the (10,3)-a network (see Figure 1.7). This chiral framework can be created by linking the tridendate 1,3,5-benzenetricarboxylate (btc) ligand to late transition metals, such as Ni^{2+} or Co^{2+}. The metal cation, M, is octahedrally coordinated and connects two btc ligands in a linear, *trans* fashion. These connections form the coordination framework. The four remaining coordination sites are available to auxiliary ligands that are not part of the framework. It is these auxiliary ligands that control the chirality, *i.e.* handedness, of the network. Rossiensky et al. have demon-strated that the chirality of the network can be influenced by the incor-poration of small chiral templating bidendate alcohols. This type of chiral

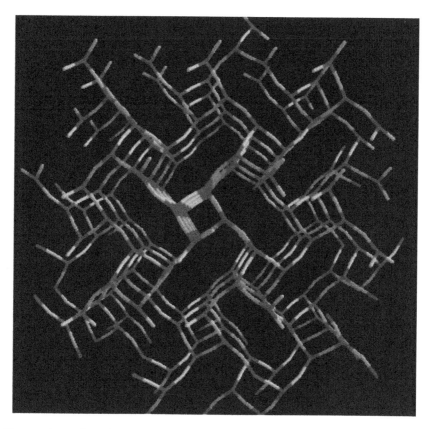

Figure 1.7 The chiral (10,3)-a M_3btc_2 network in $Ni_3btc_2(py)_6(1,2\text{-pd})_3 \cdot [(1,2\text{-pd})_{11}(H_2O)_8]$ showing the helical motif is shown. The M centres (light grey) are linear connectors and the btc centres (dark grey) produce the three connectivity. Reprinted with permission from Bradshaw *et al., J. Am. Chem. Soc.* **19**, 6106 (2004). Copyright (2004) American Chemical Society

templating was observed when the *S*-enantiomer of 1,2-propanediol (1,2-pd) was used as part of the (10,3)-a network. Compounds such as $M_3btc_2X_6Y_3 \cdot$ [guest] may be formulated. The M_3btc_2 framework forms a chiral (10,3)-a network, where X and Y represent auxiliary ligands whose chirality can control the handedness of the framework. These auxiliary ligands include nitrogen heterocycles such as pyridine (py), as well as the aforementioned bidentate alcohol, 1,2-pd. The [guest] refers to the occluded species residing in the channels. In using this strategy, Rosseinsky *et al.*[93] have been able to synthesise a variety of NCS chiral materials such as $Ni_3(btc)_2(py)_6(1,2\text{-pd})_3 \cdot [(1,2\text{-pd})_{11}(H_2O)_8]$, $Ni_3(btc)_2$

(3-pic)$_6$(1,2-pd)$_3$ · [(1,2-pd)$_9$(H$_2$O)$_{11}$],   Ni$_3$(btc)$_2$(py)$_6$(eg)$_6$ · (eg)$_x$(H$_2$O)$_y$ ($x \sim 3$, $y \sim 4$).

NCS chiral coordination frameworks have also been designed and synthesised by Lin *et al.*,[96–102] who utilised the diamondoid network (see Figure 1.8). Crystal engineering of this network was first described by Zaworotko, who noted that structures exhibiting the diamondoid network would be pre-disposed to crystallising in chiral space groups.[108] As seen in Figure 1.8, the diamondoid network consists of a three-dimensional framework of linked tetrahedra. Lin *et al.* suggested that NCS diamondoid networks could be created by connecting the tetrahedral centres with asymmetric bridging ligands. Although interpenetration could be an issue and result in a centrosymmetric framework, he suggested that this could be avoided by using an odd number of interpenetrated diamondoids bridged by asymmetric ligands. Specifically the tetrahedral metal centres are the d^{10} cations, Zn^{2+} and Cd^{2+}, that would be connected by asymmetric p-pyridinecarboxylate ligands. By using d^{10} cations, $d \rightarrow d$ transitions in the visible are avoided. In addition the p-pyridinecarboxylate ligands are rigid, imparting strength to the framework. Thus, there would be a high likelihood of creating a diamondoid framework by connecting tetrahedral (or pseudo-tetrahedral) metal centres, Zn^{2+} or Cd^{2+}, with

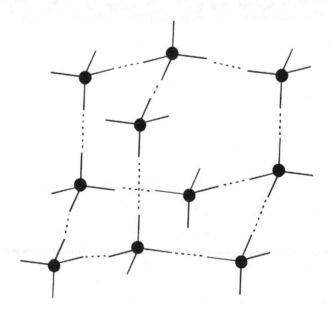

Figure 1.8   Diagram of the NCS chiral diamondoid network

asymmetric bridging ligands, p-pyridinecarboxylate. Using this strategy, Lin *et al.* have been able to successfully synthesise a variety of three-dimensional noncentrosymmetric materials exhibiting the diamondoid network. Lin *et al.* also developed strategies for creating two-dimensional NCS grid networks. With these networks, it was suggested that by connecting the metal centres with bent m-pyridinecarboxylate ligands, NCS frameworks could be created. The metal centre would be either coordinated in a *cis*-octahedral or tetrahedral manner, precluding any inversion centres. Again d^{10} metal centres are used to avoid any $d \to d$ transitions. Thus, by connecting these d^{10} metal centres, Zn^{2+} or Cd^{2+}, with asymmetric bridging ligands, m-pyridinecarboxylate, new NCS two-dimensional grids would be formed. As with the three-dimensional materials, Lin *et al.* have been able to use this strategy to create a variety of two-dimensional NCS frameworks.

Others who have synthesised chiral coordination networks include Jacobson *et al.*,[103, 104] Férey *et al.*,[105] Kim *et al.*,[106] and zur Loye *et al.*[107] Jacobson *et al.* have synthesised several chiral compounds using chiral ligands, such as 2-pyrazinecarboxylate (2-pzc) and *l*-aspartate (*l*-Asp). With both ligands, new chiral compounds were synthesised, such as $[Co_4(2\text{-pzc})_4(V_6O_{17})] \cdot xH_2O$, $[Ni_4(2\text{-pzc})_4(V_6O_{17})] \cdot xH_2O$, and $[Ni_2O(l\text{-Asp})(H_2O)_2] \cdot 4H_2O$.[103, 104] With the latter, the first chiral compound with an extended transition metal–oxide–transition metal network was synthesised. The Ni(II) cations are in octahedral coordination environments, that are edge and corner shared through oxide ligands. In addition, in an unprecedented manner, the aspartate ligands link to five Ni(II) centres. It is suggested that the steric requirements of the aspartate ligands impart chirality to the material. The use of chiral ligands to synthesise homochiral compounds has also been reported by zur Loye *et al.*[107] and Férey *et al.*[105] zur Loye *et al.* reported on the use of a chiral fluorine-based ligand, specifically 9,9-bis[(S)-2-methyl-butyl]-2,7-*bis*(4-pyridyl-ethynyl)fluorene, to synthesise a noninterpenetrating chiral square-grid polymer containing Cu(II). The grid dimensions are 25 Å × 25 Å making it one of the largest ever reported. Férey *et al.* have also reported a porous and chiral Ni(II) glutarate, $[Ni_{20}\{(C_5H_6O_4)_{20}(H_2O)_8\}] \cdot 40H_2O$.[105] In this chiral compound there is some interpenetration of the networks, but porosity is retained. The reported framework is topologically related to the (10,3)-a network discussed earlier. Finally, Seo *et al.* have used a slightly different strategy to synthesise a chiral material. They reported the synthesis of a chiral metal-organic material by using enantiopure, *i.e.* chiral, metal-organic clusters as secondary building units.[106] The chiral metal-organic cluster used was a trinuclear metal carboxylate, $[M_3(\mu_3\text{-}O)(O_2CR)_6(H_2O)_3)]^{n+}$, where M is a divalent or trivalent cation and O_2CR are organic carboxylate anions. In the Seo *et al.* report, a Zn^{2+} metal-organic cluster was used, resulting in a porous and chiral layered material.

1.4 PROPERTIES ASSOCIATED WITH NONCENTROSYMMETRIC MATERIALS

In the introduction to this chapter we briefly discussed some of the properties associated with noncentrosymmetric materials. In this section we describe these properties in more detail, as well as discussing their measurement. We will focus on the characterisation of bulk materials, as opposed to thin films or single crystals, since with the latter large (>5 mm) crystals are necessary. The techniques described can, however, be used on single crystals. In addition to having large single crystals, in many cases these crystals must be cut and polished to expose specific crystallographic faces. With new and even well-known materials, growing, cutting, and polishing such crystals is exceptionally difficult and remains an ongoing challenge. In this section of the chapter, the characterisation of second-harmonic generating, piezoelectric, pyroelectric, and ferroelectric properties in bulk noncentrosymmetric materials will be described. These phenomena have been discussed extensively in the literature[109–114] so only a brief description of each phenomenon will be given here.

- *Second-harmonic generation*

  Second-harmonic generation (SHG), or frequency doubling, is defined as the conversion of a specific wavelength of light into half its original, *i.e.* $\lambda_1 \rightarrow 1/2\,\lambda_1$, or with respect to frequency ω, $\omega_1 \rightarrow 2\omega_1$. The first report of SHG was by Franken *et al.* in 1961,[115] who reported SHG on a crystal of $\alpha\text{-}SiO_2$ using a ruby laser. Following this experimental result, a classic paper by Armstrong *et al.* was published that provided a theoretical foundation for the origin of the nonlinear optical susceptibility.[116] For several years following Franken's experimental result, large single crystals were required to measure SHG. Kurtz and Perry published, in 1968, a seminal paper that described a technique whereby SHG could be measured from polycrystalline samples.[110] It is this technique that we will describe in more detail.

- *Piezoelectricity*

Piezoelectricity, derived from the Greek *piezen*, meaning to press, was discovered in 1880 by Jacques and Pierre Curie.[117] They observed that some materials become electrically polarised when subjected to a mechanical force. Soon after, the converse effect was discovered wherein the application of a voltage resulted in a macroscopic strain. In 1910, Voigt published a standard reference detailing the electro-mechanical relationships in piezoelectric materials.[118] A thorough review of the early history of piezoelectricity can be found in Cady's seminal book.[109] Thus, with piezoelectricity, two effects are possible: direct and converse. Both direct and converse effects are used in a variety of applications. The direct effect results in generator action – the piezoelectric material converts mechanical energy to electrical energy. This generator action is used in solid-state batteries, sensing devices, and fuel lighting applications. The converse effect results in motor action – the piezoelectric material converts electrical energy to mechanical energy. This motor action is used in ultrasonic and acoustic applications, micromotor devices, and electromechanical transducers. Measurements on bulk materials utilising both direct and converse piezoelectric techniques will be described.

- *Pyroelectricity*

The pyroelectric effect may be defined as the change in spontaneous polarisation, P_s, as a function of temperature.[119] The symmetry requirements for pyroelectricity are far more restrictive compared with SHG and piezoelectricity. To exhibit a spontaneous polarisation, the material in question *must* crystallise in one of ten polar crystal classes (1, 2, 3, 4, 6, *m*, *mm2*, *3m*, *4mm*, or *6mm*). Thus, polarity is required for pyroelectric behaviour. Determining the pyroelectric coefficient may be done two ways – either measuring the pyroelectric current or the pyroelectric charge.[120] Both techniques will be described.

- *Ferroelectricity*

A ferroelectric is formally defined as a pyroelectric material that has a reversible, or 'switchable', polarisation.[112] Thus, for a material to be ferroelectric, the compound must be polar, *i.e.* must possess a perma-nent dipole moment, and must be capable of having this moment reversed in the presence of an applied voltage. The former occurs if the material crystallises in one of ten polar crystal classes (1, 2, 3, 4, 6, *m*, *mm*2, 3*m*, 4*mm*, or 6*mm*). Determining the latter is more involved. Polarisation reversal, or ferroelectric hysteresis, may be measured through a Sawyer–Tower circuit.[121] Additionally, because of the relatively large voltages needed for polarisation reversal, the material under investigation must be insulating. Another feature that is observed in some, but not all, ferroelectric materials is a dielectric anomaly at the Curie temperature. A maximum in the dielectric constant is often observed at the Curie temperature. This temperature indicates a phase-change to a centrosymmetric, nonpolar, *i.e.* nonferroelectric, often termed paraelectric, structure. We will describe measurement techniques that allow one to determine ferroelectric hysteresis curves.

This section is divided into four parts. Each part describes a specific NCS property, the history of the phenomena, and provides details on the measurement as well as an interpretation of the resulting data.

1.4.1   Second-Harmonic Generation

Second-harmonic generation (SHG), or frequency doubling, is defined as the conversion of a specific wavelength of light into half its original, *i.e.* $\lambda_1 \to 1/2\, \lambda_1$, or with respect to frequency ω, $\omega_1 \to 2\omega_1$. It was not until the invention of the laser in 1960 by Maiman[122] that sizeable nonlinear optical effects, such as SHG, could be observed. Attributable to these optical fields, the induced polarisation, P, in the material can be written as a power series: $P_i = \chi^{(1)}E + \chi^{(2)}E^2 + \ldots$ where χ is the linear electric susceptibility, with the higher order terms resulting in nonlinear effects such as SHG. These nonlinear effects are described by expanding the polarisation (equation 1.1):

$$P_i = \chi_{ij}E_j + \chi_{ijk}E_jE_k + \chi_{ijkl}E_jE_kE_l + \ldots \qquad (1.1)$$

where χ_{ij} is the electric susceptibility, with the second-order nonlinear coefficient described as χ_{ijk}. Third-order terms, χ_{ijkl}, give rise to third-harmonic generation. Only in a noncentrosymmetric environment is $\chi_{ijk} \neq 0$. Mathematically, χ_{ijk} is a third-rank tensor, and in experimental SHG measurements is replaced by d_{ijk}, where $2d_{ijk} = \chi_{ijk}$.

After the discovery of SHG in 1961,[115] large crystals, on the order of several mm, were required to investigate the phenomenon. A technique, described in 1968, allowed the determination of the SHG efficiency on polycrystalline samples. It is this technique that is described in more detail. At its most basic, the powder SHG technique requires very little instrumentation. Additionally, all of the instrumentation is commercially available. A typical set-up for powder SHG measurements is shown schematically in Figure 1.9. A low-energy laser, pulsed or continuous, is needed. Usually a commercially available Nd-YAG laser (1064 nm output) is used,[123] since any SHG will appear in the visible at 532 nm (green), and thus the experimentalist is literally able to see the SHG. The sample, a polycrystalline powder, is placed in a fused silica tube – a capillary tube or NMR tube can be used. For a 'quick and dirty measurement' that addresses the simple question of whether the material is SHG active or not, only 10–50 mg of sample is required. If more quantitative SHG information is needed, a larger amount of sample, around 1 g, is required. For more quantitative measurements, a photomultiplier tube (PMT)

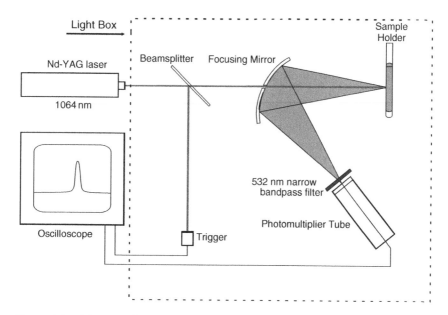

Figure 1.9 Schematic diagram of a modified Kurtz-Powder Laser System. Reprinted with permission from Ok *et al.*, *Chem. Soc. Rev.*, **35**, 710 (2006). Copyright (2006). Royal Society of Chemistry

connected to an oscilloscope is also necessary. Using the PMT and oscilloscope allows the user to collect SHG data on a standard, usually α-SiO$_2$ or urea, and compare the results with newly synthesised materials. The entire system – laser, PMT, power supplies, optics, and oscilloscope – must be placed on a flat surface. A laser table is ideal, since the various optical pieces can be attached to the table, but is not required. The total footprint of the system is 60×180 cm^2; thus only a relatively small, flat area is needed.

1.4.1.1   Measurement of SHG and Data Interpretation

Once the laser, PMT, and optics are aligned, collecting the frequency double light is reasonably straightforward. As previously stated, a small amount of the powder to be tested (10–50 mg) is placed in a fused silica tube. The SHG efficiency of the new material may be compared with standard materials. SHG properties were first measured on α-SiO$_2$; thus the material is defined to have an efficiency of 1.0 (dimensionless). Most SHG efficiencies are reported with respect to α-SiO$_2$; however, other materials can be used. BaTiO$_3$ and urea both have SHG efficiencies of $400 \times$ SiO$_2$, whereas LiNbO$_3$ has an efficiency of $600 \times$ SiO$_2$.[110, 124] Once the SHG efficiency of a polycrystalline sample of α-SiO$_2$ has been measured, it is very straightforward to roughly determine the SHG efficiency of any new material. If more accurate SHG information is required, additional experiments become necessary. Typically a larger amount of material is necessary, on the order of 1 g, and the powder needs to be sieved into particle sizes ranging between 20 μm and 120 μm. This sieving may be done using commercially available sieves. Measuring the SHG as a function of particle size, 20–120 μm, has two advantages. First, the SHG efficiency is determined more accurately. Second, type 1 phase-matching information may be determined.[110] Type 1 phase-matching, or index matching, occurs when the phase velocity of the fundamental radiation (1064 nm) equals that of the second harmonic (532 nm). If type 1 phase-matching occurs, the SHG efficiency will increase with the particle size and plateau at a maximum value. If phase-matching does not occur, the SHG efficiency will reach a maximum value and then decrease, as the particle size increases. Phase-matching (LiNbO$_3$) and nonphase-matching (α-SiO$_2$) curves are shown in the diagrams on the next page. Note that the curves are drawn to guide the eye, and are not a fit to the data.

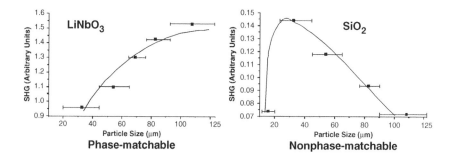

Phase-matchable   **Nonphase-matchable**

These figures also clearly illustrate that accurate SHG efficiencies can *only* be determined by measuring similar particle size ranges. For example, if the SHG of α-SiO$_2$ is measured with particles >90 μm and the unknown material is measured at a smaller particle size, the SHG efficiency of the unknown material would be overestimated. Thus, it is critical that the SHG of SiO$_2$ and the unknown material be measured at the sample particle size range, *i.e.* 45–63 μm.

Once the phase-matching capabilities of the material are known, the bulk nonlinear optical susceptibility, $<d_{eff}>$, can be estimated. The value of $<d_{eff}>$ for phase-matchable, PM, and nonphase-matchable, NPM, materials are given in equations 1.2 and 1.3, respectively.

$$<d_{eff}>_{PM} = \left\{ \frac{I^{2\omega}(A)}{I^{2\omega}(\text{LiNbO}_3)} \times 7.98 \times 10^2 \right\}^{1/2} \quad (1.2)$$

$$<d_{eff}>_{NPM} = \left\{ \frac{I^{2\omega}(A)}{I^{2\omega}(\text{SiO}_2)} \times 0.3048 \right\}^{1/2} \quad (1.3)$$

The SHG efficiency of the unknown compound (A) is either compared with LiNbO$_3$ – SHG efficiency of 600 $\times$ SiO$_2$ – or SiO$_2$ depending on the phase-matching behaviour of the unknown compound. The units for $<d_{eff}>$ are picometres per volt (pm/V).

1.4.2 Piezoelectricity

The piezoelectric phenomena also occur as both the direct and converse effect.[109] With the direct effect, an external stress, σ_{jk}, results in a change in polarisation, P_i. The direct effect is mathematically described as $P_i = d_{ijk}\sigma_{jk}$, where d_{ijk} (i, j, $k = 1$, 2, 3) is the piezoelectric charge coefficient, given in coulombs per newton (C/N). With the converse effect, an applied field, E_i, results in a strain, ϵ_{jk}. The converse effect is mathematically described as $\epsilon_{jk} = d_{ijk}E_i$, where d_{ijk} is the piezoelectric strain coefficient, given in metres per volt (m/V). With both effects, d_{ijk} is a third-rank tensor. It is important to note that the units for d_{ijk} when measuring direct or converse effects are equivalent, that is 1 C/N = 1 m/V. Often the piezoelectric equation is written as $P_i = d_{ij}\sigma_j$ ($i = 1$, 2, 3; $j = 1$, 2, ... 6), where d_{ij} is the contracted notation for d_{ijk}.[125] It is important to note that d_{ij} does not transform as a second-rank tensor. The piezoelectric constants, both charge and strain, given as d_{ij}, are usually reported as one or more terms, d_{33}, d_{31}, and / or d_{15}. With d_{33} the induced polarisation, strain, is *parallel* to the applied stress, electric field, whereas with d_{31} and d_{15} the induced polarisations, strains, are *perpendicular* to the applied stresses, electric fields. Another important variable with respect to piezoelectric devices is the electromechanical coupling factor, k. This factor describes the efficiency in the conversion of mechanical energy to electrical energy, the direct effect, or the conversion of electrical energy to mechanical energy, the converse effect. Generally large k values are desirable for efficient energy conversion; however, the coupling term does not take into account dielectric or mechanical losses.

1.4.2.1   Sample Preparation and Measurement

Unlike the SHG measurement where a loose polycrystalline powder can be used, a well-sintered ceramic is necessary for bulk piezoelectric measurements. For the measurements described herein, the dimensions of the sintered disc are a diameter of $1/2''$ and a thickness of ~ 0.5 mm. In addition, the ceramic must undergo electrical poling. With the poling technique, electrodes are applied to both sides of the sample – usually silver or gold that has been sputtered or mechanically applied. The poling process takes place above room temperature (100–300 °C), with an applied voltage (1000–2000 V) for 20–30 min. After poling, the material has a response similar to a single crystal, where the entire ceramic acts as a single unit. Although poling will not align 100% of the crystallites, the extent of alignment is sufficient to measure the piezoelectric response. It is interesting to note that it was only in 1949 that poling was discovered to be critical in changing a seemingly inert ceramic into an electromechanically active material.[126] Before this time, the assumption was that the individual crystallites in a ceramic would effectively cancel, rendering the material useless for industrial applications.

1.4.2.2 Direct Piezoelectric Measurements

The direct piezoelectric effect occurs when a mechanical force on a material results in a change in polarisation. The resultant piezoelectric charge constant, d, is a third-rank tensor, d_{ijk}, and is measured in C/N, or pC/N. Often this tensor notation is reduced to matrix notation,[125] and the d_{ijk} terms become d_{ij}, with $i = 1, 2, 3; j = 1, 2, \ldots 6$. As stated earlier, d_{ij} does not transform as a second-rank tensor. For both the direct as well as the converse effects, the d_{33} value of the material is usually reported. The subscripts specifically denote that what is being measured is a polarisation parallel to the direction of the mechanical force. Lateral, d_{31}, and shear, d_{15}, polarisations may also be determined, but these measurements are usually done on specifically cut single crystals. To measure the direct piezoelectric effect, a static or quasi-static method is used. Although this method is less precise than the resonance method,[127] the ease of use and availability of instrumentation makes the static method preferable. The static method employs a Berlincourt d_{33} Meter, for which a number of commercial systems are available.[128] The instrument is very straightforward to use. A known force is applied to the poled ceramic, as well as to a standard piezoelectric. Comparing the resultant electric signals allows one to determine the d_{33} of the sample. The value of d_{33} is simply read off the meter. These meters can measure d_{33} charge constants within a few per cent, with a range from 20 to 2000 pC/N.

1.4.2.3 Converse Piezoelectric Measurements

The d_{33} strain constant may be determined on bulk samples through converse piezoelectric measurements. As noted earlier, the strain constant is expressed in units of m/V that are equivalent to C/N. Converse piezoelectric measurements are more experimentally difficult compared with the direct measurements, but provide greater accuracy. The converse measurements use an optical technique in order to measure the small strains in the sample caused by the application of a voltage. The experimental system is composed of a high voltage amplifier and interface as well as an optical

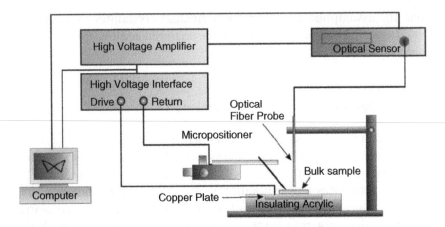

Figure 1.10 Experimental system to measure converse piezoelectric effects. Reprinted with permission from Ok *et al.*, *Chem. Soc. Rev.*, **35**, 710 (2006). Copyright (2006) Royal Society of Chemistry

sensor and probe. The optical probe remains stationary and is approximately 1 mm above the sample (see Figure 1.10).

In the native state, or zero voltage, a finite amount of light is collected in reflection from the sample. When the voltage is applied, the sample undergoes a macroscopic strain and the amount of light collected by the optical sensor changes. The change in collected light is converted to a displacement change, in m/V. Mathematically, d_{33} may then be calculated through: $d_{33} = S/E = \Delta t/V$, where S is the strain, E is the field

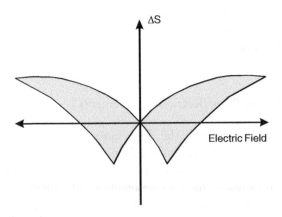

Figure 1.11 Butterfly loop observed in converse piezoelectric measurements. Reprinted with permission from Ok *et al.*, *Chem. Soc. Rev.*, **35**, 710 (2006). Copyright (2006) Royal Society of Chemistry

strength (in V/m), Δt is the change in thickness, and V is the applied voltage. A plot of strain vs electric field produces the commonly observed butterfly curves, similar to the one shown in Figure 1.11. The equation also indicates that d_{33} depends on the *change* in thickness of the sample, not the initial sample thickness. The magnitude of the piezoelectric response can vary greatly among oxide materials. For example, ZnO, $LiNbO_3$, and $LiTaO_3$ have d_{33} values of around 10 pC/N, whereas $BaTiO_3$ and PZT (lead zirconate titanate) have corresponding d_{33} coefficients of approximately 190 and between 100 and 600 pC/N for various PZT compositions.[120]

1.4.3 Pyroelectricity

The pyroelectric effect is defined as the change in spontaneous polarisation, P_s, as a function of temperature, T. The pyroelectric coefficient, p, is mathematically defined as shown in equation 1.4.

$$p = \frac{dP_s}{dT} \tag{1.4}$$

The coefficient, p, is a vector and is described in units of $\mu C/m^2/K$. Surprisingly, the effect has been known for over 2400 years, with the first account attributed to the Greek philosopher Theophrastus. He described a stone, lyngourion – probably tourmaline – that was capable of attracting straw and pieces of wood. A thorough and comprehensive description and history of the phenomenon may be found in Lang's seminal text and recent papers.[114, 119, 129] Brewster, in 1824, was the first researcher to use the term pyroelectricity.[130] Interesting, the material Brewster investigated, Rochelle salt, was studied nearly a century later by Valasek in his discovery of ferroelectricity.[131, 132] The pyroelectric effect was mainly an academic curiosity until 1938, when Ta suggested that tourmaline crystals could be used as an infrared sensor.[133] After this publication, and with the onset of the Second World War, investigation into pyroelectricity grew rapidly and remains an active area of current research among chemists, material scientists and engineers. Currently, pyroelectrics are mainly used for thermal detectors. Pyroelectric devices respond to changes in temperature and therefore can be used to detect and observe stationary or moving objects. A few of the applications for pyroelectric detectors include burglar alarms, pollution monitors, and the measurement of thermal properties of materials.

1.4.3.1   Sample Preparation and Measurement

The sample preparation for a bulk pyroelectric measurement is very similar to what is required for a bulk piezoelectric measurement, namely a well-sintered ceramic disc that has been electrically poled. Determining the pyroelectric coefficient may be divided into two groups – the measurement of the pyroelectric current and the measurement of the charge.[120] We will describe measurement techniques for both groups. In addition, the pyroelectric effect can be subdivided into primary and secondary effects. The primary effect is observed when the material is rigidly clamped under a constant strain to prevent any thermal expansion or contraction. Secondary effects occur when the material is permitted to deform, *i.e.* the material is under constant stress. Thermal expansion results in a strain that changes the spontaneous polarisation, attributable to the piezoelectric effect. Thus the secondary pyroelectric effect includes contributions caused by piezoelectricity. Exclusively measuring the pyroelectric coefficient under constant strain is experimentally very difficult. What is usually experimentally measured is the total pyroelectric effect exhibited by the material – the sum of the primary and secondary effects.

1.4.3.2   Pyroelectric Current

The most straightforward technique to measure the pyroelectric current is the direct method,[134] in which the pyroelectric material is heated uniformly at a constant rate, *i.e.* $\Delta T/\Delta t = 1-2$ °C/min. The pyroelectric coefficient is determined by measuring the pyroelectric current, given by $i(T) = (\Delta T/\Delta t)p$, where A is the sample area. Thus, a plot of $p(T)$ over a wide range of temperature can be easily obtained. More experimentally complicated methods may also be used to determine the pyroelectric current. These include radiation heating[135] and capacitive charging.[136] These methods are more accurate compared with the direct method, but experimentally more complicated.

1.4.3.3   Pyroelectric Charge

The original method for measuring the pyroelectric charge was developed in 1915.[137] This technique, known as the static method, determines the

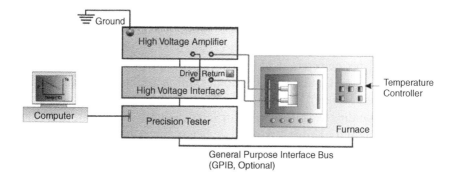

Figure 1.12 Schematic of the static method for determining the pyroelectric coefficient. Reprinted with permission from Ok *et al.*, *Chem. Soc. Rev.*, **35**, 710 (2006). Copyright (2006) Royal Society of Chemistry

charge, *i.e.* polarisation, of the material as a function of temperature. The technique works very well at discrete temperatures. The static method was improved upon by Glass, with an integration technique that allowed for larger changes in temperature.[138] The pyroelectric coefficient may be obtained by graphical differentiation. A schematic of the static method is shown in Figure 1.12. If the material under investigation is ferroelectric, *i.e.* the polarisation is reversible, the pyroelectric coefficient may be determined by measuring the temperature dependence of the remanent polarisation. The remanent polarisation is determined through a Sawyer–Tower loop (see §1.4.1.1).[121] As with the piezoelectric technique discussed earlier, graphical differentiation is used.

Clearly, ferroelectric and nonferroelectric pyroelectrics are possible, and the pyroelectric coefficient varies widely between the two groups. The pyroelectric coefficients for ferroelectrics such as $BaTiO_3$, $LiNbO_3$, and $LiTaO_3$ are −200, −83, and −176 $\mu C/m^2/K$, respectively, whereas for nonferroelectrics such as ZnO, tourmaline, and CdS the corresponding values are −9.4, −4.0, and −4.0 $\mu C/m^2/K$, respectively.[114]

1.4.4 Ferroelectricity

A ferroelectric is formally defined as a pyroelectric material that has a reversible, or 'switchable', polarisation. Ferroelectricity was discovered in *ca* 1920 by Valasek[131, 132] in Rochelle salt ($NaKC_4H_4O_6 \cdot 4H_2O$) – a material that was known at the time for its piezoelectric and pyroelectric properties. For years after this discovery, ferroelectricity was viewed simply as a scientific curiosity, and was thought to occur only rarely in materials. In 1935, the first family of ferroelectrics was discovered in KH_2PO_4 and related materials.[139, 140] The first nonhydrogen bonded ferroelectric, $BaTiO_3$, was subsequently discovered in *ca* 1945 by Wul and Goldman in the Soviet Union and von Hippel's group in the United States.[141, 142] Until this discovery it was assumed that hydrogen bonding was necessary for ferroelectricity to occur. The fact that oxides could exhibit ferroelectric behaviour ushered in a new era, and soon thereafter a number of ferroelectric oxides were discovered. A thorough and rigorous discussion that encompasses all aspects of ferroelectricity may be found in the comprehensive text by Lines and Glass.[112]

1.4.4.1   Sample Preparation and Hysteresis Loop

Similar to piezoelectric measurements, for ferroelectric measurements a well-sintered and dense (>95%) ceramic disc that has been electrically poled is necessary. The circuit design for measuring ferroelectric hysteresis curves was published in 1930.[121] Since that time there have been a few reports modifying the original design,[143, 144] but the overall concept has not changed in over 70 years. At its most basic, a linear capacitor is placed in series with the sample. An AC or DC voltage is then applied. The voltage measured across the capacitor is equivalent to the polarisation of the sample (see Figure 1.13). The measurement of spontaneous polarisations on the order of 5–50 $\mu C/cm^2$ in bulk samples requires voltages in excess of 1000 V. The circuit is

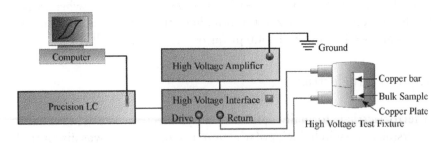

Figure 1.13 Experimental Sawyer–Tower Circuit. Reprinted with permission from Ok *et al.*, *Chem. Soc. Rev.*, **35**, 710 (2006). Copyright (2006) Royal Society of Chemistry

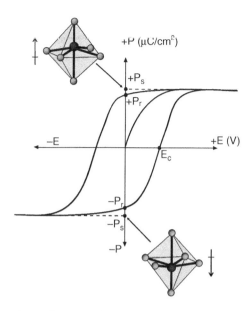

Figure 1.14 Ferroelectric hysteresis loop (polarization *vs* applied voltage). Reprinted with permission from Ok *et al.*, *Chem. Soc. Rev.*, **35**, 710 (2006). Copyright (2006) Royal Society of Chemistry

used to measure a ferroelectric hystersis loop, *i.e.* the material's switchability in the presence of an applied voltage (see Figure 1.14). The hysteresis loop is determined by measuring the polarisation of the material, in $\mu C/cm^2$, as a function of applied voltage, V. The full details of a ferroelectric hystersis loop have been extensively discussed,[145] so only a brief description will be given here. Several points of the loop are of interest, the spontaneous (P_s) and remanent (P_r) polarisations, the coercive field (E_c), and the general shape of the loop. In a ferroelectric material, when all of the dipole moments are aligned, the material is considered saturated since an increase in applied voltage will not increase the polarisation. The linear extrapolation of the curve back to the polarisation axis represents the spontaneous polarisation $(+P_s)$. As the applied voltage is reduced from its maximum positive value to zero, some dipole moments will remain aligned, and a remanent polarisation $(+P_r)$ is observed. As the applied voltage spans the range from its maximum positive to negative values, $-P_s$ and $-P_r$, will be observed. Structurally all of the dipole moments have switched from the positive to the negative – the up and down arrows in Figure 1.14. This is the 'switchability' alluded to earlier. Additional information that can be obtained from a ferroelectric hystersis loop includes the coercive field (E_c) and the shape of the loop. The coercive field is the magnitude of the external applied voltage necessary to remove all the polarisation in the material. The coercive field as well as the shape of the loop, *i.e.* the 'squareness' or sharpness, are sample preparation dependent and are influenced by grain size and homogeneity.[145] It is important to know that the claim of ferroelectricity has been made for a number of materials based on the observation of a closed hysteresis loop.[146–150] An excellent article by Scott has recently appeared that describes this 'false ferroelectricity'.[151] In this paper, he describes how some reported ferroelectric loops are simply from lossy dielectrics and have nothing to do with polarisation reversal. As previously stated, ferro-electrics may be divided into two groups, hydrogen bonding and nonhydrogen bonding. The spontaneous polarisation, P_s, values vary greatly between the two groups. With KH_2PO_4 and related materials, P_s values range from 4.0 to 6.0 $\mu C/cm^2$, whereas with oxides such as $BaTiO_3$ and $LiNbO_3$, the corresponding values are 26 and 71 $\mu C/cm^2$, respectively.[112]

1.5 OUTLOOK – MULTIFUNCTIONAL MATERIALS

The outlook for NCS inorganic materials is quite promising. In addi-tion to the aforementioned properties, these materials find uses in chiral separations and catalysis, advanced optical technologies (wave-guides and imaging) and sensors. In fact a recent book has been published that discusses advanced characterisation of polar materials – bulk and thin films.[152] These include characterisation at the nano-level, microwave dielectric properties, and microscopy. One area where inor-ganic NCS materials has seen a revival is with multifunctional, specifi-cally multi-ferroic materials.[153–155] With multi-ferroics, the definition given by Schmid will be used – "a material is considered multi-ferroic if at least two primary ferroic properties occur in the same material".[156] For our purposes, this would be magnetic ordering, anti-, ferri-, or ferro-magnetism, and ferroelectricity. The physics of these materials is beyond the scope of this chapter, but structural acentricity and polar-ity are required. A few families of multi-ferroics have been described, and we will briefly discuss each of these as well as their NCS polar nature.

1.5.1 Perovskites

With perovskite multi-ferroic materials, the most extensively studied material is $BiFeO_3$.[35, 146, 157–159] This material is reported to be ferroelectric, ferroelastic, and weakly ferromagnetic. The material exhibits trigonal symmetry, crystallising in the polar crystal class $3m$. The structure of $BiFeO_3$ consists of BiO_6 and FeO_6 octahedra that are corner-, edge-, and face-shared. The Fe^{3+} is effectively undistorted, bonded to six oxygen atoms with nearly equal Fe-O distances. The main structural distortion is from the Bi^{3+} cation, which exhibits a lone-pair. The lone-pair results in an unsymmetric coordination environment, with three 'short' Bi-O bonds ($\sim$2.34 Å) and three 'long' Bi-O bonds ($\sim$2.56 Å). Thus the BiO_6 octahedra are locally NCS and polar. It is this local polarity, attributable to the lone-pair, that is responsible for the observed ferroelectric behaviour. In other words, the polarisation associated with the lone-pair cation is being 'switched'. It is interesting to note that polarisation reversibility in lone-pair cations has only been observed with 6th period elements, $i.e.$ Tl^+, Pb^{2+}, Bi^{3+}. Ferroelectric behaviour involving 4th and 5th period lone-pair cations, $i.e.$ Se^{4+}, Sn^{2+}, Te^{4+}, I^{5+}, where the polarisation on these cations are reversed, has never been observed. It should be noted that $LiH_3(SeO_3)_2$ is ferroelectric, but the ferroelectric behaviour is attributable to hydrogen bonding and not the Se^{4+} cation.[160] We suggest that ferroelectric behaviour in 4th and 5th period lone-pair cations is extremely unlikely, if not impossible, attributable to the lone-pair being much more stereoactive compared with 6th period lone-pair cations. Polarisation reversibility in 4th and 5th period lone-pair cations would involve substantial bond breaking and/or rearrangements, that are structurally and energetically very unfavourable (see Figure 1.15).

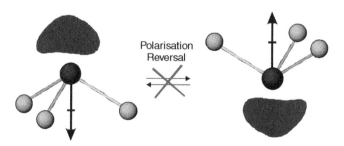

Polarisation
Reversal

Figure 1.15 Hypothetical polarisation reversibility in an MO_3E (M = lone-pair cation; E = lone-pair) polyhedron

1.5.2   Hexagonal Manganites

These materials have the $RMnO_3$ ($R =$ Sc or small, rare earth cation) stoichiometry,[161] and have been erroneously referred to as hexagonal perovskites. The compounds do not exhibit the perovskite structure. The Mn^{3+} cations are not octahedrally coordinated, rather the cation is surrounded by five oxide anions in a trigonal prismatic coordination environment. Also the 'R' cations are not 12-coordinate, as would be the case in a perovskite, but are in seven-fold coordination. The materials are multi-ferroic, with anti-ferromagnetic and ferroelectric properties.[162] The nature of the polarity and therefore the ferroelectric behaviour was only recently described.[163] Careful structural studies indicated that although the dipole moments are attributable to the R-O bonds and not the Mn-O bonds, the R-cations are not directly responsible for the ferroelectric behaviour. The noncentrosymmetry is attributable to the tilting of the MnO_5 polyhedra, which in conjunction with the dipole moments in the R-O bonds results in ferroelectric behaviour. Thus the ferroelectric behaviour in these materials is termed 'improper'[164] and occurs by a much different mechanism than $BaTiO_3$ or even $BiFeO_3$.

1.5.3   Metal Halide and Oxy-Halide Systems

Multi-ferroic behaviour has also been observed in metal halide materials, specifically $BaMF_4$-type compounds ($M =$ Mg, Mn, Fe, Co, Ni, and Zn)[165, 166] and boracites, $M_3B_7O_{13}X$ ($M =$ Cr, Mn, Fe, Co, Ni, or Cu; $X =$ Cl, Br, or I).[167–169] The former consists of puckered layers of corner-shared MF_6 octahedra that are separated by Ba^{2+} cations. The materials are iso-structural, crystallising in the orthorhombic crystal class $mm2$. Anti-ferromagnetic ordering has been observed for $M =$ Mn, Fe, Co, and Ni, but only $BaCoF_4$ and $BaNiF_4$ are ferroelectric. Similar to the hexagonal manganites, the transition metal in the $BaMF_4$ materials is in a nearly regular octahedral environment. The origin of the ferroelectricity has been recently elucidated and occurs through the rotation of the MF_6 ($M =$ Co or Ni) octahedra and polar displacements of the Ba^{2+} cation.[170] It is suggested that the reason ferroelectric behaviour occurs in the Co and Ni phases and not the Mn and Fe analogues is attributable to geometric constraints and size effects.[170] Boracite-type compounds have the stoichiometry $M_3B_7O_{13}X$ ($M =$ Cr, Mn, Fe, Co, Ni, or Cu; $X =$ Cl, Br, or I) and exhibit three-dimensional crystal structures. The structures consist of linked MO_4X, BO_4, and BO_3 polyhedra. The materials are iso-structural crystallising in the trigonal crystal class $3m$. Complex ferromagnetic ordering below 100 K has been observed in $Mn_3B_7O_{13}Cl$ and $Ni_3B_7O_{13}Cl$.[171, 172] Ferroelectric hysteresis loops have also been measured for these compounds as well,[168, 173] indicating polarisation reversibility. What is unclear is the origin of the ferroelectric behaviour. The polarity in the boracite structure is likely attributable to the MO_4X and BO_3 polyhedra. In these materials the BO_3 polyhedra are not strictly planar, but are found as a trigonal pyramid with boron at the apex. Thus, both cations are in locally acentric and polar coordination environments, whereas the BO_4 tetrahedra are only acentric. Thus, polarisation reversal, *i.e.* ferroelectricity, must occur because of the MO_4X and/or BO_3 polyhedra. To date, however, no mechanism has been confirmed.

1.6 CONCLUDING THOUGHTS

Clearly the design, synthesis, and characterisation of NCS inorganic materials will be of great importance for the foreseeable future. In this final section we briefly discuss the state of the field.

1.6.1 State of the Field

As we have described, there are a variety of strategies that have been developed to synthesise new NCS inorganic materials. The fundamental question remains – Is it possible to *a priori* rationally design and thereby synthesise a NCS inorganic material? At present, the answer is: not all the time. Of course this does not mean all is lost, and we are left with haphazardly combining reagents in the hope of synthesising a NCS material. Clearly, by using one of the strategies presented here, one is able to substantially increase the incidence of synthesising a NCS material. Let us recall the requirements for a NCS material. First the building-blocks or coordination polyhedra, i.e. MX_n where M is a metal cation, X is an anion, and n represents the number of anions surrounding the cation, must be inherently acentric. The anions cannot be related by an inversion centre associated with the cation. Thus there must be a structural distortion that requires or forces the metal cation not to be on an inversion centre. Microscopic centricity implies macroscopic centricity. Once local, microscopic, acentricity has been achieved, a second challenge must be overcome. The acentric polyhedra must be related or connected by noninversion type symmetry. Thus, acentric polyhedra are a necessary but not sufficient condition for crystallographic noncentrosymmetry. As outlined in this chapter, a host of researchers have proposed a variety of strategies for addressing both challenges. Much research, however, remains to be done, as open questions remain. Fundamentally, is it possible to control the chemical bonding interactions such that an inorganic NCS structure is *always* produced? Even within inorganic NCS structures, can chirality and polarity be controlled such that specific properties are enhanced? The design and synthesis of inorganic NCS materials has made great strides in the past decade. It is hoped that this chapter will provide an impetus for greater research in this area.

ACKNOWLEDGEMENTS

We thank the Robert A. Welch Foundation (Grant E-1457), the Texas Center for Superconductivity, the ACS PRF 47345-AC10, and the NSF (DMR-0652150) for support.

REFERENCES

[1] J.F. Nye, *Physical Properties of Crystals: Their Representation by Tensors and Matrices*, Oxford University Press, Oxford, 1985.

[2] R.A. Young and B. Post, *Acta Crystallogr.*, **15**, 337 (1962).

[3] G.S. Smith and L.E. Alexander, *Acta Crystallogr.*, **16**, 462 (1963).

[4] A. Rosenzweig and B. Morosin, *Acta Crystallogr.*, **20**, 758 (1966).

[5] C. Svensson, J. Albertsson, R. Liminga, A. Kvick and S.C. Abrahams, *J. Chem. Phys.*, **78**, 7343 (1983).

[6] I. Tordjman, R. Masse and J.C. Guitel, *Z. Kristallogr.*, **139**, 103 (1974).

[7] J.E. Geusic, H.J. Levinstein, J.J. Rubin, S. Singh and L.G. Van Uitert, *Appl. Phys. Lett.*, **11**, 269 (1967).

[8] S.C. Abrahams, W.C. Hamilton and J.M. Reddy, *J. Phys. Chem. Solids*, **27**, 1013 (1966).

[9] S.C. Abrahams, H.J. Levinstein and J.M. Reddy, *J. Phys. Chem. Solids*, **27**, 1019 (1966).

[10] E.M. Levin and H.F. McMurdie, *J. Am. Ceram. Soc.*, **32**, 99 (1949).

[11] C. Chen, B. Wu, A. Jiang and G. You, *Sci. Sin. Ser. B*, **28**, 235 (1985).

[12] S.B. Hendricks, *J. Am. Chem. Soc.*, **50**, 2455 (1928).

[13] R. Ueda, *J. Phys. Soc. Jpn.*, **3**, 328 (1948).

[14] S.C. Abrahams, *in Anomalous Scattering*, S. Ramaseshan and S.C. Abrahams, (Eds), Munksgaard, Copenhagen, 1975, p. 197.

[15] E.R. Howells, D.C. Phillips and D. Rogers, *Acta Crystallogr.*, **3**, 210 (1950).

[16] D. Rogers, E. Stanley and A.J.C. Wilson, *Acta Crystallogr.*, **8**, 383 (1955).

[17] R.E. Marsh, *Acta Crystallogr.*, **B42**, 193 (1986).

[18] R.E. Marsh, *Acta Crystallogr.*, **B51**, 897 (1995).

[19] A.L. Spek, PLATON: A Multipurpose Crystallographic Tool, Utrecht University, Utrecht, The Netherlands, 2005.

[20] G.M. Schmidt, *Pure Appl. Chem.*, **27**, 647 (1971).

[21] U. Opik and M.H.L. Pryce, *Proc. R. Soc. London*, **A238**, 425 (1957).

[22] R.F.W. Bader, *Mol. Phys.*, **3**, 137 (1960).

[23] R.F.W. Bader, *Can. J. Chem.*, **40**, 1164 (1962).

[24] R.G. Pearson, *J. Am. Chem. Soc.*, **91**, 4947 (1969).

[25] R.G. Pearson, *J. Mol. Struct.*, **103**, 25 (1983).

[26] M. Kunz and I.D. Brown, *J. Solid State Chem.*, **115**, 395 (1995).

[27] J.B. Goodenough, *Annu. Rev. Mater. Sci.*, **28**, 1 (1998).

[28] N.V. Sidgwick and H.M. Powell, *Proc. R. Soc. London, Ser. A*, **176**, 153 (1940).

[29] R.J. Gillespie and R.S. Nyholm, *Q. Rev., Chem. Soc.*, **11**, 339 (1957).

[30] L.E. Orgel, *J. Chem. Soc.*, 3815 (1959).

[31] I. Lefebvre, M. Lannoo, G. Allan, A. Ibanez, J. Fourcade and J.C. Jumas, *Phys. Rev. Lett.*, **59**, 2471 (1987).

[32] I. Lefebvre, M.A. Szymanski, J. Olivier-Fourcade and J.C. Jumas, *Phys. Rev. B*, **58**, 1896 (1998).

[33] G.W. Watson, and S.C. Parker, *J. Phys. Chem. B*, **103**, 1258 (1999).

[34] G.W. Watson, S.C. Parker and G. Kresse, *Phys. Rev. B*, **59**, 8481 (1999).

[35] R. Seshadri and N.A. Hill, *Chem. Mater.*, **13**, 2892 (2001).

[36] U.V. Waghmare, N.A. Spaldin, H.C. Kandpal and R. Seshadri, *Phys. Rev. B*, **67**, 125111 (2003).

[37] M.W. Stoltzfus, P. Woodward, R. Seshadri, J.-H. Park and B. Bursten, *Inorg. Chem.*, **46**, 3839 (2007).

[38] P.S. Halasyamani, *Chem. Mater.*, **16**, 3586 (2004).

[39] K.M. Ok, N.S.P. Bhuvanesh and P.S. Halasyamani, *Inorg. Chem.*, **40**, 1978 (2001).

[40] K.M. Ok and P.S. Halasyamani, *Chem. Mater.*, **13**, 4278 (2001).

[41] J. Goodey, J. Broussard and P.S. Halasyamani, *Chem. Mater.*, **14**, 3174 (2002).

[42] J. Goodey, K.M. Ok, J. Broussard, C. Hofmann, F. Escobedo and P.S. Halasyamani, *J. Solid State Chem.*, **175**, 3 (2003).

[43] H.-S. Ra, K.M. Ok and P.S. Halasyamani, *J. Am. Chem. Soc.*, **125**, 7764 (2003).

[44] E.O. Chi, A. Gandini, K.M. Ok, L. Zhang and P.S. Halasyamani, *Chem. Mater.*, **16**, 3616 (2004).

[45] K.M. Ok and P.S. Halasyamani, *Angew. Chem. Int. Ed.*, **43**, 5489 (2004).

[46] J.-H. Kim, J. Baek and P.S. Halasyamani, *Chem. Mater.*, **19**, 5637 (2007).

[47] J.-H. Kim, J. Baek and P.S. Halasyamani, *Chem. Mater.*, **20**, 3542 (2008).

[48] M.R. Marvel, J. Lesage, J. Baek, P.S. Halasyamani, C.L. Stern and K.R. Poeppelmeier, *J. Am. Chem. Soc.*, **129**, 13963 (2007).

[49] T. Sivakumar, H.Y. Chang, J. Baek and P.S. Halasyamani, *Chem. Mater.*, **19**, 4710 (2007).

[50] H.Y. Chang, T. Sivakumar, K.M. Ok and P.S. Halasyamani, *Inorg. Chem.*, **47**, 8511 (2008).

[51] J.R. Gutnick, E.A. Muller, A.N. Sarjeant and A.J. Norquist, *Inorg. Chem.*, **43**, 6528 (2004).

[52] E.A. Muller, R.J. Cannon, A.N. Sarjeant, K.M. Ok, P.S. Halasyamani and A.J. Norquist, *Cryst. Growth Des.*, **5**, 1913 (2005).

[53] T.R. Veltman, A.K. Stover, A.N. Sarjeant, K.M. Ok, P.S. Halasyamani and A.J. Norquist, *Inorg. Chem.*, **45**, 5529 (2006).

[54] F. Kong, S.-P. Huang, Z.-M. Sun, J.-G. Mao and W.-D. Cheng, *J. Am. Chem. Soc.*, **128**, 7750 (2006).

[55] H.-L. Jiang, S.-P. Huang, Y. Fan, J.-G. Mao and W.-D. Cheng, *Chem. Eur. J.*, **14**, 1972 (2008).

[56] R.E. Sykora, K.M. Ok, P.S. Halasyamani and T.E. Albrecht-Schmitt, *J. Am. Chem. Soc.*, **124**, 1951 (2002).

[57] R.E. Sykora, K.M. Ok, P.S. Halasyamani, D.M. Wells and T.E. Albrecht-Schmitt, *Chem. Mater.*, **14**, 2741 (2002).

[58] T.C. Shehee, R.E. Sykora, K.M. Ok, P.S. Halasyamani and T.E. Albrecht-Schmitt, *Inorg. Chem.*, **42**, 457 (2003).

[59] T.A. Sullens, R.A. Jensen, T.Y. Shvareva and T.E. Albrecht-Schmitt, *J. Am. Chem. Soc.*, **126**, 2676 (2004).

[60] R.E. Sykora and T.E. Albrecht-Schmitt, *J. Solid State Chem.*, **177**, 3729 (2004).

[61] T.Y. Shvareva, J.V. Beitz, E.C. Duin and T.E. Albrecht-Schmitt, *Chem. Mater.*, **17**, 6219 (2005).

[62] T.H. Bray, J.V. Beitz, A.C. Bean, Y. Yu and T.E. Albrecht-Schmitt, *Inorg. Chem.*, **45**, 8251 (2006).

[63] T.A. Sullens, P.M. Almond, J.A. Byrd, J.V. Beitz, T.H. Bray and T.E. Albrecht-Schmitt, *J. Solid State Chem.*, **179**, 1192 (2006).

[64] J. Goodey, K.M. Ok, J. Broussard, C. Hofmann, F.V. Escobedo and P.S. Halasyamani, *J. Solid State Chem.*, **175**, 3 (2003).

[65] S. Alvarez, D. Avnir, M. Llunell and M. Pinsky, *New J. Chem.*, **26**, 996 (2002).

[66] M. Llunell, D. Casanova, J. Cirera, J.M. Bofill, P. Alemany, S. Alvarez, M. Pinsky and D. Avnir, Shape Program, version 1.1b, University of Barcelona Barcelona, 2004.

[67] S. Alvarez, P. Alemany and D. Avnir, *Chem. Soc. Rev.*, **34**, 313 (2005).

[68] K.M. Ok, P.S. Halasyamani, D. Casanova, M. Llunell, P. Alemany and S. Alvarez, *Chem. Mater.*, **18**, 3176 (2006).

[69] P. Halasyamani, K.R. Heier, M.J. Willis, C.L. Stern and K.R. Poeppelmeier, *Z. Anorg. Allg. Chem.*, **622**, 479 (1996).

[70] P. Halasyamani, M.J. Willis, C.L. Stern and K.R. Poeppelmeier, *Inorg. Chem.*, **35**, 1367 (1996).

[71] P.S. Halasyamani and K.R. Poeppelmeier, *Chem. Mater.*, **10**, 2753 (1998).

[72] K.R. Heier, A.J. Norquist, C.G. Wilson, C.L. Stern and K.R. Poeppelmeier, *Inorg. Chem.*, **37**, 76 (1998).

[73] A.J. Norquist, K.R. Heier, C.L. Stern and K.R. Poeppelmeier, *Inorg. Chem.*, **37**, 6495 (1998).

[74] K.R. Heier, A.J. Norquist, P.S. Halasyamani, A. Duarte, C.L. Stern and K.R. Poeppelmeier, *Inorg. Chem.*, **38**, 762 (1999).

[75] M.E. Welk, A.J. Norquist, C.L. Stern and K.R. Poeppelmeier, *Inorg. Chem.*, **39**, 3946 (2000).

[76] M.E. Welk, A.J. Norquist, C.L. Stern and K.R. Poeppelmeier, *Inorg. Chem.*, **40**, 5479 (2001).

[77] M.E. Welk, A.J. Norquist, F.P. Arnold, C.L. Stern and K.R. Poeppelmeier, *Inorg. Chem.*, **41**, 5119 (2002).

[78] H.K. Izumi, J.E. Kirsch, C.L. Stern and K.R. Poeppelmeier, *Inorg. Chem.*, **44**, 884 (2005).

[79] B.C. Frazer, H.R. Danner and R. Pepinsky, *Phys. Rev.*, **100**, 745 (1955).

[80] Q. Huang and S.-J. Hwu, *Inorg. Chem.*, **42**, 655 (2003).

[81] X. Mo and S.-J. Hwu, *Inorg. Chem.*, **42**, 3978 (2003).

[82] X. Mo, E. Ferguson and S.-J. Hwu, *Inorg. Chem.*, **44**, 3121 (2005).

[83] W.L. Queen, J.P. West, S.-J. Hwu, D.G. VanDerveer, M.C. Zarzyczny and R.A. Pavlick, *Angew. Chem. Int. Ed.*, **47**, 3791 (2008).

[84] D.A. Keszler, *Curr. Opin. Solid State Mater. Sci.*, **1**, 204 (1996).

[85] P. Becker, *Adv. Mater.*, **10**, 979 (1998).

[86] D.A. Keszler, *Curr. Opin. Solid State Mater. Sci.*, **4**, 155 (1999).

[87] E. Zobetz, *Z. Krist.*, **160**, 81 (1982).

[88] S. Pan, J.P. Smit, B. Watkins, M.R. Marvel, C.L. Stern and K.R. Poeppelmeier, *J. Am. Chem. Soc.*, **128**, 11631 (2006).

[89] C.J. Kepert and M.J. Rosseinsky, *Chem. Commun.*, 31 (1998).

[90] T.J. Prior and M.J. Rosseinsky, *Chem. Commun.*, 495 (2001).

[91] T.J. Prior and M.J. Rosseinsky, *Inorg. Chem.*, **42**, 1564 (2003).

[92] M.J. Rosseinsky, *Micro. Meso. Mater.*, **73**, 15 (2004).

[93] D. Bradshaw, J.B. Claridge, E.J. Cussen, T.J. Prior and M.J. Rosseinsky, *Acc. Chem. Res.*, **38**, 273 (2005).

[94] M.J. Ingleson, J. Bacsa and M.J. Rosseinsky, *Chem. Commun.*, 3036 (2007).

[95] M.J. Ingleson, J.P. Barrio, J. Bacsa, C. Dickinson, H. Park and M.J. Rosseinsky, *Chem. Commun.*, 1287 (2008).

[96] W. Lin, O.R. Evans, R.-G. Xiong and Z. Wang, *J. Am. Chem. Soc.*, **120**, 13272 (1998).

[97] S.J. Lee and W. Lin, *J. Am. Chem. Soc.*, **124**, 4554 (2002).

[98] Y. Cui, S.J. Lee and W. Lin, *J. Am. Chem. Soc.*, **125**, 6014 (2003).

[99] Y. Cui, H.L. Ngo, P.S. White and W. Lin, *Chem. Commun.*, 994 (2003).

[100] H. Jiang and W. Lin, *J. Am. Chem. Soc.*, **125**, 8084 (2003).

[101] B. Kesanli and W. Lin, *Coord. Chem. Rev.*, **246**, 305 (2003).

[102] W. Lin, *J. Solid State Chem.*, **178**, 2486 (2005).

[103] L.M. Zheng, T. Whitfield, X. Wang and A.J. Jacobson, *Angew. Chem. Int. Ed.*, **39**, 4528 (2000).

[104] E.V. Anokhina and A.J. Jacobson, *J. Am. Chem. Soc.*, **126**, 3044 (2004).

[105] N. Guillou, C. Livage, M. Drillon and G. Férey, *Angew. Chem. Int. Ed.*, **42**, 5314 (2003).

[106] J.S. Seo, D. Whang, H. Lee, S.I. Jun, J. Oh, Y.J. Jeon and K. Kim, *Nature*, **404**, 982 (2000).

[107] N.G. Pschirer, D.M. Ciurtin, M.D. Smith, U.H.F. Bunz and H.-C. zur Loye, *Angew. Chem. Int. Ed.*, **41**, 583 (2002).

[108] M. J. Zaworotko, *Chem. Soc. Rev.*, **23**, 283 (1994).

[109] W.G. Cady, *Piezoelectricity; an Introduction to the Theory and Applications of Electromechanical Phenomena in Crystals*, Dover, New York, 1964.

[110] S.K. Kurtz and T. Perry, *J. Appl. Phys.*, **39**, 3798 (1968).

[111] B. Jaffe, W.R. Cook and H. Jaffe, *Piezoelectric Ceramics*, Academic Press, London, 1971.

[112] M.E. Lines and A.M. Glass, *Principles and Applications of Ferroelectrics and Related Materials*, Oxford University Press, Oxford, 1991.

[113] R. Schwartz, J. Ballato and G. Haerling, *in Materials Engineering*, Marcel Dekker, New York, 2004, p. 207.

[114] S.B. Lang, *Physics Today*, **58**, 31 (2005).

[115] P.A. Franken, A.E. Hill, C.W. Peters and G. Wienrich, *Phys. Rev. Lett.*, **7**, 118 (1961).

[116] J.A. Armstrong, N. Bloembergen, J. Ducuing and P.S. Pershan, *Phys. Rev.*, **127**, 1918 (1962).

[117] J. Curie and P. Curie, *Bull. Soc. Min.*, **3**, 90 (1880).

[118] V.G. Voigt, *Lehrbuch der Kristallphysik*, B.G. Teubner, Leipzig, Berlin, 1910.

[119] S.B. Lang, *Sourcebook of Pyroelectricity*, Gordon & Breach Science, London, 1974.

[120] H. Landolt, *Numerical Values and Functions from the Natural Sciences and Technology (New Series), Group 3: Crystal and Solid State Physics*, Springer Verlag, Berlin, 1979.

[121] C.B. Sawyer and C.H. Tower, *Phys. Rev.*, **35**, 269 (1930).

[122] T. Maiman, *Br. Commun. Electron.*, **7**, 674 (1960).

[123] http://www.continuumlasers.com and http://www.cohr.com/lasers. Date of access: 11.04.2010.

[124] J.P. Dougherty and S.K. Kurtz, *J. Appl. Cryst.*, **9**, 145 (1976).

[125] J.F. Nye, *Physical Properties of Crystals*, Oxford University Press, Oxford, 1957.

[126] R.B. Gray, Transducer and Method of Making Same, US Patent # 2,486,560, in US Patent # 2,486,560 (1949).

[127] R.E. Newnham, *Properties of Materials: Anisotropy, Symmetry, Structure*, Oxford University Press, Oxford, 2004.

[128] http://www.sensortech.ca and http://www.americanpiezo.com. Date of access: 11.04.2010.

[129] S.B. Lang and D.K. Das-Gupta, *in Handbook of Advanced Electronic and Photonic Materials and Devices*, H.S. Nalwa Academic Press, San Francisco, 2001, p. 1.

[130] D. Brewster, *Edinburgh J. Sci.*, **1**, 208 (1824).

[131] J. Valasek, *Phys. Rev.*, **15**, 537 (1920).

[132] J. Valasek, *Phys. Rev.*, **17**, 475 (1921).

[133] Y. Ta, *C.R. Acad. Sci.*, **207**, 1042 (1938).

[134] R.L. Byer and B. Roundy, *Ferroelectrics*, **3**, 333 (1972).

[135] A.G. Chynoweth, *J. Appl. Phys.*, **27**, 78 (1956).

[136] S.B. Lang and F. Steckel, *Rev. Sci. Instr.*, **36**, 929 (1965).

[137] W. Ackermann, *Ann. Phys.*, **351**, 197 (1915).

[138] A.M. Glass, *J. Appl. Phys.*, **40**, 4699 (1969).

[139] G. Busch, *Helv. Phys. Acta*, **10**, 261 (1937).

[140] G. Busch and P. Scherrer, *Naturwissenschaften.*, **23**, 737 (1935).

[141] B.M. Wul and I.M. Gol'dman, *Compt. Rend. Acad. Sci. URSS*, **49**, 139 (1945).

[142] A. von Hippel, R.G. Breckenridge, F.G. Chesley and L. Tisza, *Ind. Eng. Chem.*, **38**, 1097 (1946).

[143] E.D. Dias, R. Pragasam, V.R.K. Murthy and B. Viswanathan, *Rev. Sci. Instrum.*, **65**, 3025 (1994).

[144] M. Dawber, I. Farnan and J.F. Scott, *Am. J. Phys.*, **71**, 819 (2003).

[145] G.H. Haertling, *J. Am. Ceram. Soc.*, **82**, 797 (1999).

[146] A.K. Pradhan, K. Zhang, D. Hunter, J.B. Dadson, G.B. Loiutts, P. Bhattacharya, R. Katiyar, J. Zhang, D.J. Sellmyer, U.N. Roy, Y. Cui and A. Burger, *J. Appl. Phys.*, **97**, 093903 (2005).

[147] J.-S. Lee, B.S. Kang and Q.X. Jia, *Appl. Phys. Lett.*, **91**, 142901 (2007).

[148] J.Y. Park, J.H. Park, Y.K. Jeong and H.M. Jang, *Appl. Phys. Lett.*, **91**, 152903 (2007).

[149] Z.-G. Gu, X.-H. Zhou, Y.-B. Jin, R.-G. Xiong, J.-L. Zuo and X.-Z. You, *Inorg. Chem.*, **46**, 5462 (2007).

[150] D.-W. Fu, Y.-M. Song, G.-X. Wang, Q. Ye, R.-G. Xiong, T. Akutagawa, T. Nakamura, P.W.H. Chan and S.D. Huang, *J. Am. Chem. Soc.*, **129**, 5346 (2007).

[151] J.F. Scott, *J. Phys.: Condens. Matter.*, **20**, 021001 (2008).

[152] R. Waser, U. Bottger and S. Tiedke, *Polar Oxides: Properties, Characterization, and Imaging*, Wiley-VCH, Weinheim, 2005.

[153] N.A. Spaldin and M. Fiebig, *Science*, **309**, 391 (2005).

[154] M. Fiebig, *J. Phys. D: Appl. Phys.*, **38**, R123 (2005).

[155] D.I. Khomskii, *J. Magn. Magn. Mater.*, **306**, 1 (2006).

[156] H. Schmid, *Magnetoelectric Interaction Phenomena in Crystals,* Kluwer, Dordrecht, 2004.

[157] S.V. Kiselev, R.P. Ozerov and G.S. Zhdanov, *Dokl. Akad. Nauk SSSR*, **145**, 1255 (1962).

[158] S.M. Skinner, *IEEE Trans. Parts Mater. Packaging*, **6**, 68 (1970).

[159] F. Kubel and H. Schmid, *Acta Crystallogr.*, **B46**, 698 (1990).

[160] K. Vedam, Y. Okaya and R. Pepinsky, *Phys. Rev.*, **119**, 1252 (1960).

[161] E.F. Bertaut and M. Mercier, *Phys. Lett.*, **5**, 27 (1963).

[162] H. Sugie, N. Iwata and K. Kohn, *J. Phys. Soc. Jpn.*, **71**, 1558 (2002).

[163] V.B. Aken, T.T.M. Palstra, A. Filippetti and N.A. Spaldin, *Nat. Mater.*, **3**, 164 (2004).

[164] C.J. Fennie and K.M. Rabe, *Phys. Rev. B*, **72**, 100103 (2005).

[165] M. Eibschutz and H.J. Guggenheim, *Solid State Commun.*, **6**, 737 (1968).

[166] D.L. Fox, D.R. Tilley, J.F. Scott and H.J. Guggenheim, *Phys. Rev. B*, **21**, 2926 (1980).

[167] T. Ito, N. Morimoto and R. Sadanaga, *Acta Crystallogr.*, **4**, 310 (1951).

[168] E. Ascher, H. Schmid and D. Tar, *Solid State Commun.*, **2**, 45 (1964).

[169] H. Schmid, H. Rieder and E. Ascher, *Solid State Commun.*, **3**, 327 (1965).

[170] C. Ederer and N.A. Spaldin, *Phys. Rev. B*, **74**, 024102 (2006).

[171] O. Crottaz, J.P. Rivera, B. Revaz and H. Schmid, *Ferroelectrics.*, **204**, 125 (1997).

[172] S.-Y. Mao, H. Schmid, G. Triscone and J. Muller, *J. Magn. Magn. Mater.*, **195**, 65 (1999).

[173] O. Crottaz, J.-P. Rivera and H. Schmid, *J. Kor. Phys. Soc.*, **32**, S1261 (1998).

2

Geometrically Frustrated Magnetic Materials

John E. Greedan

McMaster University, Department of Chemistry, Hamilton, Ontario, Canada

2.1 INTRODUCTION

Research activity in the area of geometrically frustrated magnetic materials (GFMM) has grown rapidly over the past twenty years or so. This has not been driven by a perceived technological goal but more by the opportunity to create and study exotic states of matter such as spin glasses, spin ices and spin liquids, among others, which can be realised in such systems. Moreover, frustration is a surprisingly common thread woven through diverse scientific disciplines. It arises in protein folding, superconducting Josephson junction arrays, liquid crystals, and even in astrophysics in the 'nuclear pasta' state which arises in stellar interiors. As will be mentioned within this chapter, spin ices may be a laboratory for the study of the elusive magnetic monopole.

Magnetic materials have become the focus for much of the study of the frustration phenomenon for a number of reasons. First, magnetic materials which provide the necessary conditions for geometric frustration can be designed and synthesised with relative ease. Secondly, a very wide range of experimental techniques can be applied in the study of both the static and dynamic properties to determine the extent of spatial and

Functional Oxides   Edited by Duncan W. Bruce, Dermot O'Hare and Richard I. Walton
© 2010 John Wiley & Sons, Ltd.

temporal correlations over many decades of length and time scales. Additionally, magnetic systems lend themselves to theoretical investigation using well-defined microscopic models and thus, experiment and theory are closely linked in a symbiotic manner.

No attempt is made to be exhaustive and complete – the existing literature is too large for that to be practical. Instead, the basic principles will be outlined with minimal reference to theory and representative materials will be described with attention to structure, composition and magnetic properties characterisation. Several reviews exist which cover various stages in the development of the field of GFMM including those from the relatively early days,[1–4] while there are more recent reviews of specific materials, such as pyrochlore oxides.[5, 6]

2.2   GEOMETRIC FRUSTRATION

2.2.1   Definition and Criteria: Subversion of the Third Law

Frustration results when contradictory constraints act upon the same site in a material. Our interest here is the case when the origin of the frustration results from the intersite geometry. The canonical examples are the equilateral triangle and the regular tetrahedron shown in Figure 2.1.

If the exchange constraint is for antiferromagnetic (AF) coupling between nearest neighbours, then both 'plaquettes' are clearly frustrated. Technically, a square lattice can also be frustrated but only for the special condition that the nearest and next nearest neighbour exchange interactions are comparable. One of the key consequences of geometric frustration is that for such a material the Third Law of thermodynamics can be subverted. Figure 2.2 shows schematically the situation for a conventional magnetic material as the temperature is lowered to near zero K. According to the Third Law, the entropy, S, must approach zero as the temperature approaches zero.

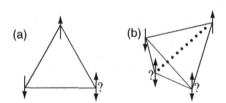

Figure 2.1   Geometric frustration illustrated by the triangle (a) and the tetrahedron (b) where the nearest neighbour exchange constraints are AF

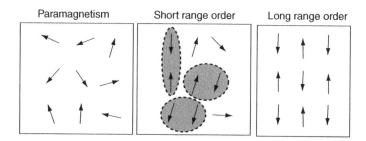

Figure 2.2 Spin correlations associated with the paramagnetic, short range ordered (SRO) and long range ordered (LRO) states

For magnetic systems an important component of the total entropy is the spin disorder entropy which is of course large at high temperatures. As the system is cooled spin–spin correlations can overcome the thermal energy and a long range ordered state becomes the ground state after passing though an intermediate state in which finite domains of short range order, called the critical region, which generally exists over a narrow thermal range, are formed. For geometrically frustrated systems, formation of a well-ordered ground state becomes very difficult, the critical regime can now extend over a wide thermal range, and in many cases, the long range ordered state is never realised. By true long range order is meant that the dimension of an ordered domain is infinite or extends throughout the entire volume of the sample.[7]

2.2.2   Magnetism Short Course

2.2.2.1   Paramagnetism and the Curie–Weiss Law

The paramagnetic state is one in which no spin–spin correlations exist on any length or time scale, *i.e.* the spin orientations are completely random as depicted in the left-hand panel of Figure 2.2. Thus, no net magnetisation exists in the absence of an applied magnetic field. Paramagnetism is a 'high temperature' state for the spins, meaning that the thermal energy, k_BT, is much greater than the spin–spin correlation energy. This latter energy is most commonly expressed in terms of the so-called Heisenberg–van Vleck–Dirac Hamiltonian,

$$H_{ex} = -2J \, S_i \cdot S_j \tag{2.1}$$

where S_i and S_j are spin operators on two sites and the magnitude of the scalar J is a measure of the interaction strength and its sign gives the relative orientation of the two spins. With the above sign convention, $J < 0$ implies antiparallel or antiferromagnetic (AF) coupling and $J > 0$ implies parallel or ferromagnetic (F) coupling. The reader is cautioned that other forms of Equation 2.1 exist with a positive sign and without the factor of two. Thus, J has units of energy and sets the energy scale for the exchange or spin–spin correlations.

As mentioned, the paramagnetic state has no net magnetisation in the absence of an applied magnetic field, so the quantity which is measured is the magnetic susceptibility, $\chi = dM/dH$, *i.e.* the magnetisation induced per unit applied magnetic field. Empirically, the measured susceptibility for most materials follows the so-called Curie–Weiss law,

$$\chi = C/(T - \theta) \qquad (2.2)$$

where C and θ are constants for a given material. Strictly speaking, the Curie–Weiss law holds only when a unique ground state is well separated from any excited states resulting from crystal field induced levels. In some cases the presence of excited states can be accomodated by adding a temperature independent term (TIP) to Equation 2.2. When there exist many excited states within an interval kT of the ground state the more complex van Vleck equation must be used.[8] The analytical form of Equation 2.2 is an hyperbola which is linearised by plotting χ^{-1} *vs* T, as shown in Figure 2.3.

C, the Curie constant, is related to the so-called effective magnetic moment, μ_{eff}, according to:

$$C = \mu_{eff}^2/8 = g^2[S(S+1)]/8 \ \mu_B = S(S+1)/2 \ \mu_B \ \{\text{if } g = 2\} \qquad (2.3)$$

where g is the Landé splitting factor, S is the spin quantum number and μ_B is the Bohr magneton. Thus, C is a quantitative measure of S for a

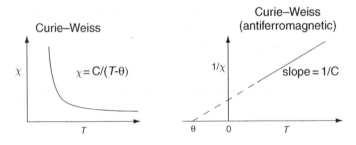

Figure 2.3 The Curie–Weiss law

magnetic system. On the other hand, θ can be shown, within the context of the so-called Mean or Molecular Field Theory (MFT), to be related to the exchange energy through:[9]

$$\theta = \frac{2S(S+1)}{3k} \sum_{m=1}^{N'} z_m J_m \tag{2.4}$$

where z_m is the number of neighbours of a spin and the sum N extends over all possible sets of neighbours from nearest neighbours ($m = 1$) outward. Thus, θ is the weighted, algebraic sum of all of the exchange interactions in a magnetic material. Its sign determines the net exchange constraint, $\theta < 0$ implies AF coupling and $\theta > 0$ F coupling given the sign convention of 2.1, and its magnitude, $k_B|\theta|$, is a measure of the magnetic exchange energy. Again, as a caution it should be noted that there may be other contributions to θ in addition to exchange. These often arise in rare earth ions due to complex electronic structures induced by the crystal field but can occur for d-group ions as well. In rare earths with a large magnetic moment and weak exchange interactions the magnetic dipole–dipole interaction can be significant as well.

2.2.2.2   The Frustration Index and Mean Field Theory

As the temperature is lowered such that $k_B T \sim k_B|\theta|$, i.e. $|\theta|/T \sim 1$, the paramagnetic state should become unstable with respect to some ordered state. According to MFT, for ferromagnetism a transition to a long range ordered state occurs for $T_c = \theta$, where T_c is the so-called 'critical' temperature or the Curie temperature in older literature. In practice one finds that θ/T ranges between $\sim$1.1–1.4 at most. For antiferromagnets the situation is more complex but, empirically, long range AF order is usually established for $|\theta|/T_c \sim 2$–3. The most extreme case known for conventional AF materials, according to MFT, is $\theta/T \sim 5$.[9] This observation led to the use of the frustration index, $f \sim |\theta|/T_c$ as a simple rule of thumb.[2] In general the observation that $f >> 5$ or so is taken as evidence for the importance of magnetic frustration in a real material.

2.2.2.3   The Role of the Spin Quantum Number

Quantum mechanics allows for statistical fluctuations among the quantum states for a given system. For magnetic systems the effect of such fluctuations becomes more pronounced as S decreases. For example for large S, say $5/2$ or $7/2$, the state multiplicity, $2S(S + 1)$, is also large and a fluctuation from say $M_z = -5/2$ to $M_z = -3/2$ has a minor effect on the nature of a ground state as the net spin direction is not changed. However, for $S = \frac{1}{2}$ there exist only two states, $M_z = +/-\frac{1}{2}$ and a fluctuation will result in a change in spin direction. Thus, low spin states, $S = \frac{1}{2}$ and 1 are often called quantum spin states, as the effects of quantum mechanical spin state fluctuations on ground magnetic configurations are the largest. This is a separate issue from geometric frustration but leads to an exacerbation of such effects for low S-based magnetic materials.

2.2.3 Frustrated Lattices – The Big Four

As already shown in Figure 2.1, the equilateral triangle and the regular tetrahedron provide the basis for geometric frustration. In real materials these discrete plaquettes are condensed to form extended arrays with periodic symmetry called lattices. Condensation can involve sharing either corners or edges in two or three dimensions and this results in the four 'canonical' frustrated lattices shown in Figure 2.4. These are the edge-sharing triangular lattice, the corner-sharing triangular lattice (called the kagomé lattice after a Japanese basket weaving pattern), the edge-sharing tetrahedral lattice which is the familiar face-centred-cubic lattice and the corner-sharing tetrahedral lattice which is known as the pyrochlore lattice. This set of four is not exhaustive but does account for a substantial majority of all known geometrically frustrated lattices in real materials.

Other lattices of considerable interest are found in the garnet and langesite structures for example and can be regarded as analogous to the kagomé lattice but with distortions or extensions to three dimensions. Other structures, such as the so-called 'hyper-kagomé' and kagomé-staircase which also feature three-dimensional arrays of corner-sharing triangles, will be described in more detail in later sections.

2.2.4 Ground States of Frustrated Systems: Consequences of Macroscopic Degeneracy

As already mentioned, frustrated magnets appear to defy the Third Law in that, often, a long range ordered magnetic ground state is not stable but

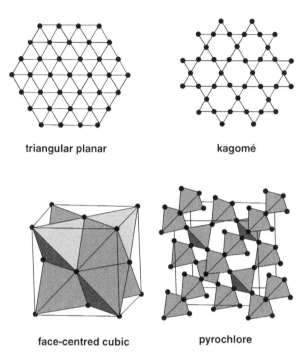

triangular planar   **kagomé**

face-centred cubic   **pyrochlore**

Figure 2.4 The four principal frustrated lattices: triangular planar (TP), kagomé, face-centred cubic, pyrochlore

instead more exotic states such as spin glasses, spin liquids or the curious spin ice state may be found. The difficulty in satisfying the Third Law follows directly from the singular feature of frustrated magnetic materials – an enormous ground state degeneracy. Following Moessner,[10] the full ground state degeneracy is given below:

$$D = N[n(q - 2) - q]/2 \qquad (2.5)$$

where D is the total degeneracy, N is the number of 'plaquettes' (triangles or tetrahedra), n is the number of spin components per site ($n = 1$ for Ising, 2 for XY and 3 for Heisenberg spins) and q is the number of spins per plaquette. The most extreme example is the Heisenberg, pyrochlore case, where $n = 3$ and $q = 4$, for which $D = N$. That is, the degeneracy is equal to the actual number of tetrahedra in the sample which can of course approach 10^{22}! Thus, the selection of a single ground state from such macroscopic degeneracy is profoundly difficult.

Detailed theory in the form of MFT and computations using Monte Carlo simulations have shown for the pyrochlore lattice with nearest neighbour AF spin constraints, that a long range ordered state cannot exist above $T = 0\,K$.[11, 12] In general additional perturbations such as further neighbour interactions, inclusion of magnetic dipolar interactions, site disorder, application of pressure or external magnetic fields, *etc.* are needed to lift the N-fold degeneracy of the ground state. The gamut of ground states observed include complex long range ordered states, *i.e.* with bizarre ordering wave vectors, spin glasses, spin liquids (also called co-operative paramagnets) and spin ices, as already mentioned. Recent theory has shown for example that a spin glass state can be stabilised on the pyrochlore lattice with a minimal level of disorder.[13]

A brief description of some of these states will now be presented.

2.2.4.1 Complex Long Range Order

Often, this takes the form of noncollinear spin ordering. The best known example is that for the triangular lattice where the so-called 120° structure is often found. This is shown in Figure 2.5 where a chiral degeneracy is indicated. There is one report of the corresponding '109°' structure, Figure 2.6, in a pyrochlore lattice material, the metastable, pyr-FeF$_3$.[14] Other more extreme examples will be discussed in sections which follow.

2.2.4.2 Spin Glass State

Spin glasses exhibit, as the name suggests, a random, spin frozen ground state which is depicted in Figure 2.7. Here, the spins are random in space

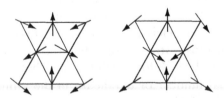

Figure 2.5 The noncollinear 120° magnetic structure for a TP lattice. The two forms are degenerate chiral pairs.[3] Reprinted with permission from Gaulin, 1994 [3]. Copyright (1994) Springer Science + Business Media

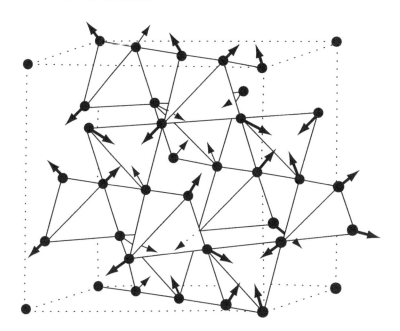

Figure 2.6 The 109° magnetic structure. Note that chirality is present, as for each tetrahedron, the four spins can be pointing either all in or all out.[14] Reprinted from Férey *et al.*, 1986 [14]. Copyright (1986) Elsevier

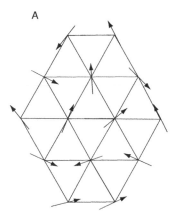

Figure 2.7 The spin glass state showing spins frozen in time with random orientations

and the spin dynamics occur over very long time scales, *i.e.* the spins appear to be instantaneously frozen. The analogy with the glassy state of matter is quite close. It is widely held that the spin glass state requires both frustration and positional disorder to be stabilised.[15]

Archetypal spin glasses are formed by dissolving low concentrations of magnetic elements, such as Fe or Mn, in diamagnetic metallic matrices, such as Au or Cu. Positional disorder in such cases is obvious and the magnetic frustration is provided by the nature of the spin exchange in metals, the so-called RKKY (Ruderman, Kittel, Kasuya, Yosida) interaction, for which the sign of the exchange constraint changes as a function of distance.[15] Spin glass states also are found in insulators such as diluted magnetic salts, e.g. $Eu_{1-x}Sr_xS$, where a spin glass state obtains for $x\sim0.5$. In this case, competing F and AF, nearest neighbour (nn) and next nearest neighbour (nnn), respectively, provide magnetic site frustration.[16] It should be emphasised that the transition to the spin glass state is a true cooperative phenomenon which occurs at a sharp temperature, T_f or T_g, and displays so-called critical behaviour in which the thermodynamic properties such as magnetic susceptibility or heat capacity diverge as T_g is approached according to a set of universal 'critical exponents'. This is in distinct contrast to a superparamagnet in which domains of essentially noninteracting coupled spins undergo blocking behaviour as the temperature is decreased, i.e. the spin dynamics also slows considerably but there is no true phase transition. A spin glass and a superparamagnet can be distinguished using data on spin dynamics obtained from a.c. susceptibility studies.[15] Thus, the spin glass ground state should not occur in systems which lack significant positional disorder. Nonetheless, it is found for nominally well-ordered geometrically frustrated materials such as the pyrochlore, $Y_2Mo_2O_7$ and a number of B-site ordered double perovskites, $A_2BB'O_6$, in which the magnetic B' site ions form a frustrated f.c.c. lattice. These examples will be discussed in some of the following sections.

2.2.4.3   Spin Liquid or Cooperative Paramagnetic State

In this rather more rare and exotic case, the spins never undergo a phase transition to a frozen state in spite of evidence for the presence of strong spin exchange.[17,18] One model for the spin liquid is the 'Resonating Valence Bond' (RVB) ground state due to Anderson but clearly inspired by Linus Pauling.[18] Anderson's original proposal involved a triangular planar lattice decorated with $S = \frac{1}{2}$ spins. He envisaged a ground state consisting of a superposition of components which involve AF coupled spin dimers on adjacent sites. Each contributing state would involve dimers between different pairs of neighbours. An attempt to illustrate this is shown in Figure 2.8. Thus, a critical feature of the spin liquid state, at least in the

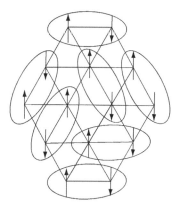

Figure 2.8   A representation of the RVB spin liquid state on a TP lattice. The ground state is a superposition of states in which each spin is AF coupled to another to form $S = 0$ dimers

RVB picture, is that overall, a spin singlet ground state will result. As well, another characteristic is the presence of significant spin dynamics down to the lowest temperatures. There are relatively few accepted examples of the spin liquid state, the pyrochlore oxide $Tb_2Ti_2O_7$ being one.[19]

2.2.4.4   Spin Ice State

This is also a fairly rare ground state and one with very strict requirements. Unlike the cases already outlined, where nn AF exchange is dominant, the spin ice state actually forms when the nn constraint is F but the spins must be subject to a dominant axial anisotropy. The only known examples occur in pyrochlore oxides involving rare earth ions, such as $Dy_2Ti_2O_7$ or $Ho_2Ti_2O_7$. The spin ice configuration is shown in Figure 2.9 for a single tetrahedron. The dominant magnetic anisotropy constrains the magnetic moments to lie along axes which originate at the four corners of the tetrahedra and pass through the centre of the opposite base. If the nn constraint is F, this results in a highly degenerate ground state involving moment configurations of 'two in, two out' on each tetrahedron.

It was noticed by Harris and Bramwell[20] that this situation is topologically equivalent to that for hexagonal water ice in which each oxygen atom is surrounded by two short covalent bonds and two longer hydrogen bonds, also shown in Figure 2.9. This configuration, 'two in, two out' or 'two short, two long' has become known as the 'ice rules'. The ground state degeneracy is calculated by noting that of the $2^4 = 16$ possible spin

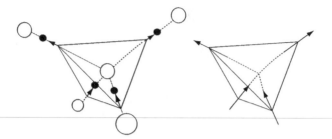

Figure 2.9 The spin ice ground state (right) and the analogy with the proton configuration about an oxygen atom in water ice (left). In spin ice the four moments are constrained to lie along <111> directions of the tetrahedron and have the two in, two out configuration. In water ice two protons (small, solid circles) are covalently bonded at short O–H distances and two are hydrogen bonded at longer O–H distances.[6] Reprinted with permission from Gardner *et al.*, 2010 [6]. Copyright (2010) American Physical Society

configurations for four spins on a tetrahedron, 6 satisfy the condition that the vector sum of the moments equals zero. For a system with N spins and $N/2$ tetrahedra, the total number of microstates, $\Omega = (2^4)^{N/2} = 4^N$. But each spin is shared by one other tetrahedron, so the 16 configurations per tetrahedron are not independent. Following Pauling's arguments for the water ice problem, $2^2 = 4$ spins are allocated for each tetrahedron and only 6/16 of these satisfy the zero total moment condition. Thus, the true degeneracy, $\Omega_0 = [2^2(6/16)]^{N/2}$. Then, the total entropy is $S = k_B \ln \Omega_0 = Nk_B/2\ln(3/2)$. This excess entropy had been observed in water ice by Giauque and by Ramirez *et al.* and later by many others in the pyrochlore oxide $Ho_2Ti_2O_7$, as will be described in Section 2.3.4.1.[21,22]

2.3   REAL MATERIALS

2.3.1   The Triangular Planar (TP) Lattice

2.3.1.1   Ideal (Undistorted)Triangular Planes

While the TP lattice is realised in many inorganic materials, the transition metal di- and trihalides such as CrX_3 or NiX_2, where X is a Cl, Br or I, come to mind, for most of these the nn, intraplanar exchange is F and frustration does not play a role. In part this is due to the predominance of intraplanar 90° M–X–M superexchange which is often F. The salts, VBr_2 and VCl_2, do appear to have intraplanar AF interactions.[23] Other examples of geometrically frustrated systems occur in so-called ordered NaCl oxide materials, such as AMO_2 where M is a TP metal ion and A = Li, Ag, Cu or Na. These compounds crystallise in $R\bar{3}m$ and the A and M ions occupy alternating perfect TP layers along the stacking direction which is <111> of the cubic NaCl unit cell as shown in Figure 2.10.

Early work was motivated in part by Anderson's RVB model, which has been described earlier, with $S = \frac{1}{2}$ ions as a particular focus. For example it was hoped that $NaTiO_2$ and $LiNiO_2$ would provide realisations of the elusive spin liquid ground state. Unfortunately, neither material has proved to be suitable for different reasons. $NaTiO_2$ appears to be metallic[24] and for $LiNiO_2$, while the detailed nature of its ground state is still debated, the dominant in plane exchange appears to be F and the most stoichiometric samples behave as spin glasses.[25] One of the best examples of a GFMM from this series is $LiCrO_2$ which is insulating and does show a large frustration index, $f = 10$. However, this $S = 3/2$ system shows AF long range order (LRO) below 62 K but with a very complex magnetic structure consisting of a double k-vector 120° structure.[26]

$AgCrO_2$ and $CuCrO_2$ also exhibit complex magnetic order with wave vectors $\mathbf{k} = (0.327\,0.327\,0)$ and $\mathbf{k} = (1/3\,1/3\,\xi)$, respectively, i.e. modulated 120° structures.

In addition the magnetic Bragg peaks exceed the resolution width, indicating that LRO is not truly established in three dimensions.[27]

In Table 2.1 we collect some relevant data for selected TP systems.

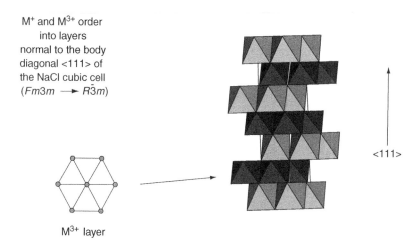

M^+ and M^{3+} order into layers normal to the body diagonal <111> of the NaCl cubic cell ($Fm3m \longrightarrow R\bar{3}m$)

<111>

M^{3+} layer

Figure 2.10 The layered structure of the ordered NaCl materials, $M^+M'^{3+}O_2$ with $R\bar{3}m$ symmetry. The ideal TP topology of the magnetic, M^{3+}, layers is shown

Table 2.1 Selected magnetic data for some $ACrO_2$ and VX_2 layered triangular antiferromagnets

Compound	T_N (K)	θ_C (K)	$|\theta_C|/T_N$	Ref.
VCl_2	36	−437	12	23
VBr_2	29	−335	12	23
$LiCrO_2$	62	−600	10	26
$AgCrO_2$	34	168	5	27
$CuCrO_2$	27	−199	7	27

While several of these have respectable frustration indices, it should be recalled that these are also low dimensional magnetic materials with relatively strong intraplanar exchange but weak interplanar exchange. This low dimensionality will also inhibit LRO and will contribute to a large f value.

2.3.1.2 Distorted Triangular Planes

A few compounds exist with slightly distorted triangular planes. Good examples are the anhydrous alums, $AM(SO_4)_2$ where A = K, Rb or Cs and M is a trivalent transition metal ion. The structure is shown in Figure 2.11. The slightly distorted triangular planar layers are well separated by the large Group 1 ions. The actual space group is $C2/m$ and the triangular edges are of rather different length, 5.15 Å and 4.82 Å and the angles are 64° and 57°, typically. For most cases, M = Fe with S = 5/2, but a Ti^{3+} S = ½ member is also known, $KTi(SO_4)_2$. The M = Fe phases order between 4 K and 8 K and show modest f values of about 7. The Ti material on the other hand does not order above 1.2 K and shows $f > 10$, which should provide motivation for further studies.[28,29]

An additional set of distorted TP magnets is provided by another form of site-ordered NaCl oxide, namely, A_5MO_6. Two examples which have been studied in some detail are Li_4MgReO_6 and Li_5OsO_6.[30, 31] Both involve S = ½ ions, Re^{6+} in the former and Os^{7+} in the latter. The crystal structure, $C2/m$, is shown in Figure 2.12. There are actually two sets of triangular planes within the ab and ac planes, as seen in the figure inset. The various exchange pathways are indicated on Figure 2.13 and relevant interatomic distances for the two compounds are given in Table 2.2.

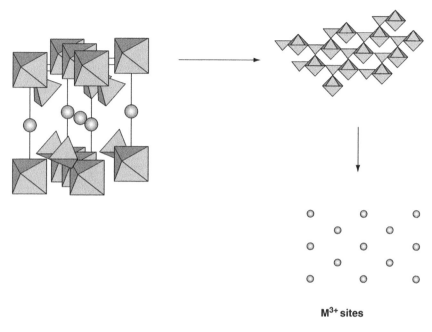

Figure 2.11 The layered structure of the anhydrous alums, $AM(SO_4)_2$. The M sites are shown as octahedra, the $[SO_4]^{2-}$ groups as tetrahedra and the A^+ cations as spheres. The topology of the M sites is also shown

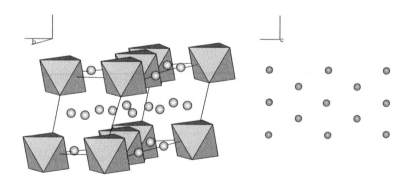

Figure 2.12 The unit cell of the $C2/m$ ordered NaCl materials of type A_5MO_6. The MO_6 octahedra are shown in dark grey and the topology of an M-site layer is shown on the right

While these are iso-structural and iso-spin materials, their properties are remarkably different as indicated in Table 2.3. Li_4MgReO_6 shows a spin frozen ground state below $T_f \sim 12$ K with no long range order and a large frustration index, $f \sim 14$, while Li_5OsO_6 orders AF below ~ 40 K with essentially no evidence for frustration, $f \sim 1$.

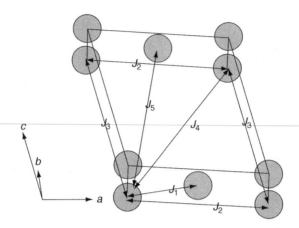

Figure 2.13 Possible exchange pathways for a A_5MO_6 material.[31] Reprinted with permission from Derakhshan *et al.* [31]. Copyright (2008) American Physical Society

Table 2.2 Interatomic distances for various exchange pathways in Li_5OsO_6 and Li_4MgReO_6. See Figure 2.13 for identification of the pathways[30,31]

M–M pathway(Å)	Li_5OsO_6	Li_4MgReO_6
J_1	5.065	5.092
J_2	5.047	5.098
J_3	5.008	5.082
J_4	5.783	5.851
J_5	6.495	6.554

Table 2.3 Comparison of relevant magnetic properties between Li_5OsO_6 and Li_4MgReO_6

Property	Li_5OsO_6	Li_4MgReO_6
μ_{eff} (μ_B)	0.92	1.14
θ_C (K)	-34	-166
$T_{N,f}$	40	~ 12
Ground state	AF LRO	Spin glass
F	~ 1	~ 14

A partial explanation for this surprising difference is provided by a combination of an analysis of the role of magnetic site coordination polyhedron in determining the electronic structure and application of the Spin Dimer Model.[32] The octahedron about Os^{7+} in Li_5OsO_6 is subject to a much stronger psuedo-tetragonal compression, 4 Os – O bonds at 1.908 Å

Table 2.4 Relative values for the exchange pathways $J_1 - J_5$ (Figure 2.13) calculated on the Spin Dimer Model for occupation of d_{xy} only and equal occupation of d_{xy}, d_{xz} and d_{yz} for Li_5OsO_6. The values for Li_4MgReO_6 follow the same pattern

Pathway	d_{xy} only	d_{xy}, d_{xz}, d_{yz} equally
J_1	0.022	0.04
J_2	5×10^{-5}	0.96
J_3	1×10^{-5}	1.00
J_4	1.00	0.43
J_5	0.019	0.02

and 2 at 1.846 Å, while for the Re^{6+} site in Li_4MgReO_6, there are 4 Re – O bonds at 1.962 Å and 2 at 1.932 Å. A tetragonal compression splits the $t_{2g}{}^1$ configuration such that the d_{xy} orbital lies lowest with the d_{xz} and d_{yz} orbitals nearly degenerate. The relative magnitudes of $J_1 - J_5$ differ greatly depending on which crystal field states are occupied. In Table 2.4 two extreme cases are considered, the occupation of only the d_{xy} orbital and equal occupation of d_{xy}, d_{xz} and d_{yz}.

Note that for d_{xy} only occupation, one finds a dominant J_4, suggesting low-dimensional magnetism rather than geometric frustration. Evidence for low-dimensional magnetism is found in Li_5OsO_6 and the stronger pseudo-tetragonal compression suggests that the d_{xy} only model is a decent approximation in this case. On the other hand, if the three t_{2g} orbitals are equally occupied, J_2, J_3 and J_4 are now comparable and these form a triangular lattice which would favour geometric frustration, suggesting that this a good approximation for Li_4MgReO_6 with its smaller tetragonal compression. The presence of positional disorder between Li^+ and Mg^{2+} combines with geometric frustration to provide the conditions for the spin glass ground state. One object lesson from the above exercise is that for low symmetry systems such as these, the actual geometry of the magnetic ion sites is less important than the relative strengths of the interconnecting superexchange pathways.

2.3.2 The Kagomé Lattice

The term, kagomé, originates from a traditional Japanese basket weave pattern but in fact the motif is used widely in Japanese culture. Currently, kagomé lattice materials also represent one of the most active research areas. This is due in part to the theoretical tractability of the two-dimensional kagomé lattice and also to expectations that exotic ground states should be found, especially with quantum, $S = \frac{1}{2}$ and $S = 1$ spins.[33,34]

We will refer to these materials as 2DKAF (two-dimensional kagomé antiferromagnets).

2.3.2.1 Ideal (Undistorted) Kagomé Lattices

These are fairly rare and there are only two families of materials which have been studied with any intensity. The first are the jarosites, a sulfate/hydroxide mineral class with composition, $AM_3(SO_4)_2(OH)_6$. The A cation can be more or less any large singly charged ion such as Na^+, K^+, Rb^+, Ag^+, Tl^+, $[NH_4]^+$ or $[H_3O]^+$. The majority of magnetic jarosites involve $M = Fe^{3+}$ but some are known for $M = Cr^{3+}$. The crystal structure, $R\overline{3}m$, is shown in Figure 2.14 where it can be seen clearly that the M-layers do form a geometrically perfect kagomé lattice. Jarosites are generally prepared using hydrothermal methods and are plagued by vacancies within the M layers due to protonation of the $[OH]^-$ sites. Thus, full coverage of the M lattice is rarely found unless special measures are taken. A synthetic breakthrough was reported recently by Grohol *et al.*[35] by which materials

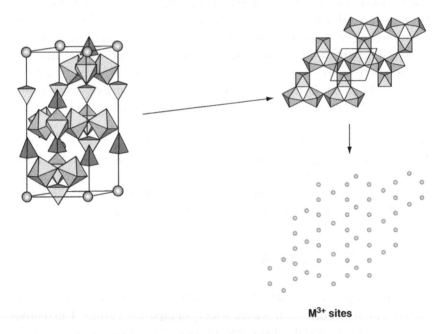

M³⁺ sites

Figure 2.14 The jarosite, $AM_3(SO_4)_2(OH)_6$, structure. The MO_6 units are shown as octahedra, the $[SO_4]^{2-}$ anions as tetrahedra and the A^+ cations as spheres. The H atoms associated with the OH^- anions are omitted. The top right shows one layer of MO_6 octahedra and the bottom right the perfect kagomé net of M^{3+} sites

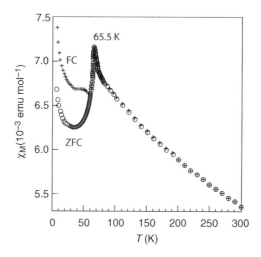

Figure 2.15 The DC magnetic susceptibility of $KFe_3(SO_4)_2(OH)_6$ with full occupation of the kagomé net by Fe^{3+}. Note the zero field cooled (ZFC)/field cooled (FC) divergence below $T_N = 65.5$ K.[35] Reprinted with permission from Grohol *et al.*, 2003 [35]. Copyright (2003) American Physical Society

with fully occupied M sites, M = Fe, were obtained routinely. A typical result is shown in Figure 2.15 and a summary is presented in Table 2.5. Note that in addition to the obvious susceptibility maximum, signalling T_N, a ZFC/FC divergence below T_N, noted T_D in the table is a common property of the jarosite family. The ordering temperatures, T_N, found for these stoichiometric jarosites are considerably higher than those reported on samples with large vacancy concentrations, as would be expected.[4]

The intraplanar spin structure of the iron jarosites is described by the so-called $\mathbf{k} = (0\,0)$ structure, Figure 2.16, left, rather than the $\mathbf{k} = (\sqrt{1/3}\ \sqrt{1/3})$ structure, right, in spite of both classical and quantum theories for the kagomé lattice which predict the latter.[36]

Perhaps the most interesting iron jarosite is that for $A = [H_3O]^+$ which is reported to show ~98% coverage of the M sites, even in earlier work.[37,38]

Table 2.5 Summary of results for Fe^{3+} jarosites, $AFe_3(SO_4)_2(OH)_6$[35]

A^+	θ_C (K)	T_N (K)	T_D (K)	f
Na	−825	61.7	~58	14
K	−828	65.4	~53	13
Rb	−829	64.4	~53	13
NH$_4$	−812	61.8	~53	13

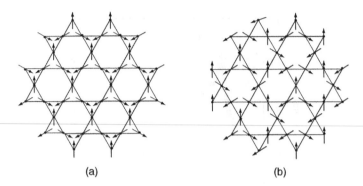

Figure 2.16 Two possible magnetic structures for the kagomé lattice. (a) $k = (0\,0)$. (b) $k = (\sqrt{3} \times \sqrt{3})$.[4] Reprinted with permission from Greedan, 2001 [4]. Copyright (2001) Royal Society of Chemistry

While there is an apparent magnetic anomaly near 13 K, neutron diffraction shows a lack of long range magnetic order down to 1.9 K. The 13 K anomaly is identified as a spin glass like freezing and has been investigated by muon spin relaxation methods.[39] Neutron diffraction at the lowest temperature on polycrystalline samples shows that the magnetic scattering has the asymmetric Warren line shape, characteristic of two-dimensional spin correlations, with a short correlation length of ~19 Å, Figure 2.17.[37, 38]

Remarkably, dilution of the Fe site with Al^{3+} results in long range order evidenced by resolution limited Bragg peaks.[40] This was interpreted, originally, as an 'order by disorder' process, based on the

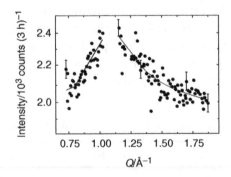

Figure 2.17 Diffuse magnetic scattering for $(D_3O)Fe_3(SO_4)_2(OH)_6$ at 1.5 K. The fit is to the Warren line shape function which describes two-dimensional spin correlations.[37] Reprinted with permission from Wills *et al.*, 1998 [37]. Copyright (1998) EDP Sciences

assumption that all jarosites required vacancies on the Fe sites to show order. As mentioned, recent work has shown this to be incorrect and a re-thinking of these doping results for the $A = [H_3O]^+$ material is perhaps in order.

The crystal chemistry of jarosites does extend to other M^{3+} ions, especially Cr^{3+} which is $S = 3/2$ compared with $S = 5/2$ for Fe^{3+}. Unfortunately, there are apparently no detailed studies of any stoichiometric Cr jarosite. $KCr_3(SO_4)_2(OH)_6$ with 76% Cr sites occupied is reported to show AFLRO only below 1.8 K with an ordered Cr^{3+} moment of only $\sim 1\mu_B$ which is 1/3 of the expected value.[41] With θ_C values ranging from -55 K to -70 K for this material, $f \sim 31$ to 39, indicating high levels of frustration. It should be noted that the values for T_N are much lower than those for Fe jarosites with similar site coverage, by more than an order of magnitude. More effort should be directed to the preparation and study of fully stoichiometric Cr jarosites. Unfortunately, jarosites with $S = \frac{1}{2}$ or $S = 1$ ions, such as Ti^{3+} or V^{3+}, are unknown.

The second family of undistorted kagomé lattice materials, the so-called paratacamites, $Zn_{4-x}Cu_x(OH)_6Cl_2$, especially the $x = 3$ phase, which is known as herbertsmithite, has been the object of intense study quite recently. The crystal structure ($R\bar{3}m$) is shown in Figure 2.18. Of course the magnetic ion in herbertsmithite, $Zn_3Cu(OH)_6Cl_2$, is Cu^{2+}, $S = \frac{1}{2}$, providing a quantum spin on the kagomé lattice. Initial reports that $\theta_C = -300$ K coupled with no sign of LRO down to 50 mK

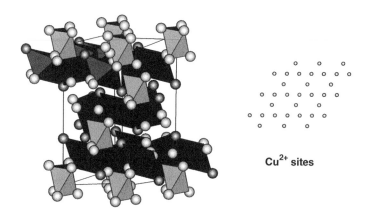

Figure 2.18 The structure of herbertsmithite, $ZnCu_3(OH)_6Cl_2$. The CuO_6 octahedra are shown as dark grey, the ZnO_6 octahedra as light grey. The Cl^- and O atoms as dark and light grey, respectively. The H atoms have been omitted. Note that the Cu containing layers are connected by edge-sharing ZnO_6 octahedra. The perfect kagomé net of Cu^{2+} ions is shown on the right

induced much enthusiasm, indicating perhaps that a spin liquid ground state had indeed been found.[42, 43]

Initially, it was thought that the Cu lattice was indeed a perfect kagomé with no disorder. This was a not unreasonable assumption, as, in solution or molecular chemistry Zn^{2+} and Cu^{2+} generally exhibit very different stereochemistries. Unfortunately, it quickly became apparent from a battery of studies, including NMR and neutron diffraction among others, that there was significant Zn–Cu intersite exchange of about 10%.[44–46] Nonetheless, spin dynamics consistent with a spin liquid-like ground state are reported. However, ^{17}O NMR data seem to indicate that the ground state may not be a spin singlet as would be expected for the RVB model. Heat capacity and neutron inelastic scattering data also support a spin liquid ground state without a true spin (singlet-triplet) gap.[45,46]

Very recently, a metastable polymorph with the same composition, $ZnCu_3(OH)_6Cl_2$, called kapellasite has been synthesised and characterised structurally and with preliminary magnetic measurements.[47,48] Kapellasite crystallises in $P\bar{3}m1$ and differs from herbertsmithite in that the connection between the Cu^{2+} kagomé planes involves only weak Cl–H–O hydrogen bonds, see Figure 2.19. In herbertsmithite, the interlayer connection involves edge-sharing ZnO_6 octahedra and Zn/Cu mixing compromises the two dimensionality of the system. Thus, it is argued that kapellasite is potentially a better approximation to the 2DKAF. Magnetic susceptibility data suggest a nonmagnetic ground state for this material as $T \rightarrow 0\,K$. Clearly, much more work is needed here.

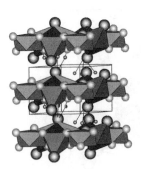

Figure 2.19 The structure of kapellasite, a second form of $ZnCu_3(OH)_6Cl_2$. The colouring scheme is the same as that for Figure 2.18 (herbertsmithite) except that the H atoms are shown as very small spheres. Note that the Zn and Cu octahedra are accommodated within the same layer by edge-sharing and that the layers are now connected only by Cl–H–O hydrogen bonds. The Cu sites form a perfect kagomé net as in Herbertsmithite

2.3.2.2 Distorted Kagomé Lattices

The mineral volborthite, $Cu_3V_2O_7(OH)_2 \cdot 2H_2O$, has long been considered a good approximation to the 2DKAF. The monoclinic crystal structure C2/m now lacks a threefold axis which requires some distortion of the kagomé net, Figure 2.20, with Cu–Cu distances 3.03 and 2.98 Å and angles of 62 and 58° but the layer is strictly planar.[49]

The relatively long interlayer separation of >7 Å aids in the realisation of 2DKAF behaviour. The first investigation of volborthite in the context of 2DKAF was in 2001 where $\theta_C = -115$ K was reported but no long range order was detected.[50] A broad maximum occurs in the DC susceptibility at ~ 20 K with a Curie tail extending to lower temperatures but no ZFC/FC divergence, Figure 2.21. Heat capacity and ^{51}V NMR found no sign of either AFLRO nor a spin singlet–triplet gap to 1.8 K, indicating $f > 64$. Measurements were extended to the mK regime using d.c. SQUID and ^{51}V NMR which found an anomaly below 1.4 K which was characterised as a spin freezing which, however, involves only $\sim 40\%$ of the Cu spins. Another 20% were thought to be involved in AFSRO of the $\sqrt{3} \times \sqrt{3}$ type.[51] Volborthite and herbertsmithite have been compared recently.[52]

A more recently reported kagomé model is the complex fluoride material, $Rb_2Cu_3SnF_{12}$, whose structure ($R\bar{3}$) is shown in Figure 2.22.[53] This material was discovered, apparently, as result of a search among compounds of similar composition, such as $Cs_2Cu_3ZrF_{12}$ which is isostructural at room temperature but undergoes structural phase transitions at lower temperatures. The Rb/Sn phase retains the $R\bar{3}$ structure to the lowest temperatures studied. Note that the kagomé layers are now slightly

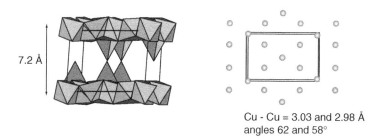

7.2 Å

Cu - Cu = 3.03 and 2.98 Å
angles 62 and 58°

Figure 2.20 The structure of volborthite, $Cu_3V_2O_7(OH)_2 \cdot 2H_2O$. The CuO_6 octahedra are shown in dark grey and the V tetrahedra in light grey. The waters of hydration are omitted as are the H atoms associated with the $[OH]^-$ anions. The Cu lattice is shown on the right and the distortions from a perfect kagomé topology are indicated

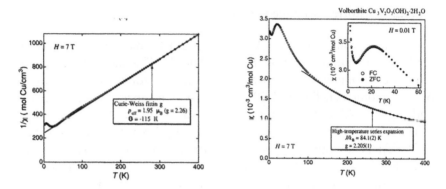

Figure 2.21 Some magnetic properties of volborthite. Left, the Curie–Weiss fit showing $\theta_c = -115$ K and right, the broad maximum at ~ 20 K.[50] Reprinted with permission from Hiroi et al., 2001 [50]. Copyright (2001) Physical Society of Japan

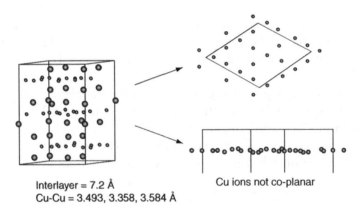

Interlayer = 7.2 Å
Cu-Cu = 3.493, 3.358, 3.584 Å

Cu ions not co-planar

Figure 2.22 The structure of $Rb_2Cu_3SnF_{12}$. The Rb^+ ions are shown as large grey spheres and the Cu^{2+} ions as small, dark grey spheres. The F^- and Sn^{4+} ions are omitted. The diagrams on the right show the Cu layers in projection (top) and a side view (bottom) which illustrates the nonco-planarity. The variations in Cu–Cu distances are given

nonco-planar with Cu-Cu distances (Å) of 3.493, 3.358 and 3.584 but with the angles also highly distorted from 60°. One result of these distortions is that there are now four nn exchange pathways to consider within the puckered planes, Figure 2.23. For this material the d.c. susceptibility data can be well understood in terms of a disordered spin singlet ground state with a spin gap, unlike volborthite or herbertsmithite.

A final distorted kagomé structure type is presented by the langasite family, with composition AGa_5SiO_{14} where $A =$ lanthanide such as Pr

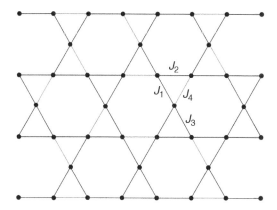

Figure 2.23 The four exchange pathways within the distorted kagomé layer in $Rb_2Cu_3SnF_{12}$.[53] Reprinted with permission from Morita *et al.*, 2008 [53]. Copyright (2008) Physical Society of Japan

or Nd. The langasite structure type, $Ca_3Ga_2Ge_3O_{14}$, is quite complex and can be described in terms of the various cation site coordinations as $A_3BC_3D_2O_{14}$, where A is eightfold square antiprismatic, B octahedral and C and D tetrahedral.[54] The A site can be occupied by lanthanide ions or Group II ions. The space group is $P321$, thus, the crystals are noncentric and piezoelectric and the materials have applications in for example surface acoustic wave (SAW) devices which has been the primary interest until recently.[55] It was noted in 2006 that the A cation net has some characteristics of the kagomé lattice, Figure 2.24, in which the triangles are perfect but the lattice is distorted from the ideal.[56]

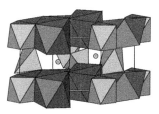
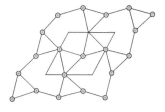

perfect triangle but distorted net

Nd-Nd = 4.192 Å (intraplanar)
Nd-Nd = 5.062 Å (interplanar)

Figure 2.24 The langesite structure, $Nd(Pr)_3Ga_5SiO_{14}$. The Nd(Pr) square antiprismatic sites are dark grey, the Ga octahedral and tetrahedral sites are light grey. The Si atom is a small sphere. The distorted kagomé net of Nd(Pr) ions is shown on the right

Initial susceptibility data on single crystals of the Nd compound indicated a large, negative θ_C value, -52 K, that a strong uni-axial magneto-crystalline anisotropy was present and that no sign of magnetic order was found to 1.6 K.[56] Now, θ_C values for most lanthanide insulators contain contributions from exchange, dipolar and so-called crystal field effects, with the latter generally dominant. It is thus important to attempt to separate these contributions by, for example, measuring magnetically very dilute samples to isolate the single ion crystal field component. Nonetheless, initially, the Nd langasite was regarded as highly frustrated. Inelastic neutron scattering data on the Nd material were interpreted as arising from a magnetically disordered state with strong correlations such as expected from a spin liquid.[57] Subsequent spin dynamics measurements using muon spin relaxation and [69, 71]Ga NMR indicated rapidly fluctuating spins down to 60 mK, supporting the spin liquid model.[58] However, a more recent set of studies pointed out several inconsistencies among the data obtained by various probes and through analysis of single ion effects arising from the crystal field manifold structure, showed that much of the spin dynamics was not due to frustration, at least for the Nd phase.[59] That is, single-ion effects appear to mask any evidence of magnetic frustration within the temperature range covered in most of the experiments so far. There has been relatively less work reported on the Pr langasite. This phase shows a very small $\theta_C = -2.6$ K, in contrast to the Nd analogue, but still no AFLRO down to 35 mK. Nonetheless, there is a broad maximum in the heat capacity near 6 K but a T^2 law is found at low temperatures. Such a power law is often taken as arising from strong two-dimensional spin correlations. Spin excitation spectra are also consistent with a highly degenerate ground state but, surprisingly, no magnetic diffuse scattering has been observed.[60] It was speculated that the ground state could be quite exotic, some type of spin nematic state. This system merits deeper investigation.

2.3.2.3 Three-Dimensional Kagomé-Like Lattices

Often, the modifier, kagomé, is used as either a prefix or a suffix to indicate a lattice connectivity involving the corner-sharing of triangles even when the lattice involves three spatial dimensions. Thus, one sees 'kagomé-staircase' and 'hyper-kagomé' applied to magnetic materials. In other cases the corner-sharing triangular motif is described by the mineral name of the structure type, such as the garnets. All of these will be discussed in the following paragraphs.

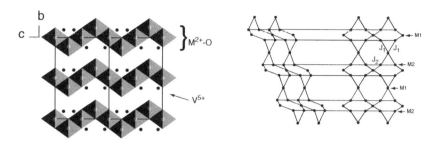

Figure 2.25 The structure of the kagomé staircase materials $Ni_3V_2O_8$ and $Co_3V_2O_8$ showing zigzag layers edge-shared octahedra of M^{2+} ions. (Left) The Co or Ni sites. (Right) The two crystallographically different cation sites, M1 and M2, are indicated.[62] Reprinted with permission from Rogado *et al.*, 2002 [62]. Copyright (2002) Elsevier Ltd

The kagomé-staircase materials are the vanadates $Ni_3V_2O_8$ and $Co_3V_2O_8$, first described by Sauerbrie *et al.* in *Cmca*.[61] The magnetic lattice is shown in Figure 2.25 which resembles a Dali-esque kagomé lattice arrayed on an apparent staircase.[62] For both materials the triangles are isosceles with distances (Å) of 2.94 and 2.97 for the Ni case. The spin states are $S = 1$ (Ni) and 3/2 (Co). It is debatable whether these materials are truly frustrated in the conventional sense as $\theta_C = -30\,K$ (Ni) and even $+14\,K$ (Co). The main fascination with these compounds is the remarkably complex magnetic ordering process which involves for the Ni phase, four apparent phase transitions as the temperature is decreased and at least two for the Co material, Figure 2.26. These have been studied in detail and the resulting phase diagram is indeed complex with incommensurate phases giving way to commensurate phases as T decreases.[63,64] A full understanding requires competing nn and nnn interactions along with anisotropy terms and it is clear that the frustrated magnetic site lattice topology plays a key role. Very recently, a coupling between the magnetic and dielectric properties has been reported.[65]

To date there is only one material which has been called a hyper-kagomé, namely the unusual oxide, $Na_4Ir_3O_8$ which involves the $S = \frac{1}{2}$ ion Ir^{4+} which has the low spin configuration, t_{2g}^5. From a structural point of view, this compound could be discussed with the spinels but it is also convenient to do so here. The reported space group(s) are the noncentric and chiral pair $P4_132$ and $P4_332$. Na^+ ions occupy two types of octahedral sites, differing from the classic spinel, where tetrahedral sites are occupied. Three Na^+ ions per formula unit reside solely on one such site, while the remaining Na^+ shares the other

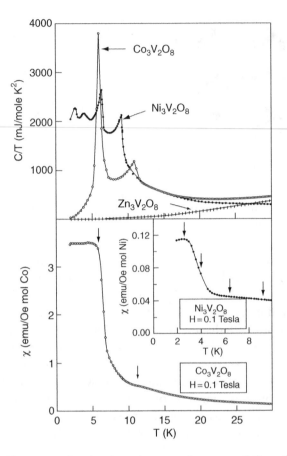

Figure 2.26 Heat capacity (top) and magnetic susceptibility (bottom) for $Ni_3V_2O_8$ and $Co_3V_2O_8$. Note the four phase transitions for the Ni material.[63] Reprinted with permission from Rogado *et al.*, 2002 [62]. Copyright (2002) Elsevier Ltd

octahedral site in an ordered manner with the three Ir^{4+} ions. This site ordering leads to a three-dimensional corner-sharing triangular magnetic lattice, as shown in Figure 2.27. There is little experimental work published on this compound but the measured $\theta_C = -650\,K$ and there is no sign of magnetic order down to 1.8 K, giving $f > 360$.[66] The heat capacity shows a broad anomaly near 30 K but no sharp feature nor does the magnetic susceptibility. At low temperatures the heat capacity follows a T^2 law. These results along with the invariance of the heat capacity with applied fields up to 12 T have lead to speculation that the ground state is spin liquid like. Recent theoretical work has cast some doubt on this interpretation.[67] It was shown that when spin orbit coupling

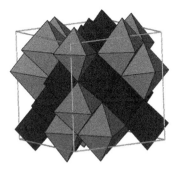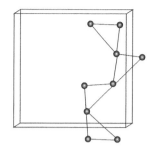

Figure 2.27 The structure of the so-called 'hyper-kagomé' material, $Na_4Ir_3O_8$. The magnetic Ir^{4+} (t_{2g}^5) sites are shown as dark grey octahedra and the Na sites as light grey octahedra. Part of the Ir^{4+} lattice, illustrating the corner-sharing triangular connectivity, is shown on the right

and the allowed Dzyaloshinskii–Moriya (DM) interactions are taken into account, the Ir – O – Ir superexchange is actually nonfrustrated and a spin liquid state is unlikely. For other sets of parameters, a spin liquid state could be recovered. Clearly, studies on this new material are in their infancy and much more work is needed to clarify the true situation.

Finally, the garnets can be discussed here as the connectivity of the magnetic lattice is indeed based on corner-sharing triangles in three dimensions. The garnet composition is $A_3B_2C_3O_{12}$, and the site coordinations are eightfold dodecaheral for A, octahedral for B and tetrahedral for C. While garnet materials have been studied for decades, the compounds of interest here are those for which only the A-site is magnetic and in fact one material dominates, $Gd_3Ga_5O_{12}$ or GGG as it is often denoted. At one time GGG was used as a substrate material for the so-called bubble memory technology and high quality single crystal samples are readily available. The A-site lattice is shown in Figure 2.28. There exist two interpenetrating corner-sharing triangular lattices.

Wolf *et al.* were the first to realise the frustrated nature of this material, noting that while $\theta_C = -2.3$ K, no long range order was observed down to 0.35 K.[68,69] This was considered surprising as gallium garnets with A = Dy, Nd, Sm and Er order just below 1 K.[70, 71] In zero applied field, the heat capacity of GGG shows a broad maximum at about 0.8 K, Figure 2.29 which could be reproduced using three exchange constants, $J_1(nn) = -0.107$, $J_2(nnn) = 0.003$ K and $J_3(nnnn) = -0.010$ K. Soon, reports of long range order induced by applied fields appeared[72] and eventually, a phase diagram was deduced, Figure 2.30, in which the AFLRO phase is seen exist below 0.4 K within a range of fields.[73] Also, at low fields, a spin freezing was found with $T_f \sim 0.15$ K.[74]

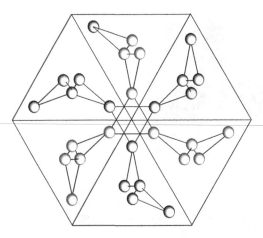

Figure 2.28 The Gd sites in GGG, $Gd_3Ga_5O_{12}$ projected along the (111) axis. There are two interpenetrating Gd lattices, each involving corner-sharing triangles shown as dark and light grey spheres.[75] Reprinted with permission from Petrenko *et al.*, 1998 [75]. Copyright (1998) American Physical Society

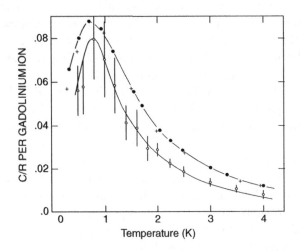

Figure 2.29 Heat capacity of GGG in zero applied field. The solid line and points are the experimental results. The solid line and open circles represent a fit using three exchange constants (see text).[69] Reprinted with permission from Kinney and Wolf, 1979 [69]. Copyright (1979) American Institute of Physics

The first neutron diffraction measurements, using a sample enriched with non-absorbing ^{160}Gd, showed a remarkable result, Figure 2.31, in that relatively sharp Bragg peaks of magnetic origin appeared below the reported T_f from susceptibility measurements which are superimposed on

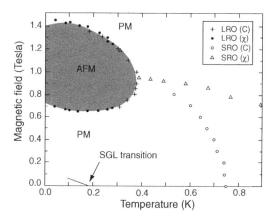

Figure 2.30 The magnetic phase diagram for GGG showing the field induced AFLRO state.[73] Reprinted with permission from Schiffer *et al.*, 1994 [73]. Copyright (1994) American Physical Society

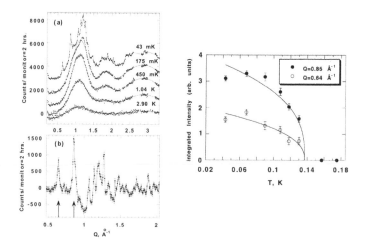

Figure 2.31 Neutron diffraction data for GGG using ^{160}Gd. Note the appearance of sharp Bragg peaks between 175 mK and 43 mK (top). The bottom panel shows the result of subtracting the 175 mK data from 43 mK. The temperature dependence of two strong Bragg peaks, (right) gives $T_c \sim 140$ mK.[75] Reprinted with permission from Petrenko *et al.*, 1998 [75]. Copyright (1998) American Physical Society

a diffuse background.[75] The magnetic peaks are incommensurate with the chemical lattice and their width exceeds the resolution limit of the instrument, indicating that the order is short range with an estimated correlation length of ~ 100 Å. From integrated intensities it was suggested that $\sim$85% of the Gd spins are in a liquid-like state at 43 mK. There is still no consensus regarding the reason for the coexistence of the two components to the ground state of GGG. One suggestion is that lattice defects serve to nucleate the AFSRO domains. Direct studies of spin dynamics using μSR show no evidence for spin freezing to 25 mK, *i.e.* the spins remain dynamic with a finite fluctuation rate.[76, 77] Neutron diffraction in applied fields shows remarkable behaviour with the appearance of both F and AF type reflections.[78] This is clearly a complex system which is still not well understood. A very recent paper suggests a reconciliation of some of the conflicting data and concepts.[79]

2.3.3   The Face-Centred Cubic Lattice

2.3.3.1   B-Site Ordered Double Perovskites

The face-centred cubic (f.c.c.) lattice is one of the four archetypal frustrated lattices and has received the least attention. The most important class of oxide materials here are the *B*-site ordered double perovskites, $A_2BB'O_6$, which have been of recent interest in other contexts, such as colossal magneto-resistance materials, *e.g.* Sr_2FeMoO_6, and related compounds. The structural relationships between the standard perovskite structure and the *B*-site ordered double perovskites is displayed in Figure 2.32. Both the *B* and *B'* sites form f.c.c. lattices. When only one is magnetic, geometric frustration results. Criteria for the expectation of *B*-site ordering in perovskites have been presented some time ago, and are shown in Figure 2.33 in terms of a phase diagram with axes expressing the differences in formal charge and ionic radius.[80]

It should be emphasised that these are only guidelines and an actual determination of the extent of site mixing is needed for each specific case. The literature on this class of perovskites is considerable and the review here will be selective rather than exhaustive.

First, as with the basic perovskites, the *B*-site ordered systems can adopt a variety of space groups which can be understood in terms of coordinated rotations of the octahedra.[81] A group-subgroup 'family tree' is shown in Figure 2.34. Of the several possible space groups, three are particularly common, *Fm3m*, *I4/m* and *P2$_1$/n* with a few reports of $I\bar{1}$ (the nonstandard settings are usually adopted in the latter cases to conform to the commonly recognised setting for the perovskite

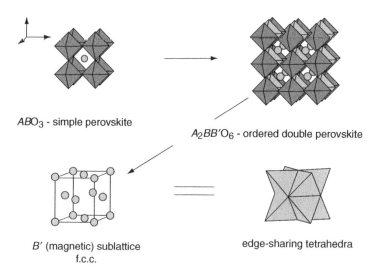

Figure 2.32 The relationship between the simple perovskite and *B*-site ordered double perovskite structures. The *B'* magnetic lattice is f.c.c.

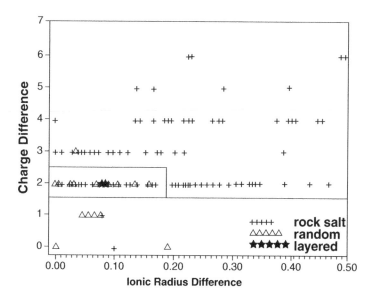

Figure 2.33 A phase diagram indicating the range of stability of *B*-site ordered double perovskites, denoted as +++++ rock salt, in terms of differences in ionic radii and formal charge.[80] Reprinted with permission from Anderson *et al.*, 1993 [80]. Copyright (1993) Elsevier

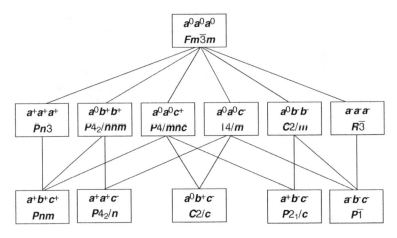

Figure 2.34 Group theory 'family tree' for B-site ordered double perovskites.[81] Reprinted with permission from Howard *et al.*, 2003 [81]. Copyright (2003) IUCr

structure). One expects the descent in symmetry to be a function of a modified Goldschmidt tolerance factor,

$$t = (A - O)/\sqrt{2}(<B, B'-O>)$$

where the denominator is the average between the B–O and B'–O distances. That is, for $t \sim 1$ the $Fm3m$ structure (also known as the elpasolite structure) is anticipated, while, as t decreases, $I4/m$ and then $P2_1/n$ are often found. The perovskite structure can accommodate a wide variety of ions for the A and B sites which allows systematic study of the roles of spin quantum number, S, and space group symmetry on the magnetic properties. For example S can range from 7/2 to 1/2, *i.e.* from the classical to the quantum limit and the range of space groups has already been detailed. Of course the space group determines the local site symmetry and together with the chemical bonding with the oxide ligands, the crystal field acting on the magnetic ion. For heavier elements, spin-orbit coupling will also be an important factor. All magnetic interactions in these materials occur by a super-super exchange pathway(SSE), B'–O–B–O–B', and the choice of the A cation determines the angles involved. Clearly, there is much scope for design of materials to test hypotheses. Selected data are presented in Table 2.6. Some comments are in order. Of course $S = 5/2$ examples come from the $3d$ series and these appear to be only marginally frustrated using the f criterion.

Table 2.6 Some selected magnetic data for $A_2BB'O_6$ perovskites

S	Space group	Compound	θ_C (K)	$T_{c,f}$ (K)	f	B' site symm.	Grd. state
5/2	$Fm3m$	Ba_2MnWO_6	−64	9	7	$m3m$	AFLRO
5/2	$P2_1/n$	Sr_2MnWO_6	−30	10	3	$\overline{1}$	AFLRO
3/2	$Fm3m$	Ba_2YRuO_6	−571	36	16	$m3m$	AFLRO
3/2	$P2_1/n$	La_2LiRuO_6	−170	30	6	$\overline{1}$	AFLRO
1	$Fm3m$	Ba_2YReO_6	−480	40	12	$m3m$	SG
1	$I4/m$	Sr_2NiWO_6	−175	54	3	$4/m$	AFLRO
1/2	$Fm3m$	Ba_2LiOsO_6	−40	8	5	$m3m$	AFLRO
1/2	$Fm3m$	Ba_2NaOsO_6	−10	6.8	1.5	$m3m$	FLRO
1/2	$Fm3m$	Ba_2YMoO_6	−91	2<	>45	$m3m$	?
1/2	$I4/m$	Sr_2MgReO_6	−426	45	9	$4/m$	SG
1/2	$P2_1/n$	Sr_2CaReO_6	−443	14	32	$\overline{1}$	SG

Note that θ_C is larger by a factor of two for cubic Ba_2MnWO_6[82] relative to monoclinic Sr_2MnWO_6.[83, 84] which reflects the more favourable 180° SSE of the former compared with angles of 162° for the latter. Nonetheless, neutron diffraction data show evidence for frustration in the form of diffuse magnetic scattering which persists well above T_c, particularly for Ba_2MnWO_6.[82]

When a $4d$ or $5d$ element is the magnetic ion, θ can be very large values, near −600 K in some cases, which reflects the greater spatial extent of these orbitals relative to those for 3d ions. Note also, that as S decreases, the role of frustration increases. The $S = 3/2$ materials both show much higher f values than the isostructural $S = 5/2$ compounds.[85, 86] Thus far, materials with $S \geq 3/2$ always show AFLRO. This changes considerably for quantum spins. For $S = 1$ Sr_2NiWO_6 shows AFLRO[83] but Ba_2YReO_6[87] does not order. The $S = 1/2$ double perovskites show a remarkable range of properties. The lower symmetry compounds, Sr_2CaReO_6 and Sr_2MgReO_6, do not order to 2 K.[88, 89] Curiously, both show a spin frozen ground state in spite of the apparent lack of significant levels of positional disorder. The cubic phases, Ba_2LiOsO_6, Ba_2NaOsO_6 and Ba_2YMoO_6 could not differ more. Ba_2LiOsO_6 is weakly frustrated and shows AFLRO.[90] Ba_2NaOsO_6[90, 91] is quite unusual, showing nearly no frustration, $f \sim 1$, and FLRO, in spite of the negative θ_C. Both the effective and ordered moments are very small compared with the values expected for spin only $S = \frac{1}{2}$, ~0.6 μ_B vs 1.73 μ_B and 0.2 μ_B vs 1 μ_B, respectively. The large single ion spin orbit coupling expected for the 5d^1 configuration of Os^{7+} is implicated here. There have been two efforts to understand the low moments and the ferromagnetism on theoretical grounds.[92, 93] Finally, Ba_2YMoO_6 appears to be strongly frustrated,

$f > 45$, and shows no LRO down to 2 K.[94] Clearly the magnetic ground state in these $S = 1/2$ materials is sensitive to a large number of factors such as exchange, spin orbit coupling, orbital ordering and others. It is anticipated that a more detailed study of this class of double perovskites will yield interesting new insights and results.

2.3.4 The Pyrochlores and Spinels

2.3.4.1 Pyrochlores

This is by far the largest class of GFMM studied to date. As a major review of the pyrochlore oxides has been published very recently,[6] the following discussion will again be highly selective. First, the pyrochlore composition is $A_2B_2O_7$ where A is usually a trivalent lanthanide and B a tetravalent transition element. There is somewhat less scope for substitutional chemistry, relative to the double perovskites for example, but fortunately, a large number of pyrochlore oxides exist as shown in Figure 2.35. Earlier reviews list more examples.[95, 96]

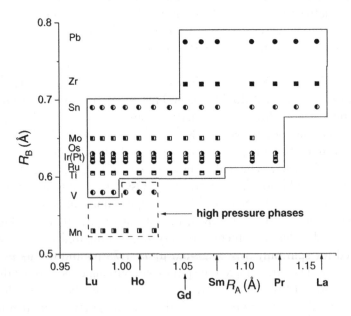

Figure 2.35 A 'structure field' stability map for selected pyrochlore oxides

Before proceeding further, it is instructive to recall that the pyrochlore structure is an ordered variant of the cubic fluorite structure. For oxides, the fluorite formula is MO_2 and the space group is $Fm3m$, so the composition of one unit cell is M_4O_8 and the unit cell constant is ~ 5 Å. The transformation to pyrochlore involves replacing the single M^{4+} ion with two ions of quite different radius and charge, A^{3+} and B^{4+}. Each occupies half of the M^{4+} f.c.c. lattice in an ordered manner. Oxide vacancies are also created for charge balance and these order along with the remaining oxide ions to accommodate the different coordination demands of the larger A^{3+} and smaller B^{4+} ions. As a result of the cation, anion and vacancy orderings, the pyrochore unit cell is doubled relative to fluorite in all three directions, giving a ~ 10 Å and a unit cell composition of $A_{16}B_{16}O_{56}\square_8$ ($\square$ denotes a vacant oxide site).

Since the pyrochlore unit cell contains 88 atoms, a full diagram, showing all of the atoms, is not always informative. Instead Figure 2.36 displays the coordination polyhedra of the A and B sites, separately. Note that the B-site is octahedrally coordinated and the octahedra share corners, resulting in a rather rigid three-dimensional lattice.

The crystallographic description for pyrochlores is given in Table 2.7. The space group is $Fd3m$ with A in $16d$, B in $16c$ and the two oxide ions in $48f$ and $8b$ and the formula can be rewritten as $A_2B_2O_6O'$. As the International Tables give two origins for this space group, there exist four possible settings for the structure and all have been used in the earlier

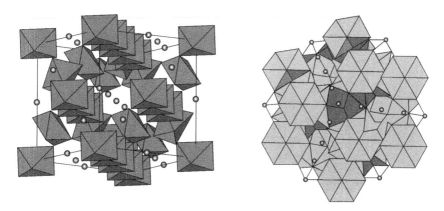

Figure 2.36 (Left) The corner-sharing octahedral sublattice of the B-site ions (BO_6) in the pyrochlore structure. The A ions are shown as light spheres. (Right) The A-site hexagonal bipyramidal polyhedra (AO_8) in the pyrochlore structure. The B-site ions are shown as light spheres

Table 2.7 Crystallographic description of the pyrochlore structure, $A_2B_2O_6O'$ in space group (No. 227) with origin at 16c

Atom	Wyckoff position	Point symmetry	Minimal coordinates
A	16d	$\overline{3}m(D_{3d})$	1/2, 1/2, 1/2
B	16c	$\overline{3}m(D_{3d})$	0,0,0
O	48f	$mm(C_{2v})$	x, 1/8, 1/8
O'	8b	$\overline{4}3m(T_d)$	3/8, 3/8, 3/8

literature but that given in Table 2.7 with 16c at the origin is now taken as standard.

There is only one adjustable structural parameter, associated with the O(48f) site, which is generally found within the narrow range 0.31 to 0.33. This parameter controls the B–O–B angle between corner-sharing BO_6 octahedra. The main issue here is the topology of the A and B sites both of which consist of a three-dimensional network of corner-sharing tetrahedra, Figure 2.37, and thus, geometric frustration is associated with both sites. As well, the local coordination of the A and B sites is relevant for the magnetism. While the B site is a nearly perfect octahedron, with six equivalent B–O distances and a small trigonal compression, the A site is eightfold and is best described as an hexagonal bipyramid, $AO_6O'_2$. The A-site coordination polyhedron is remarkable and is shown in Figure 2.38, where the O_6 ring forms a chair conformation and the two A–O' bonds are normal to the average plane of the O_6 ring. These A–O' bonds are among the shortest ($\sim$2.2 Å, where the sum of the normal A^{3+} and O^{2-} radii are $\sim$2.4–2.5 Å) known in lanthanide oxide chemistry and impart a strong axial symmetry to the crystal field at the A site. The A–O' bonds are directed along <111> directions within the unit

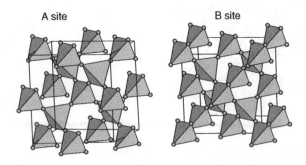

Figure 2.37 The topology of the A (16d) and B (16c) sites in the pyrochlore structure. Both form a three-dimensional array of corner-sharing tetrahedra

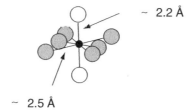

Figure 2.38 The coordination polyhedron (hexagonal bipyramid) for the A site in pyrochlore. The A ion is the small, black sphere, the O ions are the grey spheres and the O′ ions are the white spheres

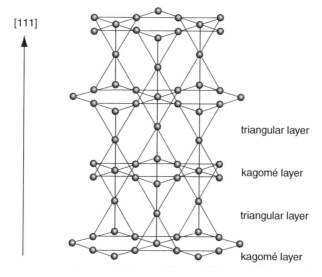

Figure 2.39 A view of the B (or A) sites in the pyrochlore lattice as a stacking of alternating kagomé and triangular planar layers normal to the <111> direction.[6] Reprinted with permission from Gardner *et al.*, 2010 [6]. Copyright (2010) American Physical Society

cell. With respect to each tetrahedron, the <111> directions connect apices to the midpoint of the opposite triangular face. Finally, the $16c$ or $16d$ sites can be viewed as a stacking of alternating kagomé and triangular planar layers along <111> directions, Figure 2.39.

The discussion will deal with three categories of magnetic pyrochlores, A-site only magnetic, B-site only magnetic and both A and B sites magnetic.

(a) A-site magnetic For this class of magnetic materials in which the magnetic ion is a trivalent lanthanide, the $4f$ electrons are very well

localised which implies that insulators only are found, orbital angular momentum is unquenched, indeed, J is a good quantum number, that superexchange interactions will be very weak, on the order of 10^{-1} to 1 K and that magnetic dipolar interactions can compete with superexchange. As orbital angular momentum is prominent, interactions with the local ligand environment – the crystal field – will lead to strong magnetocrystalline anisotropy, *i.e.* the moments will be pinned along specific crystallographic directions.

By far the greatest volume of work has been reported for the rare earth titanate pyrochlores, *i.e.* for $B = Ti^{4+}$, but there has been more recent interest in stannate phases as well. This is due in part to the relative ease in the growth of large, high quality single crystals of most members of the titanate series. This series begins with $A = Sm$ but this ion is a strong neutron absorber and the single ion properties are notoriously complex.

In addition the $A = Eu$ and Tm materials both show magnetic singlet ground states at low temperature, *i.e.* are not magnetic in the temperature range of interest. Table 2.8 lists what is known to date. The range of properties and ground states shows remarkable variety as one sees complex AFLRO, spin liquid and spin ice ground states represented.

Gd^{3+} with $4f^7$ is a half-filled $4f$ shell and therefore an S-state ion, $^8S_{7/2}$, where the crystal field will be largely irrelevant and this is born out in the isotropic behaviour. As already emphasised, when dealing with the lanthanides, interpretation of the θ_C value requires caution due to the fact that there are three potential contributions from exchange, dipolar interactions and the crystal field. Dilution experiments are the best way to separate the collective effects, exchange and dipolar, from the crystal field. For example, $\theta_C = -9.6$ K for pure $Gd_2Ti_2O_7$ while a nearly zero value of -0.9 K is found for $(Gd_{0.02}Y_{0.98})_2Ti_2O_7$.[97] On the other hand, for $Tb_2Ti_2O_7$ the

Table 2.8   Selected magnetic properties of the $A_2Ti_2O_7$ pyrochlores

A	θ_C (K)[a]	T_c (K)	Anisotropy[b]	Ground state
Gd	−10	1.0, 0.7	Isotropic	Complex AFLRO
Tb	−19	–	Easy axis	Spin liquid
Dy	1.2	–	Easy axis	Spin ice
Ho	1.9	–	Easy axis	Spin ice
Er	−24	1.2	Easy plane	Complex AFLRO
Yb	0.71	–	Easy plane	Spin liquid?

[a] Caution should be taken to separate crystal field and dipolar contributions from exchange, see text.
[b] This is with respect to the <111> axis directions.

pure sample shows $\theta_C = -19\,K$ while for $(Tb_{0.02}Y_{0.98})_2Ti_2O_7$ one finds $\theta_C = -6\,K$ which is the crystal field contribution, leaving $-13\,K$ for exchange plus dipolar.[19] As the Gd titanate orders only below $1\,K$ and Tb titanate does not order at all, both clearly qualify as highly frustrated systems.

Returning to the Gd case, reports from heat capacity experiments both in zero and applied magnetic fields, indicated a complex low temperature behaviour with at least two transitions in zero field, at $\sim 0.9\,K$ and $\sim 0.6\,K$.[98] Neutron diffraction data, taken on a $^{160}Gd_2Ti_2O_7$ sample disclosed a very complex magnetic ordering, with wave vector $\mathbf{k} = (1/2\ 1/2\ 1/2)$, Figure 2.40.[99, 100] Solution of the magnetic structure required an examination of both the Bragg magnetic peaks and the magnetic diffuse scattering. This is the structure between $1\,K$ and $0.7\,K$ and for each tetrahedron only three of the four spins are ordered. As the temperature is reduced to as low as $50\,mK$, still only $\sim 27\%$ of these remaining spins actually order, so there is still partial spin dynamics to the lowest temperatures. This magnetic structure is not the one calculated on a model of isotropic classical spins on a pyrochlore lattice with dipolar interactions.[101]

With such a delicate zero field ground state, it is perhaps not surprising that application of even modest magnetic fields results in a complex phase diagram.[98, 102] See Gardner, Gingras and Greedan for more details.[6]

$Tb_2Ti_2O_7$ is generally regarded as a robust example of a spin liquid. The major elements for this characterisation are the absence of evidence

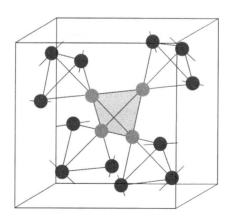

Figure 2.40 The '4-k' magnetic structure found for $Gd_2Ti_2O_7$. The dark spheres show sites which order below $0.9\,K$. The light spheres are the sites which order only partially below the second transition at $0.7\,K$.[100] Reprinted with permission from Stewart *et al.*, 2004 [100]. Copyright (2004) IOP Publishing Ltd

of LRO from any probe – neutrons, specific heat, μSR, *etc.* down to very low temperatures, *e.g.* 50 mK and direct evidence for considerable spin dynamics down to 17 mK.[103,104] Also, elastic neutron scattering on single crystals shows a 'chequerboard' pattern which is consistent with spin–spin correlations on a nearest neighbour, 3.5 Å, length scale. Nonetheless, in spite of intense efforts, a detailed understanding of the origins of this spin liquid behaviour does not yet exist. The system appears to be more Ising- than Heisenberg-like and theory would predict ordering, as is actually found for isostructural $Tb_2Sn_2O_7$ to be discussed later. The experimental situation has become more confused with the observation of structural peak broadening using high resolution X-ray scattering[105] and recent theory suggests that the low lying excited crystal field state at 18 K above the ground state may play a major role in influencing the spin dynamics.[106] A remarkable feature of $Tb_2Ti_2O_7$ is that AFLRO can be induced by the application of pressure, Figure 2.41.[107]

The spin ice phases, $Dy_2Ti_2O_7$ and $Ho_2Ti_2O_7$, have received considerable attention. As mentioned in Section 2.2.4. these materials are called spin ices due to the analogy with water ice, in particular the experimental observation of the excess heat capacity as shown in Figure 2.42. Here it is important to note that both materials show a positive θ_C indicative of net F interactions and that the anisotropy is of the easy axis type, the two necessary requirements for the formation of the spin ice state. Interestingly, it has been shown that the origin of the F interaction is actually the magnetic dipolar interaction, not superexchange.[108] The former depends on distance as r^{-3} and is thus of very long range. It is thus paradoxical that a nn exchange constraint results from a long range interaction. An attempt at paradox resolution has been made.[109] Further development of these ideas has led to the remakable suggestion that magnetic monopoles are involved and that this very elusive particle might actually be detectable in spin ice materials![110] Very recent experiments claim the observation of magnetic monopoles using diffuse neutron scattering in both $Dy_2Ti_2O_7$ and $Ho_2Ti_2O_7$.[111, 112]

For both ions the ground crystal field state is well isolated from excited states, by >250 K for Ho and >100 K for Dy.[113, 114] In addition the ground states are nearly pure $|+/-M_J(max)>$, $|+/-8>$ for Ho and $|+/-15/2>$ for Dy, which is consistent with the strong axial anisotropy and rather different from the Tb material where the ground state is not pure $|+/-M_J(max)>$ and the lowest excited state is only 18 K above.

The spin ices undergo a type of spin flop transition in applied magnetic fields in which the '2 in 2 out' ground state is transformed to a '3 in 1 out'

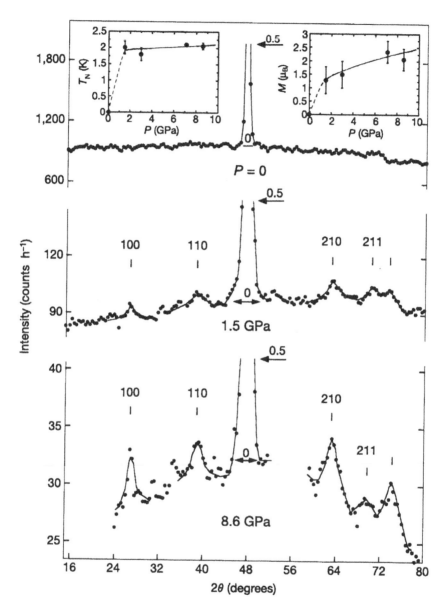

Figure 2.41 Long range order induced in the spin liquid $Tb_2Ti_2O_7$ by application of hydrostatic pressure. New Bragg peaks appear in the neutron diffraction pattern at 1.4 K with increasing pressure.[107] Reprinted with permission from Mirabeau *et al.*, 2002 [107]. Copyright (2002) Macmillan Publishers Ltd

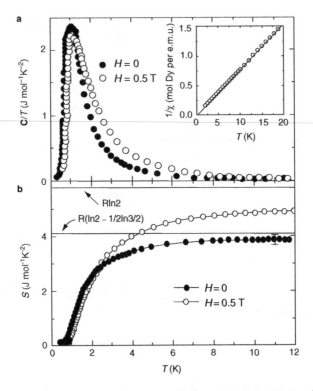

Figure 2.42 Heat capacity (a) and entropy (b) for spin ice $Dy_2Ti_2O_7$ showing the agreement with the Pauling prediction of $S = R(\ln2 - 1/2\ln3/2)$.[22] Reprinted with permission from Ramirez *et al.*, 1999 [22]. Copyright (1999) Macmillan Publishers Ltd

state at rather modest fields of 1 T applied along <111>. Interestingly, application of pressure in the GPa range to spin ices does not induce long range magnetic order unlike the case for the spin liquid, $Tb_2Ti_2O_7$.[115]

At this stage the so-called 'stuffed spin ice' materials are relevant. These are pyrochlores with excess *A* ions substituted on the *B* sites, such as $Ho_{2+x}Ti_{2-x}O_{7-x/2}$.[116] These phases can be regarded as intermediates between the pyrochlore and C-type rare earth oxide structures. As as result of the substitution the rare earth sublattice will contain some edge sharing tetrahedra as the 16*d* and 16*c* sites taken together form a f.c.c. lattice. From the perspective of magnetic frustration, it is remarkable that the spin ice behaviour of the parent pyrochlores such as $Ho_2Ti_2O_7$ is not destroyed.[117, 118]

$Er_2Ti_2O_7$ is one of two easy plane titanates and with a large negative θ, it is not a spin ice candidate. Very early heat capacity data indicated LRO at ~1.2 K.[119] A comprehensive study by Champion *et al.*[120]

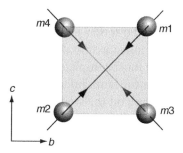

Figure 2.43   The spin configuration for $Er_2Ti_2O_7$ found from a spherical neutron polarimetry study. The moments point along the edges of each tetrahedron.[121] Reprinted with permission from Poole *et al.*, 2007 [121]. Copyright (2007) IOP Publishing Ltd

confirmed LRO *via* neutron diffraction and found a crystal field spectrum with a ground state consistent with an easy plane and separated by ~75 K from excited states. Solution of the magnetic structure, Figure 2.43, was difficult and required application of the relatively new technique of spherical neutron polarimetry.[121]

The remaining titanate, $Yb_2Ti_2O_7$ has proved to be enigmatic. While the earliest reports suggested LRO below 0.2 K from a sharp heat capacity maximum,[117] recent studies of spin dynamics using both Mössbauer and μSR methods disclosed instead a discontinuous change in the spin fluctuation rate at that temperature, Figure 2.44.[120] Subsequent neutron

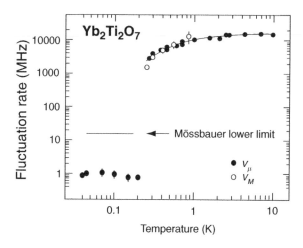

Figure 2.44   The remarkable first-order like collapse of the spin fluctuation rate in $Yb_2Ti_2O_7$.[122] Reprinted with permission from Bonville *et al.*, 2002 [122]. Copyright (2002) American Physical Society

diffraction studies seemed to indicate ferromagnetism[123] but polarised neutron diffraction data ruled out that possibility.[124] The current picture is one of a spin liquid as significant spin dynamics are seen below 0.2 K.

The $A_2Sn_2O_7$ series is the most complete among pyrochlore oxides as A extends from La to Lu, the entire lanthanide series. While single crystals are difficult to grow, the literature on the stannate pyrochlores has expanded rapidly. In general the stannates form with a slightly larger unit cell constant that the corresponding titanates. For example a for $Gd_2Sn_2O_7$ is 10.460 Å compared with 10.185 Å for $Gd_2Ti_2O_7$, an increase of about 3%. Table 2.9 shows relevant magnetic data for this series and there are many surprises when compared with the isostructural titanates.

The $A = Pr$ phase, which does not exist for the $B = Ti$ series has been assigned a spin ice ground state from studies using a variety of probes such as a.c. susceptibility, neutron scattering and heat capacity.[125–127] Note the easy axis anisotropy and slightly positive θ_C which match with the canonical spin ices for the heavy rare earths, $A = Dy$ and Ho. However, some unusual very low temperature spin dynamics suggest unique properties for this new spin ice.

$Nd_2Sn_2O_7$, also new with the stannates, appears to be a conventional AF material.[119,128]

$Gd_2Sn_2O_7$ shows surprising differences from the corresponding titanate. First, there is only one transition temperature at 1.0 K [128, 129] and the magnetic structure is much simpler, showing the so-called $k = (000)$ state predicted by Palmer and Chalker[101] for the Heisenberg (isotropic) spin system on the pyrochlore lattice, Figure 2.45.[130] This remarkable difference in magnetic structures for two isostructural materials with such a small difference in cell volume is difficult to understand. A proposal suggesting that third neighbour interactions stabilise the complex ground state in $Gd_2Ti_2O_7$ has been advanced.

Table 2.9 Selected magnetic properties of the $A_2Sn_2O_7$ pyrochlores

A	θ_C (K)	T_c (K)	Anisotropy	Ground state
Pr	0.32	–	Easy axis	Spin ice
Nd	−0.2	0.9		AFLRO
Gd	−6.6	1.0	Isotropic	AFLRO
Tb	−11	0.9	Easy axis	AFLRO?
Dy	1.7	–	Easy axis	Spin ice
Ho	1.8	–	Easy axis	Spin ice
Er	−14	0.02 <	Easy plane	?
Yb				?

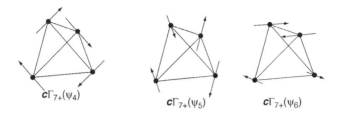

$c\Gamma_{7+}(\psi_4)$ $c\Gamma_{7+}(\psi_5)$ $c\Gamma_{7+}(\psi_6)$

Figure 2.45 Possible spin configurations for $Gd_2Sn_2O_7$ found from neutron diffraction and predicted by Palmer and Chalker. The three configurations are indistinguishable from powder data.[130] Reprinted with permission from Wills *et al.*, 2006 [130]. Copyright (2006) IOP Publishing Ltd

While $Tb_2Ti_2O_7$ falls clearly into the spin liquid camp, the situation for the corresponding stannate is less well understood. The earliest studies showed that spin liquid behaviour gave way to a complex ordering which occurred over two temperature ranges at 1.3 K and 0.9 K. An apparently static magnetic structure with both F and AF character, a kind of ordered spin ice, was determined.[131] On the other hand, μSR data obtained on a sample showing prominent magnetic Bragg peaks and a clear phase transition from specific heat at 0.88 K showed no sign of static spins![132, 133] The persistent spin dynamics are apparent to other probes such as neutron spin echo.[132] These observations are certainly difficult to reconcile. Models involving very large, ordered spin domains relaxing very rapidly were proposed. However, a very recent study using externally implanted muons seem to indicate a strong static component to the ground state, after all.[127] Discussions about this material are likely to continue.

Both $Dy_2Sn_2O_7$ and $Ho_2Sn_2O_7$ appear to be spin ice materials in parallel with their titanate counterparts, although only the $A = Ho$ material has been studied in detail.[135, 136]

$Er_2Sn_2O_7$, unlike the corresponding titanate, shows no sign of LRO from either bulk susceptibility or spin dynamics studies such as μSR.[129, 137]

Very little is known about $Yb_2Sn_2O_7$.

Before leaving the A-site magnetic pyrochlores, some comments on the origin of the magnetic anisotropy, which plays a large role in determining the nature of the ground state, is appropriate. Of course the origin of magnetic anisotropy is in the interaction of the crystal field (CF) with the relevant electronic eigenstates of the ion in question. Recall that for the lanthanides, $J = L + S$ is a good quantum number and the magnetic eigen states with be formed from the various free ion $|M_J\rangle$ states. The subject of CF in rare earth ions has a long history and will not be discussed in great detail here. One formalism which is useful for our purposes is due to

Stevens called the 'operator equivalent' method.[138] In this approach the CF Hamiltonian for an ion in $\bar{3}m$ symmetry, appropriate to the A site in pyrochlores is given as:

$$H_{CF} = B_2^0 O_2^0 + B_4^0 O_4^0 + B_4^3 O_4^3 + B_6^0 O_6^0 + B_6^3 O_6^3 + B_6^6 O_6^6$$

where the B_n^m are numerical factors and the O_n^m are the operator equivalents. For example, $O_2^0 = 3J_z^2 - J(J + 1)$ and $O_6^6 = J_+^6 + J_-^6$. In general when the O_n^m operate on a wave function the result is a linear combination of $|M_J>$ states. The exception is the O_2^0 operator which involves only J_z which cannot mix states of different $|M_J>$. In fact the $B_2^0 O_2^0$ term arises due to axial symmetry and it has already been noted that the A-site in pyrochlores has a very pronounced axial symmetry due to the very short A–O' bond. If we assume that the $B_2^0 O_2^0$ term plays a dominant role in determining the CF, then the observed pattern of anisotropy can be understood. The factor $B_2^0 = A_2^0 <r^2> \alpha_J(1-\sigma_2)$, where A_2^0 is a lattice point charge sum which is invariant for a given crystal structure and $<r^2>$ and σ_2 are the expectation value of r^2 (radius) of the $4f$ electrons and a screening constant. α_J is the critical factor which is derived from the Wigner–Eckhart theorem and is the only factor to change sign as a function of the A-site ion.

A_2^0 is known to be positive for pyrochlores and $<r^2>$ and σ_2 are also positive so the sign of B_2^0 is determined entirely by α_J. If $B_2^0 > 0$, the $B_2^0 O_2^0$ operates on the free ion wave function to give a ground state containing only $|M_J(min) >$ which implies that the moments will lie normal to the quantisation axis (<111> for the A-site) and if $B_2^0 < 0$, the ground state will consist of $|M_J(max)>$, *i.e.* the moments will lie parallel to the quantisation (<111>) axis. Table 2.10 shows how the magnetocrystalline anisotropy correlates with the sign of α_J. While this simple approach cannot predict accurately the full CF ground state wave

Table 2.10 Correlation of the sign of the 'Stevens' factor, α_J[138] with the observed magnetocrystalline anisotropy for the $A_2Ti_2(Sn)_2O_7$ pyrochlores

A	Sign of α_J	Ground state $|M_J >$ for dominant $B_2^0 O_2^0$	Predicted anisotropy	Observed anisotropy
Pr	−	$|4 >$ (max)	Axis	Axis
Tb	−	$|9 >$ (max)	Axis	Axis
Dy	−	$|+/−15/2 >$ (max)	Axis	Axis
Ho	−	$|8 >$ (max)	Axis	Axis
Er	+	$|+/− ½ >$ (min)	Plane	Plane
Yb	+	$|+/− ½ >$ (min)	Plane	Plane

function, as the other terms cannot be ignored for quantitative work, it never fails to predict the correct anisotropy and it should be noted that the actual ground state wave functions found for $A =$ Dy and Ho are very nearly pure $|+/-15/2 >$ and $|8 >$, respectively.

(b) B-site magnetic   Here, four materials will be discussed, $Y_2Mo_2O_7$, $Y_2Ru_2O_7$, $Y_2Ir_2O_7$ and $Y_2Mn_2O_7$.

$Y_2Mo_2O_7$ has played a critical role in the development of the GFMM field since the initial report of its unexpected magnetic properties in 1986.[139] Mo^{4+} is an $S = 1$ ion and it had been shown by Hubert to have a fairly large, negative $\theta_C = -200$ K.[140] Figure 2.46 (top left) shows the

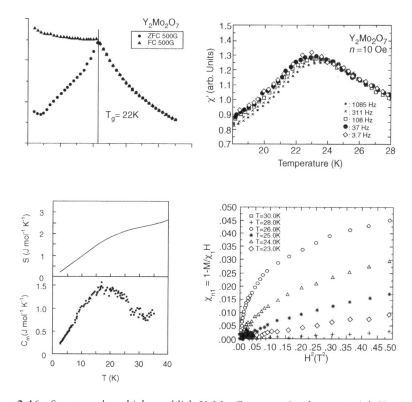

Figure 2.46 Some results which establish $Y_2Mo_2O_7$ as a spin glass material. Top left: d.c. susceptibility showing a ZFC/FC cusp at $T_f = 22$ K. Top right: a.c. susceptibility showing frequency dependence.[141] Bottom left: Heat capacity indicating the absence of a sharp lambda anomaly and the linear temperature dependence at low temperatures.[142] Bottom right: The build up of the nonlinear susceptibility as T approaches T_f.[143] Reprinted with permission from Miyoshi *et al.*, 2000 [141]. Copyright (2000) Physical Society of Japan

DC magnetic susceptibility for this material at low temperatures where a clear ZFC/FC divergence is evident at $T_f = 22$ K which is one of the signatures of a spin glass. Subsequent intensive study, including AC susceptibility (top right), specific heat (bottom left) and the so-called nonlinear susceptibility (bottom right) have shown that $Y_2Mo_2O_7$ behaves as an ideal, canonical spin glass.[141–143]

As already discussed, stabilisation of the spin glass state is thought to require both frustration and positional disorder. Yet, the crystal structure could be refined very well on a perfectly ordered model from neutron powder diffraction data.[144] Subsequently, reports of disorder have appeared from local probe studies such as EXAFS and ^{89}Y NMR.[145,146]

The EXAFS data were interpreted in terms of a very large variance, σ^2, in the Mo–Mo distances (but not the Y–Y, Mo–O or Y–O distances) which implied a static disorder in the Mo–Mo bond distance with $\sigma = 0.16$ Å. ^{89}Y wide line NMR studies showed that a multitude of resonances appeared when the sample was cooled below about 200 K, the Weiss temperature, θ_C, for the compound. This led to speculation regarding the role of geometric frustration in driving this disorder and searches for a phase transition. Very recently, the technique of neutron pair distribution function (NPDF) analysis was applied to $Y_2Mo_2O_7$.[147] This approach has the advantage that both the average and local structures can be analysed from essentially the same neutron diffraction data set. Data suitable for NPDF are recorded to very large momentum transfers, $Q > 40$ Å^{-1}, while the same data can be truncated at say $Q \sim 9$–10 Å^{-1} for conventional Rietveld refinement. First, now with data of much better quality than the initial study, the average structure could still be refined, very well assuming the perfectly ordered model. However, the anisotropic atomic displacement parameters (ADP) were very large relative to those for isostructural $Y_2Sn_2O_7$, especially for the 48f O site, and showed anomalous behaviour upon lowering the temperature. However, no evidence for a true crystallographic phase transition was observed from 300 K to 15 K, which is well below the T_f of 22 K. NPDF analysis involves fitting the real space pair wise distribution function, $G(r)$, obtained from a Fourier transform of the high Q diffraction data. The $G(r)$ will show peaks for each bond pair with relative intensities given by $N_{ab}b_ab_b$ where N_{ab} is the number of neighbours contributing to a given bond between atoms a and b at distance r and b_a and b_b are the neutron scattering lengths of the atoms. The $G(r)$ for $Y_2Mo_2O_7$ at 300 K is shown in Figure 2.47 (top). Even without fitting it is clear the the main issue is with the Y–O1 peak (denoted as Y–O1), it is much weaker than anticipated, and not with the Mo–Mo peak with is not split or broadened to the extent demanded by the interpretation of the EXAFS results. Note that disorder associated with the Y – O1 pair is consistent with the ^{89}Y NMR data. Models involving either a split O1 site or anisotropic atomic displacement parameters for all sites give a satisfactory fit to the $G(r)$, Figure 2.47 (bottom).

Analysis of the effect of O disorder on the Mo – O – Mo superexchange interaction, using spin dimer analysis, shows a significant perturbation to the average structure value. This observation fits well with the Saunders–Chalker model for the appearance of the spin glass state for systems with

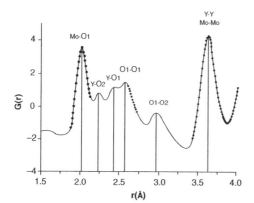

atom Pair	number	distance (Å)
Mo - O1	6	2.018
Y - O2	2	2.214
Y - O1	6	2.455
O1 - O1	2	2.613
O1 - O2	3	2.938
Mo - Mo Y - Y	12	3.616

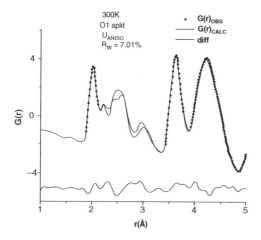

Figure 2.47 Top: The experimental $G(r)$ for $Y_2Mo_2O_7$. Note that the Y – O1 peak is much weaker than the Mo – O1 peak while the number of neighbours is the same. Also, the Mo – Mo peak is neither split nor broadened as had been predicted from analysis of EXAFS data. Bottom: A fit the the $G(r)$ assuming a split O1 site.[147] Reprinted with permission from Greedan *et al.*, 2009 [147]. Copyright (2009) American Physical Society

subtle disorder.[13] Thus, a long standing puzzle seems to be understood at least in part. The larger issue of local disorder and its role in pyrochlore oxides in general is a rather open one at this time.

$Y_2Ru_2O_7$ is also $S = 1$, coming from a $4d^4$, t_{2g}^4 low spin configuration. Thus, it is quite surprising that this material shows clear AFLRO at ~ 77 K as evidenced by d.c. susceptibility and heat capacity data, Figure 2.48, and confirmed by neutron diffraction in the form of rather well-resolved Bragg peaks.[148, 149] As $\theta_C \sim -1100$ K (about five times that for $Y_2Mo_2O_7$), this is still a highly frustrated system, $f \sim 14$, but LRO occurs. Studies of spin dynamics by inelastic neutron scattering indicate that strong AF correlations are present even at 300 K and this system is regarded as one in which a transition occurs from a liquid-like 'cooperative paramagnetic' state to LRO. The contrast with $Y_2Mo_2O_7$ is striking. Nothing is known of the local order or disorder in the ruthenate material.

Potentially, the most interesting B-site only material is $Y_2Ir_2O_7$ which involves a $5d^5$, t_{2g}^5 low spin configuration, i.e. $S = 1/2$. There exists much theory to the effect that a $S = 1/2$ spin system on a pyrochlore lattice should not order under nn AF constraints.[150] $Y_2Ir_2O_7$ is a Mott insulator but

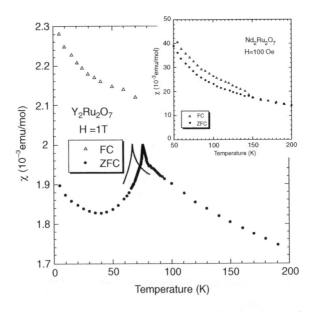

Figure 2.48 DC magnetic susceptibility for $Y_2Ru_2O_7$ showing a clear ordering transition at ~ 75 K.[148] Reprinted with permission from Ito et al., 2001 [148]. Copyright (2001) Elsevier Ltd

appears to undergo a magnetic transition at ~150 K.[151, 152] Originally, no heat capacity anomaly corresponding to the magnetic feature was observed, but later experiments did show a clear lambda type anomaly in the $Sm_2Ir_2O_7$ phase.[153] These materials present significant synthetic challenges. At this stage it is fair to state that the situation for the ground state of $Y_2Ir_2O_7$ is unclear, much more work, especially spin dynamics studies, is needed. In the context of $S = 1/2$ pyrochlores, $Lu_2V_2O_7$ should be mentioned. This material is actually a ferromagnetic insulator with $T_c = 74$ K.[154] While it is not, thus, frustrated in the normal sense, it exhibits some unusual properties such as orbital ordering[155] and a negative magnetoresistance.[156]

The final compound in the B-site only series is $Y_2Mn_2O_7$ based on Mn^{4+} which has configuration $3d^3$ (t_{2g}^3) with $S = 3/2$.

The manganate(4+) pyrochlores, unlike the perovskites of the same ion are not stable at ambient pressure, Figure 2.35. The various synthetic procedures used have been summarised.[6] $Y_2Mn_2O_7$ would appear to be an odd example of frustration as $\theta_C = +$ 42(1) K. Yet, DC susceptibility seems to indicate $T_c = 15$ K. In many ways this compound is a very atypical ferromagnet. First, the ratio $\theta_C/T_c \sim 3$ which is well above the usual value of 1.1–1.4 found for most true ferromagnets. As well, the AC susceptibility shows an edge at ~15 K but also a broad maximum at ~7 K which has the frequency dependence of a spin glass. Neutron diffraction showed only rather broad magnetic peaks. Heat capacity data on different samples show somewhat different results, Figure 2.49. While those of Reimers et al.[157] show no anomaly near 15 K, a very small lambda peak is seen in those of Shimakawa.[158] In both cases very substantial excess heat capacity well above 15 K is evident in comparison with the diamagnetic lattice match, $Y_2Sn_2O_7$. Small angle neutron scattering data (SANS) also indicate anomalous behaviour when compared with the true ferromagnet, $Tl_2Mn_2O_7$. For a ferromagnet SANS scattering peaks at T_c and this is seen for the Tl manganate while the data for the Y phase shows only a weak anomaly at the suspected T_c followed by a sharp rise as the temperature is lowered.[6] Is $Y_2Mn_2O_7$ a frustrated ferromagnet? This neglected material merits further attention.

(c) A and B sites magnetic   The most widely studied pyrochlores in this category are those with $B =$ Mo, Ru and recently, Ir.

A remarkable crossover from metallic to semiconducting behaviour is found for the molybdates as the radius of the rare earth ion is decreased. For $A =$ Nd, Sm and Gd the oxides are metallic and ferromagnetic

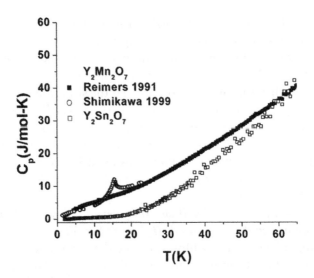

Figure 2.49 Comparison of heat capacity data for $Y_2Mn_2O_7$ from different sources and that of the diamagnetic $Y_2Sn_2O_7$.[6] Reprinted with permission from Gardner *et al.*, 2010 [6]. Copyright (2010) American Physical Society

Mo–Mo coupling is observed with long range ferromagnetic order, while for A = Tb and smaller rare earths, semiconducting and antiferromagnetic Mo–Mo correlations pertain with resulting spin glass like behaviour.[159, 160] The origin of this metal/semiconductor crossover is not fully understood. One proposal suggests a type of double exchange mechanism involving e_g and a_{1g} sub bands which result from splitting the Mo t_{2g} band due to the axial crystal field at the Mo site.[161] Another invokes a band crossing involving the t_{2g} band and high lying rare earth p-bands which decrease markedly in energy with a decrease in lattice constant.[162] There is some spectroscopic evidence to support both models but no definitive picture has emerged. It should be noted that the Mo–O–Mo angle which is a key parameter in the Solovyev approach appears to change discontinuously between the metallic series A = Nd, Sm and Gd and the semiconducting series A = Tb, Ho, Y and Er, Figure 2.50.

While there is no doubt that a crossover occurs, there exists some contradictory evidence regarding the exact placement of the metal/semiconductor boundary. Data taken on polycrystalline samples prepared under a strictly controlled oxygen partial pressure by the use of a buffer gas place the boundary between Gd and Tb.[160] Reports based on single crystals show that the A = Gd phase is nonmetallic.[162, 163] Further studies have demonstrated that the single crystals are oxygen deficient

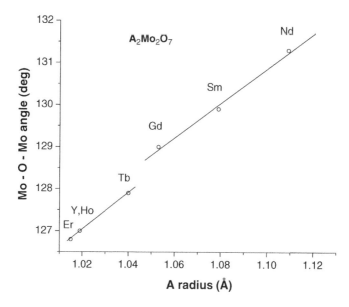

Figure 2.50   A discontinuity in the Mo – O – Mo angle in the $A_2Mo_2O_7$ pyrochlores between A = Tb and Gd, *i.e.* at the semiconductor/metal boundary

and that the transport properties are strongly correlated with the lattice constant which increases sharply with increasing oxygen deficiency, especially for $Gd_2Mo_2O_7$.[6] The conclusion is that $Gd_2Mo_2O_7$ samples close to oxygen stoichiometry are indeed metallic.

Among the three metallic members of this series, $Nd_2Mo_2O_7$ has received by far the greatest attention. The magnetic structure of this phase was determined to consist of two sublattices which were only weakly coupled.[164] The Mo lattice magnetic structure consists of a slightly noncollinear ferromagnetic array which orders at $\sim$93 K for the sample studied but the Nd lattice, which begins to order below $\sim$20 K, shows the familiar two in, two out configuration seen in the spin ice materials. Overall, the intersublattice coupling is approximately antiferromagnetic with ordered moments reduced from spin only ($\sim$1.2 μ_B for Mo) or free ion values ($\sim$1.5 μ_B for Nd). The noncollinear nature of the Nd spins has been implicated in the observation of the unprecedented temperature dependence of the anomalous Hall effect (AHE). In ferromagnets, the Hall effect or the transverse conductivity, σ_{xy}, has two contributions, one proportional to the applied magnetic field and one proportional to the spontaneous sample magnetisation. The component dependent on the sample is called the AHE. In normal ferromagnets, the AHE tends to zero as T approaches zero. In $Nd_2Mo_2O_7$ the AHE actually increases as $T \to 0$ and retains a finite value. The origin of this unusual AHE has been assigned to the chiral nature of the Nd spin configuration (recall that the two in, two out configuration is sixfold degenerate).[165, 166] However, this interpretation has been questioned and a lively debate has ensued which is, as of yet, unresolved.[167, 168]

$Sm_2Mo_2O_7$ has been also been suggested as an AHE material and shows a giant magneto resistance.[169] More recently, it has been proposed as a candidate metallic spin ice.[170]

$Gd_2Mo_2O_7$ has also been investigated by neutron diffraction using ^{160}Gd substitution.[171] The Mo sublattice magnetisation sets in just below $T_c \sim 83$ K and saturates rapidly at a low value of ~ 1.0 μ_B while the Gd sublattice magnetises gradually with decreasing temperature, reaching a saturation value of only 5.7 μ_B, well below the 7.0 μ_B expected for a $S = 7/2$ ion. Overall, the orientation of the two sublattices is parallel or ferromagnetic. The low moments have been explained in terms of strong fluctuations seen in μSR experiments. Remarkably, the LRO ground state is destroyed by application of pressures above 2.5 GPa. This may be due to the opening of a Mott-Hubbard gap in the Mo d band with pressure but at present this is speculative.

Of the semiconducting molybdates, only $Tb_2Mo_2O_7$ has been studied in much detail. This material shows a spin glass freezing at $T_f = 25$ K, very similar to $Y_2Mo_2O_7$ and very strong diffuse features in the neutron diffraction pattern.[172] Neutron inelastic scattering shows directly the spin freezing at 25 K[173] but other probes of spin dynamics, such as μSR indicate that Tb spin fluctuations persist down to very low temperatures, 50 mK.[174]

Members of the series $A_2Ru_2O_7$, which extends from $A = $ Pr to Yb, are all insulators, in contrast to the molybdates and in all cases the Ru sublattice orders AF with no indication of spin glass behaviour. There does exist a strong correlation between the A cation radius and T_c which increases from 75 K for $A = $ Y to ~ 160 K for $A = $ Pr.[148] From neutron diffraction on the $A = $ Er phase the Ru^{4+} moment saturates very near to the free ion value of 2 μ_B, again in sharp contrast to the $A = $ Mo materials.[175] Early reports suggested a possible spin ice state for $Dy_2Ru_2O_7$ and $Ho_2Ru_2O_7$.[176] However, neutron diffraction and specific heat data show, conclusively, that that the Ho moments order at 1.4 K and that the magnetic entropy is not that expected for spin ice.[177, 178]

On the other hand the $A_2Ir_2O_7$ materials resemble much more closely the molybdates in that a metal/insulator cross over occurs as a function of decreasing A cation radius.[151]

In this initial work the metal/insulator boundary was set between Eu and Gd. As already mentioned, magnetic transitions are seen, not only in the $A = Y$ phase but in the metallic $A = Sm$ material as well.[152] Further work showed that sample preparation methods are important and that IrO_2 can be lost in the synthesis. Methods involving a 10% excess of IrO_2 in sealed Pt tubes resulted in well crystallised samples which showed metal/insulator transitions with decreasing temperature for $A = Nd$, Sm and Eu but that $Pr_2Ir_2O_7$ remains metallic to the lowest temperatures, Figure 2.51.[153] In contrast to most other series members, the $A = Pr$ phase remains paramagnetic to 0.3 K.[179] The lack of spin freezing has been interpreted in terms of a metallic spin liquid state with even some Kondo behaviour.[180]

2.3.4.2 Spinels

The spinel structure, AB_2O_4, is also described in $Fd3m$ and in normal spinels, the octahedrally coordinated B-site ion occupies $16d$, with the

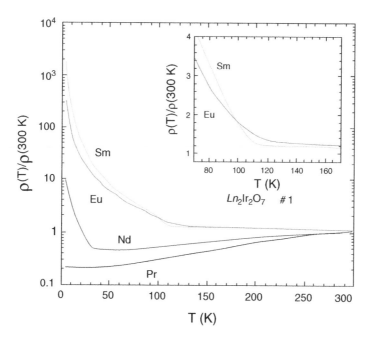

Figure 2.51 Evidence for a metal-insulator transition with temperature for $Nd_2Ir_2O_7$, $Sm_2Ir_2O_7$ and $Eu_2Ir_2O_7$. Note that $Pr_2Ir_2O_7$ remains metallic to the lowest temperatures.[153] Reprinted with permission from Matsuhira *et al.*, 2007 [153]. Copyright (2007) Physical Society of Japan

tetrahedral A ion on $8a$ and the single O^{2-} ion at $32e$. While the B sublattice is clearly frustrated, the A sublattice is that of diamond and is also frustrated in an interesting manner. As with pyrochlores, materials with only the B-site and only the A-site magnetic are of interest.

(a) B-site magnetic spinels This case comprises the larger class of frustrated spinels. It is important to recall that while the B-site sublattice topology is identical between pyrochlores and spinels, the connectivity of the BO_6 octahedra is very different, with corner-sharing for the former and edge-sharing for the latter. Thus, the $B–B$ cation distances will be systematically shorter in spinels than in pyrochlores. Among the spinels of interest are $ZnFe_2O_4$, $GeNi_2O_4$, $GeCo_2O_4$, $ZnCr_2O_4$, $LiMn_2O_4$, $Li_2Mn_2O_4$ and λ-MnO_2.

$ZnFe_2O_4$, also known as the mineral franklinite, has been studied for some time. It is generally regarded as a normal spinel but, apparently, rapid cooling can introduce some level of inversion. Early studies of magnetic properties found that long range order sets in below 10 K and that the magnetic structure is complex and probably incommensurate with the chemical cell although all magnetic peaks could be indexed with $k = (0\ 0\ ½)$.[181] Later, a very detailed study involving neutron diffraction, μSR and Mössbauer spectroscopy was undertaken on a sample with negligible inversion.[182] The same magnetic structure was found but the spin dynamics could also be monitored. The ND data showed the development of magnetic SRO above 80 K, more than eight times T_N, with a correlation length of $\sim$30 Å. Remarkably, the SRO persisted to 4.2 K, well below T_N and the SRO component remained dynamic rather than frozen with a relaxation rate $>1\ \mu$s, which is the same order as seen in some pyrochlore spin liquids.

$GeCo_2O_4$ and $GeNi_2O_4$ have been known for many years and among the first studies are those from Bertaut *et al.*[183,184] At first glance these materials would not seem to be frustrated according to the usual criterion. For example θ_c for $GeCo_2O_4$, is widely reported to be fairly large and positive, values range from 57 to 81 K with a strong dependence on the temperature range chosen for the Curie–Weiss fit.[185,186] Yet, this material shows order AF below 21 K. The corresponding $GeNi_2O_4$ has θ_c ranging from –4 to –6 K, at least suggesting net AF interactions but still smaller than T_N of $\sim$ 11K. In both cases the temperature range investigated extends to $\sim$ 400 K, the upper limit for most SQUID devices currently available. For the Co compound it has been shown that low lying crystal field excited states cannot be ignored and thus, application of the Curie–Weiss law is not justified within the usual temperature range measured.[186] This is the same situation which is obtained with many lanthanide ions, such as Nd^{3+}, which has been documented in an earlier section. In fact the original report for $GeNi_2O_4$ shows susceptibility data taken up to 1500 K and a true Curie–Weiss regime does not appear to be obtained unless $T > 900$ K! The θ_C value derived from these data is ~ -200 K.[183] These examples illustrate the danger of applying the Curie–Weiss law at too low temperatures.

Results from heat capacity and neutron scattering indicate the frustrated nature of both compounds. For $GeNi_2O_4$ there are actually two transitions at 12.08 K and 11.43 K from heat capacity data. Above the T_N region the magnetic entropy extends to ~ 400 K, indicating an extensive regime of AF SRO, Figure 2.52. The entropy associated with each transition is only 2–3% of the expected $R\ln 3$ for an $S = 1$ state. Both of these observations are consistent with a highly frustrated magnetic system. Neutron diffraction data show a magnetic structure with $k = (1/2\ 1/2\ 1/2)$ based on the chemical cell. The original model proposed by Bertaut seems to be valid.[183, 187] Here the view of the pyrochlore lattice as alternating sheets of kagomé (K) and triangular planar (TP) topology stacked along the (111) direction is valuable, Figure 2.39, and the model involves F kagomé and TP spin

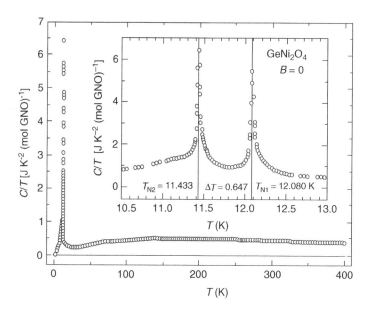

Figure 2.52 Heat capacity data for $GeNi_2O_4$ showing two sharp maxima at low temperatures and a large magnetic contribution extending to very high temperatures.[186] Reprinted with permission from Lashley *et al.*, 2008 [186]. Copyright (2008) American Physical Society

orientations which alternate such that adjacent K planes are AF coupled and the same for adjacent TP planes, *i.e.* K(+)TP(+)K(−)TP(−), *etc.* Moment directions within the K planes are normal to (111) and those within the TP planes can be either parallel or perpendicular to the accuracy of the powder data. The two transition temperatures for GeNi$_2$O$_4$ have been interpreted as sequential ordering of the K and TP layers, however, there is only a single transition for GeCo$_2$O$_4$ which orders at 20.6 K. An analysis of the exchange interactions in both materials out to the fourth nearest neighbours argues strongly for a large role for frustration in both of these spinels.[187]

It is worth noting that both Ge-based spinels remain cubic below the ordering temperatures and down to 1.5 K. This is in contrast to ZnCr$_2$O$_4$. An S = 3/2 material, ZnCr$_2$O$_4$, is clearly frustrated by the standard criterion with reported θ_C = −330 K to −390 K while AFLRO sets in only below T_N = 9.5 K to 12.5 K, giving f = 31–35.[188–190] There is general agreement, currently, that T_N = 12.5 K. The most remarkable feature of this compound which has drawn considerable attention is the first order cubic to tetragonal crystallographic phase transition which accompanies and, as has been argued, drives the AFLRO.[190] The symmetry of the tetragonal, low temperature phase is $I\bar{4}2m$ but the details of the actual magnetic structure are still elusive, as it is quite complex, involving four ordering wave vectors which are commensurate with the chemical cell.[191, 192] This coupled crystallographic/magnetic phase transition has features analogous to both spin Peierls and Jahn-Teller phenomena. Another remarkable observation concerns the presence of hexagonal spin clusters in the cubic phase at 15 K, just above the lattice/magnetic phase transition.[193] These spin clusters have been dubbed 'protectorates' and it is argued that such structures are probably present in many GFAF materials at low temperature.

Interestingly, CdCr$_2$O$_4$ undergoes a similar coupled structural/magnetic phase transition at a slightly lower temperature of 7.8 K but both the crystal symmetry, $I4_1/amd$, and the magnetic structure, incommensurate, differ from those of the Zn phase.[191] As well, doping experiments show that the Néel state of ZnCr$_2$O$_4$ is remarkably sensitive to very low levels of Cd doping, just 3% is sufficient to destroy the Néel state and induce spin glass behaviour.[194] A very recent study suggests that the crystallographic symmetry of the low temperature state in ZnCr$_2$O$_4$ is actually orthorhombic.[195] These are clearly complex systems and a full understanding of the ground state will require further efforts. It is interesting to note that GeCo$_2$O$_4$ which is also S = 3/2, shows little of the complexity of the corresponding chromate spinel.

The series of Mn based spinels, $LiMn_2O_4$, λ-MnO_2 and $Li_2Mn_2O_4$ also show the effects of geometric frustration. Both λ-MnO_2 and $Li_2Mn_2O_4$ are metastable phases which are derived from $LiMn_2O_4$ by adding or removing Li at room temperature either by 'soft chemical' or electrochemical methods. The relationship between these materials and the thermodynamically stable Mn oxides are shown in Figure 2.53.

$LiMn_2O_4$, the 'parent' compound in this series is of course well known as a potential cathode material in Li ion batteries. It undergoes a first order crystallographic phase transition from $Fd3m$ to $Fddd$ below ~283 K where the unit cell is now much larger, $a = 24.750$ Å, $b = 24.801$ Å and $c = 8.1903$ Å which is $a(LT) \sim b(LT) \sim 3a(HT)$ and $c(LT) \sim c(HT)$, where (LT) and (HT) stand for the low temperature and high temperature cells, respectively. The LT cell volume expands sixfold.[196,197] Again, the magnetic properties do not suggest a high degree of frustration, $\theta_c = -300$ K and magnetic order occurs below 65 K or $f \sim 5$.[198] The crystallographic phase transition involves a partial charge ordering of the Mn^{3+} and Mn^{4+} ions both from powder neutron and single crystal X-ray analysis. Another report claims more or less perfect charge ordering but no detailed structure solution is presented in support.[200] The magnetic structure is very complex and has not been solved – there is a suggestion that the magnetic cell may contain 1152 Mn sites![198] Results of neutron diffraction studies of the magnetic correlations differ widely. Wills $et\ al.$ showed a coexistence between AFLRO and diffuse magnetic scattering down to 10 K.[198]

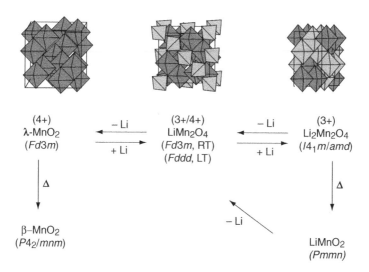

Figure 2.53   The soft chemical pathways to the metastable $Li_2Mn_2O_4$ and λMnO_2 and the relationships to the thermodynamically stable $LiMnO_2$ and β-MnO_2

A second study at about the same time found no magnetic Bragg peaks below 65 K.[199] A third study on the sample claimed to exhibit perfect charge ordering showed mainly Bragg peaks and less diffuse scattering.[200] Later, a detailed study using polarised neutrons on the sample of Wills *et al.* gave clear evidence for the coexistence of Bragg and diffuse magnetic scattering down to 1.5 K. In fact the relative contributions of the two components were nearly equal. As well, inelastic scattering indicated that most of the moments contributing to the diffuse scattering were static but that up to 20% were still dynamic even at 1.5 K.[201]

The situation for λ-MnO_2 is somewhat simpler. The earliest known study of magnetism on a sample in which the Li was removed by acid leaching showed $\theta_C = -104(4)$ K.[202] The unit cell constant for this sample is $a = 8.0407(7)$ Å which indicates a high level of Li removal. A sample with Li_xMnO_2 where $x = 0.11(8)$ gave $a = 8.155$ Å.[203] An AF transition was found at $T_N = 32$ K, so $f\sim3$, suggesting a low level of frustration. Now, this material is isostructural with $ZnCr_2O_4$ and has the same $S = 3/2$ spin state but there was no evidence for a structural phase transition coincident with the magnetic transition. Neutron diffraction data did indicate significant diffuse magnetic scattering persisting up to ~100 K, more than $3 \times T_N$, a characteristic of frustrated systems. As well, the magnetic structure is very complex and is described by $\mathbf{k} = (\frac{1}{2}\,\frac{1}{2}\,\frac{1}{2})$, the same ordering wave vector found for the spinels $GeNi_2O_4$ and $GeCo_2O_4$. Nonetheless, the pattern of magnetic intensities is very different. In the latter two spinels the $\frac{1}{2}\,\frac{1}{2}\,\frac{1}{2}$ magnetic reflection is by far the most intense, for example the intensity ratio $(\frac{1}{2}\,\frac{1}{2}\,\frac{1}{2})/(\frac{3}{2}\,\frac{1}{2}\,\frac{1}{2})$ is ~8 but for λ-MnO_2, it is ~1.4. Two rather complex magnetic structures (128 spins per magnetic cell) have been proposed for λ-MnO_2, shown in Figure 2.54, both giving a reasonable fit to the data.[202, 204]

The final member of this series, $Li_2Mn_2O_4$, is perhaps the most interesting. Due to the presence of the Jahn-Teller ion, Mn^{3+}, the symmetry is reduced to tetragonal, $I4_1/amd$ with cell constants $a = 5.649(3)$ Å and $c = 9.198(5)$ Å, that is, $a_t \sim a_c/\sqrt{2}$ and $c_t \sim a_c$. As a result of the distortion, the Mn–Mn distances are now $2 \times 2.842(2)$ Å and 4×3.046 Å, compared with six equal Mn–Mn distances for cubic $LiMn_2O_4$ of 2.915 Å.

The susceptibility data are dominated by a broad maximum just below 200 K and there is no discernible Curie–Weiss regime out to 600 K. Below about 50 K there is an apparent magnetic phase transition.[205] Neutron diffraction data show the remarkable result that the magnetic scattering peaks have the Warren line shape, characteristic of two-dimensional scattering, Figure 2.55. There is no sign of a crossover to three-dimensional LRO down to 1.6 K. A two-dimensional correlation length derived from

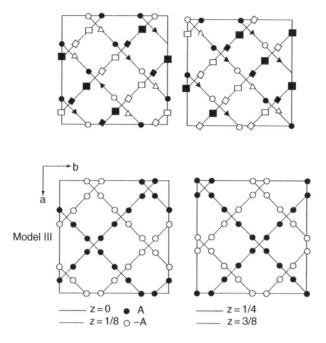

Figure 2.54 Possible magnetic structures for λ-MnO_2 shown in projection on a cubic face of the doubled pyrochlore unit cell. Top:[204] The filled and empty symbols represent up and down spins, respectively. For the meaning of the squares, triangles and circles, consult Morgan *et al.*[204] Bottom:[202]. Reprinted with permission from Morgan *et al.*, 2003 [204]. Copyright (2003) American Physical Society

analysis of the Warren peaks indicates a maximum of about 90 Å and that the apparent 'transition' is very abrupt, Figure 2.55. From the positions of the two observable Warren peaks, a tentative magnetic structure, the so-called $\sqrt{3} \times \sqrt{3}$ structure involving only the kagomé layer spins, was proposed. A subsequent μSR study supports this assignment and in addition finds that significant spin dynamics persist to low temperatures, the origin of which is likely the spins on the TP layers which appear to remain uncoupled.[203]

(b) *A-site magnetic spinels* This is quite a new area where, with only a few exceptions, papers have appeared very recently. The first systematic study of *A*-site magnetic spinels appears to be that of Roth.[207] Of course an important issue with these materials is the extent of inversion. This was addressed by Roth who determined inversion levels using both X-ray and neutron diffraction in AAl_2O_4 spinels where $A^{2+} = $ Mn, Fe and Co to

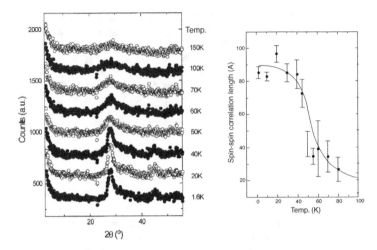

Figure 2.55 Left: The appearance of asymmetric Warren line shape peaks with decreasing temperature for $Li_2Mn_2O_4$, indicative of two-dimensional spin correlations. Right: The temperature dependence of the spin–spin correlation length showing an abrupt, first-order like increase below 60 K.[205] Reprinted with permission from Wills *et al.*, 1999 [205]. Copyright (1999) American Chemical Society

range from 5 to 15% A ions on the B site. It was commented that the degree of inversion is a strong function of the heat treatment program. Most of the sparse literature concerns these three aluminates. A sample of $FeAl_2O_4$ with about 17% Fe on the B site was determined to be a spin glass with $T_f \sim 13$ K from AC and DC susceptibility data.[208] Tristan *et al.* appear to be the among the first to regard the A-site spinels as GFAF systems.[209] Figure 2.56 shows the A-site connectivity for up to third nearest neighbours where the triangular motif is obviously ubiquitous. Table 2.11 shows the relevant structural and magnetic data for the three well-studied aluminate spinels obtained from X-ray diffraction,

Figure 2.56 The connectivity of A-site ions in spinel, AB_2O_4 out to third nearest neighbours. Note the prevalence of triangles.[209] Reprinted with permission from Tristan *et al.*, 2005 [209]. Copyright (2005) American Physical Society

Table 2.11   Relevant structural and magnetic parameters for the AAl_2O_4 spinels

Property	$MnAl_2O_4$	$FeAl_2O_4$	$CoAl_2O_4$
S	5/2	2	3/2
θ_C (K)	$-143(5)$	$-130(1)$	$-104(2)$
T_f, T_c (K)	40(0.5)	12(0.5)	4.8(2)
f	4	11	22
a_0 (Å)	8.2246(3)	8.1572(3)	8.1078(3)
% A on B site	6(2)	8(2)	8(2)

susceptibility and heat capacity studies. While the $A = $ Mn phase has a modest f value and does order at the relatively high temperature of 40 K, the other two aluminates are highly frustrated. The low temperature specific heat data follow a T^2 law instead of the linear dependence found for canonical spin glasses which has led to speculation that a more liquid-like state is involved.

Subsequent neutron diffraction studies showed a collinear magnetic structure for the $A = $ Mn phase but only broad, diffuse scattering for $A = $ Co. The ordered moment for Mn is only 3.7 μ_B, much reduced from $5\mu_B$ expected for $S = 5/2$.[210] Another group reported slightly different results for $CoAl_2O_4$ with a lower $\theta_c = -89$ K and a $T^{2.5}$ power law dependence in the specific heat.[211] This sample was claimed to show an inversion of only 4%.

Interestingly, frustration effects are greatly diminished in the ARh_2O_4 series where the inversion levels are essentially zero due to the enormous octahedral site preference of $Rh^{3+}(t_{2g}^6)$. For example $CoRh_2O_4$ has $\theta_C = -31$ K with $T_N = 30$ K, giving $f \sim 1$. Thus, even small levels of inversion may be sufficient to destroy long-range order, recall the situation of $Y_2Mo_2O_7$, and give rise to spin glass behaviour. Nonetheless, the huge difference in θ_C for the rhodate and aluminate Co spinels is difficult to understand. Clearly, there remain a number of unresolved issues with these materials. A detailed theory of A-site spinels has recently appeared.[212]

2.3.5   Other Frustrated Lattices

In addition to the well-studied GFAF materials with kagomé and pyrochlore lattices, a number of new lattice topologies have been discovered recently. Only two examples will be described in some detail, $BaM_{10}O_{15}$, where $M = $ V and Cr and Li_3MgRuO_6 – a new type of ordered NaCl material.

The $BaM_{10}O_{15}$ compounds can be regarded as variants of V_2O_3 ($V_{10}O_{15}$) in which large Ba^{2+} ions substitute on the O^{2-} sublattice. In V_2O_3 with the corundum structure, V^{3+} ions occupy 2/3 of the octahedral sites. In $BaM_{10}O_{15}$ the presence of Ba^{2+} ions in the close-packed O^{2-} layers, which form an ordered pattern, direct the occupation of the octahedral sites which are now 5/8 filled by $2M^{2+}$ and $8M^{3+}$ ions per formula unit. The M-site topology now involves M_{10} clusters which are comprised of edge-sharing tetrahedra interconnected by corner sharing. Relevant structural details are shown in Figure 2.57. Note that these materials contain the V^{2+} and Cr^{2+} oxidation states which are very rare in oxides. There is no convincing crystallographic evidence for charge ordering between the M^{2+} and M^{3+} ions.

The $M = Cr$ material does not show LRO down to 5 K but instead a spin-glassy signature in the DC susceptibility with $T_f \sim 25$ K and a broad. diffuse peak develops in the neutron diffraction pattern below 120 K. From the width of the diffuse peak a correlation length of ~ 3 Å is derived which is the Cr-Cr nearest neighbour distance.[213]

$BaV_{10}O_{15}$ has been studied in greater detail. Magnetic susceptibility data show a very large, negative $\theta_C = -1156(15)$ K. There are two phase transitions at low temperature, a first-order structural transition at 135 K, $Cmca$ to $Pbca$, and a magnetic transition at 40 K. Thus, this is highly frustrated material as f ~ 30.[214] Neutron diffraction data confirm the onset of AFLRO below 40 K but the magnetic structure is complex, requiring two ordering wave vectors, $k = (0\,0\,0)$ and $k = (1/2\,0\,0)$. Heat capacity results show that only $\sim 11\%$ of the expected entropy is removed at T_c and that there is evidence for SRO extending well below T_c. Remarkably, doping with only 5% Ti quenches the AFLRO completely, leaving only a spin glass like divergence at 13 K, yet the Curie–Weiss

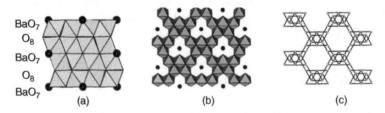

Figure 2.57 Structure of $BaV(Cr)_{10}O_{15}$. (a) Stacking of close packed layers indicating the substitution of Ba^{2+} for O^{2-}. Ba^{2+} ions are shown as spheres. (b) The pattern of occupation of the octahedral sites with the Ba^{2+} ions again shown as spheres. (c) One half of the cation sublattice which features $V(Cr)_{10}$ clusters formed by edge-sharing tetrahedra

law still holds giving $\theta_C = -1100\,K$ or $f\sim85$.[215] Clearly, the magnetic ground state of this material is very sensitive to small perturbations.

Ordered NaCl oxides have been discussed in an earlier section. In addition to the $C2/m$ ordering scheme which results in a TP lattice, there exists another variation described by $Fddd$. The first $Fddd$ magnetic material, $Li_3Mg_2RuO_6$, has been described recently.[213] The Ru^{5+} $(S = 3/2)$ sublattice is shown in Figure 2.58 which is seen to consist of ribbons of edge-sharing triangles running parallel to (110) directions which are linked along (001) by corner sharing. The Curie–Weiss law holds above $100\,K$ giving $\theta_C = -109\,K$ and AFLRO sets in below $17\,K$, giving $f\sim6$. Heat capacity data show a excess magnetic contribution well above T_c extending to at least $50\,K$ or $3 \times T_c$. Further signs of frustration appear in the magnetic structure which is very complex and not yet solved. Two ordering wave vectors, $\mathbf{k} = (1/2\,0\,1/3)$ and $\mathbf{k} = (1/2\,1/2\,0)$, are needed to index the magnetic unit cell which is nine times the volume of the chemical cell and contains 96 Ru spins. A spin dimer analysis suggests that the dominant exchange interactions are in the ab plane which would imply a TP model which would be low dimensional. Yet, the

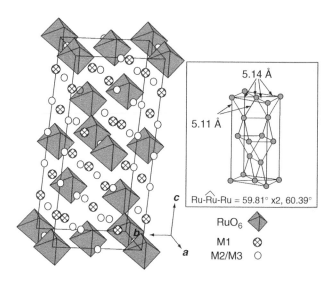

Figure 2.58 The structure of $Li_3Mg_2RuO_6$. The RuO_6 octahedra are shown in grey and the open and hatched spheres represent the Li rich and Mg rich positions, respectively. The Ru sublattice, inset, is seen to involve ribbons of edge-sharing triangles along <110> directions which share corners along <001>.[216] Reprinted with permission from Derakhshan et al., 2008 [216]. Copyright (2008) American Chemical Society

diffuse magnetic scattering at 18 K, just above T_c, shows a symmetrical Lorenztian rather than the asymmetric Warren line shape indicating that the SRO is three-dimensional. More work is needed to understand this material and efforts to synthesise other *Fddd* compounds with quantum spins would be very welcome.

2.4　CONCLUDING REMARKS

I hope that it is clear that geometrically frustrated magnetic materials represent a fascinating and rapidly expanding research topic. Since 2000 there have been four international meetings focusing on highly frustrated magnetic materials, in Waterloo, Grenoble, Osaka and Braunschweig with the next planned for Baltimore in 2010. It is an area in which progress is driven by interdisciplinary cooperation. Inorganic materials/ solid state chemists are ideally suited to the discovery, preparation of high quality samples, crystal growth, structure determination (both average and local structures) and basic properties characterisation. Condensed matter physics experimentalists are needed for application of more sophisticated characterisation methods such as neutron inelastic scattering, muon spin relaxation, neutron spin echo and nuclear magnetic resonance, among others. The traditional role of the theorist in providing explanations for observations and pointing the way for new experiments is of critical importance. Many of the key discoveries in this field, such as the spin glass rare earth molybdate pyrochlores, the spin ice and spin liquid states in titanate pyrochlores, preparation and growth of stoichiometric jarosites, herbertsmithite and kapellasite have resulted from such wide ranging interdisciplinary collaborations.

New ideas have emerged. It is remarkable that the study of spin ice connects to both the structure of water ice and the potential involvement and observation of the elusive magnetic monopole! The spin liquid provides an example of an extremely delicate ground state which is highly susceptible to perturbations. For example long range order can be induced in $Tb_2Ti_2O_7$ either by increasing the lattice constant through replacement of the smaller Ti^{4+} with the larger Sn^{4+} or by decreasing the lattice constant by application of hydrostatic pressure.

In fact there are numerous examples of materials which are remarkably similar in crystal structure and spin quantum number, yet, show a high level of divergence in the nature of the ground state. The case of the $S = \frac{1}{2}$ ordered NaCl compounds, Li_5OsO_6 (AFLRO) and Li_4MgReO_6 (spin glass), illustrates the role of local environment on electronic structure and the dimensionality of the magnetic exchange. Among the $S = 1$ pyrochlores, $Y_2Mo_2O_7$ is a spin glass with very subtle disorder while $Y_2Ru_2O_7$, which has a frustration index five times larger, shows AFLRO. The cubic B-site ordered double perovskites with $S = \frac{1}{2}$ exhibit F (Ba_2NaOsO_6), AF (Ba_2LiOsO_6) and apparently nonordered (Ba_2YMoO_6) ground states. For the $S = 3/2$ B-site spinels, $ZnCr_2O_4$ distorts before ordering while $GeCo_2O_4$ remains cubic below T_c.

Thus, these materials are likely to continue to fascinate researchers in many disciplines for some time to come.

REFERENCES

[1] A.P. Ramirez, *Ann. Rev. Mater. Sci.*, **24**, 453 (1994).

[2] P. Schiffer and A.P. Ramierz, *Commun. Condens. Matter Phys.*, **10**, 21 (1996).

[3] B.D. Gaulin, *Hyperfine Interactions*, **85**, 159 (1994).

[4] J.E. Greedan, *J. Mater Chem.*, **11**, 37 (2001).

[5] J.E. Greedan, *J. Alloys Compd.*, **408**, 444 (2006).

[6] J.S. Gardner, M.J.P. Gingras and J.E. Greedan, *Rev. Mod. Phys.*, **82**, 53 (2010).

[7] H.E. Stanley, *Introduction to Phase Transitions and Critical Phenomena*, Oxford University Press, New York and Oxford, 1971.

[8] F.E. Mabbs and D.J. Machin, *Magnetism and Transition Metal Complexes*, Chapman & Hall, London, 1973, p. 12.

[9] J.S. Smart, *Effective Field Theories of Magnetism*, W. B. Saunders, Philadelphia and London, 1996, p. 77.

[10] R. Moessner, *Can. J. Phys.*, **79**, 1283 (2001).

[11] J.N. Reimers, A. J. Berlinsky and A.-C. Shi, *Phys. Rev. B*, **43**, 865 (1991).

[12] J.N. Reimers, *Phys. Rev. B*, **45**, 7287 (1992).

[13] T.E. Saunders and J.T. Chalker, *Phys. Rev. Lett.*, **98**, 157201 (2007).

[14] G. Férey, R. De Pape, M. LeBlanc and J. Pannetier, *Rev. Chim. Miner.*, **23**, 474 (1986).

[15] J.A. Mydosh, *Spin Glasses: An Experimental Introduction*, Taylor & Francis, London and Washington, DC 1993.

[16] K. Binder, W. Kinzel, H. Maletta and D. Stauffer, *J. Magn. Magn. Mater.*, **15–18**, 189 (1980).

[17] J. Villain, *Z. Phys. B*, **33**, 31 (1979).

[18] P.W. Anderson, *Mater. Res. Bull.*, **8**, 153 (1973); P.W. Anderson, *Science*, **235**, 1196 (1987).

[19] M.J.P. Gingras, B.C. den Hertog, M. Faucher, J.S. Gardner, L.J. Chang, B.D. Gaulin and N.P. Raju, *Phys. Rev. B*, **62**, 6496 (2000).

[20] M.J. Harris, S.T. Bramwell, D.F. McMorrow, T. Zeiske and K.W. Godfrey, *Phys. Rev. Lett.*, **79**, 2554 (1997); M.J. Harris, S.T. Bramwell, P.C. Holdsworth and D.Champion, *Phys. Rev. Lett.*, **81**, 4496 (1998).

[21] W.F. Giauque and J.W. Stout, *J. Am. Chem. Soc.*, **58**, 1144 (1936).

[22] A.P. Ramirez, A. Hayashi, R.J. Cava, R. Siddharthan and B.S. Shastry, *Nature*, **399**, 333 (1999).

[23] K. Hirakawa, H. Ikeda, H. Kadowaki and K. Ubukoshi, *J. Phys. Soc. Jpn.*, **52**, 2882 (1983); M. Neil, C. Cros, G. LeFlem, M. Pouchard and P. Hagenmuller, *Physica*, **86–88B**, 702 (1977).

[24] S.J. Clarke, A.J. Fowkes, A. Harrison, R.M. Ibberson and M.J. Rosseinsky, *Chem. Mater.*, **10**, 372 (1998).

[25] J.N. Reimers, J.R. Dahn, J.E. Greedan, C.V. Stager, G. Liu, I.J. Davidson and U. von Sacken, *J. Solid State Chem.*, **102**, 542 (1993); A.-L. Barra, G. Chouteau, Stepanov, A. Rougier and C. Delmas, *Eur. Phys. J.*, **B7**, 551 (1999).

[26] J.L. Soubeyroux, D. Fruchart, G. Le Flem and C. Delmas, *J. Magn. Magn. Mater.* **14**, 159 (1979); H. Kadowaki, H. Takei and K. Motoya, *J. Phys.: Condens. Matter*, **7** 6869 (1995).

[27] J.-P. Doumerc, A. Wichainchai, A. Ammar, M. Pouchard and P. Hagenmuller, *Mater. Res. Bull.*, **21**, 745 (1986); Y. Oohara, S. Mitsuda, H. Yoshizawa, N. Yaguchi, H. Kuriyama, T. Asano and M. Mekata, *J. Phys. Soc. Jpn.*, **63**, 847 (1994).

[28] S.T. Bramwell, S.G. Carling, C.J. Harding, K.D.M. Harris, B.M. Karuiki, L. Nixon and I.P. Parkin, *J. Phys.: Condens. Matter*, **8**, L123 (1996).

[29] H. Serrano-Gonzales, S.T. Bramwell, K.D.M. Harris, B.M. Karuiki, L. Nixon and I.P. Parkin, *Phys. Rev. B*, **59**, 14451 (1999).

[30] M. Bieringer, J.E. Greedan and G.M. Luke, *Phys. Rev. B*, **62**, 6521 (2000).

[31] S. Derakhshan, J.E. Greedan and L.M.D. Cranswick, *Phys. Rev. B*, **77**, 014408 (2008).

[32] M.H. Whangbo, H.J. Koo and D.J. Dai, *J. Solid State Chem.*, **176**, 417 (2003).

[33] P. Lecheminant, B. Bernu, C. Lhuillier, L. Pierre and P. Sindzingre, *Phys. Rev. B*, **56**, 2521 (1997).

[34] Ch. Waldtmann, H.-U. Everts, B. Bernu, C. Lhuillier, P. Sindzingre, P. Lecheminant and L. Pierre, *Eur. Phys. J. B*, **2**, 501 (1998).

[35] D. Grohol, D.C. Nocera and D. Papoutsakis, *Phys. Rev. B*, **67**, 064401 (2003).

[36] A.B. Harris, C. Kallin and A.J. Berlinsky, *Phys. Rev. B*, **45**, 2899 (1992).

[37] A.S. Wills, A. Harrison, S.A.M. Mentink, T.E. Mason and Z. Tun, *Europhys. Lett.*, **42**, 325 (1998).

[38] G.S. Oakley, D. Visser, J. Frunzke, K. H. Anderson, A. S.Wills and A. Harrison, *Physica B*, **267–8**, 142 (1999).

[39] A. Harrison, K.M. Kojima, A.S. Wills, F. Fudamato, M.I. Larkin, G.M. Luke, B. Nachumi, Y.J. Uemura, D. Visser and J.S. Lord, *Physica B*, **289–90**, 217 (2000).

[40] A.S. Wills, A. Harrison, C. Ritter and R.I. Smith, *Phys. Rev. B*, **61**, 6156 (2000).

[41] S.-H. Lee, C. Broholm, M.F. Collins, L. Heller, A.P. Ramirez, Ch. Kloc, E. Bucher, R.W. Erwin and N. Lacevic, *Phys. Rev. B*, **56**, 8091 (1997).

[42] M. Shores, E. Nytko, B. Bartlett and D.G. Nocera, *J. Am. Chem. Soc.*, **127**, 13462 (2005).

[43] J.S. Helton, M. Matan, M.P. Shores, E.A. Nytko, B.M. Bartlett, Y. Yoshida, Y. Takano, A. Suslov, Y. Qui, J.-H. Chung, D.G. Nocera and Y.S. Lee, *Phys. Rev. Lett.*, **98**, 107204 (2007).

[44] A. Olariu, P. Mendels, F. Bert, F. Duc, J.C. Trombe, M.A. de Vries and A. Harrison, *Phys. Rev. Lett.*, **100**, 087202 (2008).

[45] M.A. deVries, K.V. Kamenev, W.A. Kocklelmann, J. Sanchez-Benitiz and A. Harrison, *Phys. Rev. Lett.*, **100**, 157205 (2008).

[46] S.-H. Lee, H. Kikuchi, Y. Qui, B. Lake, Q. Huang, K. Habicht and K. Kiefer, *Nat. Mater.*, **6**, 853 (2007).

[47] W. Krause, H.-J Bernhardt, R.S.W. Braithwaite, U. Kolitsch and R. Pritchard, *Miner. Mag.*, **70**, 329 (2006).

[48] R.H. Coleman, C. Ritter and A.S. Wills, *Chem. Mater.*, **20**, 6897 (2008).

[49] M.A. Lafontaine, A. Le Bail and G. Férey, *J. Solid State Chem.*, **85**, 220 (1990).

[50] Z. Hiroi, M. Hanawa, N. Kobayashi, M. Nohara, H. Tagaki, Y. Kato and M. Takagawa, *J. Phys. Soc. Jpn.*, **70**, 3377 (2001).

[51] F. Bert, D. Bono, P. Mendels, F. Ladieu, F. Duc, J.-C. Trombe and P. Millet, *Phys. Rev. Lett.*, **95**, 087203 (2005).

[52] Z. Hiroi, H. Yoshida, Y. Okamoto and M. Takigawa, *J. Phys. Conf. Ser.*, **145**, 012002 (2009).

[53] K. Morita, M. Yano, T. Ono, H. Tanaka, K. Fujii, Uekusa, Y. Narumi and K. Kindo, *J. Phys. Soc. Jpn.*, **77**, 043707 (2008).

[54] S. Park and D. Keszler, *Solid State Sci.*, **4**, 799 (2002).

[55] K. Shimamura, H. Takeda, T. Kohno and T. Fikuda, *J. Cryst. Growth*, **163**, 388 (1996).

[56] P. Bordet, I. Gelard, K. Marty, A. Ibanez, J. Robert, V. Simonet, B. Canals, R. Ballou and P. Lejay, *J. Phys.: Condens. Matter*, **18**, 5147 (2006).

[57] J. Robert, V. Simonet, B. Canals, R. Ballou, P. Bordet, P. Lejay and A. Stunault, *Phys. Rev. Lett.*, **96**, 197205 (2006).

[58] A. Zorko, F. Bert, P. Mendels, P. Bordet, P. Lejay and J. Robert, *Phys. Rev. Lett.*, **100**, 147201 (2008).

[59] V. Simonet, R. Ballou, J. Robert, B. Canals, F. Hippert, P. Leiay, P. Fouquet, J. Ollivier and D. Braithwaite, *Phys. Rev. Lett.*, **100**, 237204 (2008).

[60] H.D. Zhou, C. R. Wiebe, Y.-J. Jo, L. Balicas, R.R. Urbano, L.L. Lumatta, J.S. Brooks, P.L. Kuhns, A.P. Reyes, Y. Qui, J.R.D. Copley and J.S. Gardner, *Phys. Rev. Lett.*, **102**, 067203 (2009).

[61] E.E. Sauerbrie, R. Faggiani and C. Calvo, *Acta. Cryst. B*, **29**, 2304 (1973).

[62] N. Rogado, G. Lawes, D.A. Huse, A.P. Ramirez and R.J. Cava, *Solid State Commun.*, **124**, 229 (2002).

[63] G. Lawes, M. Kenzelmann, N. Rogado, K.H. Kim, G.A. Jorge, R.J. Cava, A.H. Aharony, O. Entin-Wohlman, A.B. Harris, T. Yildirim, Q.Z. Huang, S. Park, C. Broholm and A.P. Ramrez, *Phys. Rev. Lett.*, **93**, 247201 (2004).

[64] M. Kenzelmann, A.B. Harris, A.H. Aharony, O. Entin-Wohlman, T. Yildirim, Q.Z. Huang, S. Park, G. Lawes, C. Broholm, N. Rogado, R.J. Cava, K.H. Kim, F.A. Jorge and A.P. Ramirez, *Phys. Rev. B*, **74**, 014429 (2006).

[65] G. Lawes. M. Kenselmann and C. Broholm, *J. Phys.: Condens. Matter.*, **20**, 434205 (2008).

[66] Y. Okamoto, M. Nohara, H. Aruga-Katori and H. Tagaki, *Phys. Rev. Lett.*, **99**, 137207(2007).

[67] G. Chen and L. Balents, *Phys. Rev. B*, **78**, 094403 (2008).

[68] W.P. Wolf, M. Ball, M.T. Hutchings, M.J.M. Leask and A.F.G. Wyatt, *J. Phys. Soc. Jpn.*, **17** (Suppl. B1), 443 (1962).

[69] W.I. Kinney and W.P. Wolf, *J. Appl. Phys.*, **50**, 2115 (1979).

[70] J. Fillipi, J.C. Lasjaunias, A. Ravex, F. Tcheou and J. Rossat-Mignod, *Solid State Commun.*, **23**, 613 (1971).

[71] D.J. Onn, H. Meyer and J.P. Remeika, *Phys. Rev.*, **156**, 663 (1967).

[72] S. Hov, H. Bratsberg and A.T. Skjeltorp, *J. Magn. Magn. Mater.*, **15–18**, 455 (1980); A.P. Ramirez and R.N. Kleiman, *J. Appl. Phys.*, **69**, 5252 (1991).
[73] P. Schiffer, A.P. Ramirez, D.A. Huse and A.J. Valentino, *Phys. Rev. Lett.*, **73**, 2500 (1994).
[74] P. Schiffer, A.P. Ramirez, D.A. Huse, P.L. Gammel, U. Yaron, D.J. Bishop and A.J. Valentino, *Phys. Rev. Lett.*, **74**, 2379 (1995).
[75] O.A. Petrenko, C. Ritter, M. Yethiraj and D. McK. Paul, *Phys. Rev. Lett.*, **80**, 4570 (1998).
[76] S.R. Dunsiger, J.S. Gardner, J.A. Chakalian, A.L. Cornelius, M. Jaime, R.F. Kiefl, R. Movshovich, W.A. McFarlane, R.I. Miller, J.E. Sonier and B.D. Gaulin, *Phys. Rev. Lett.*, **85**, 3504 (2000).
[77] I.M. Marshall, S.J. Blundell, F.L. Pratt, A. Husman, C.A. Steer, A.I. Coldea, W. Hayes and R.C.C. Ward, *J. Phys.: Condens. Matter*, **14**, L157 (2002).
[78] O.A. Petrenko, G. Balakrishnan, D. McK. Paul, M. Yethiraj and J. Klenke, *Appl. Phys.*, **A74** (Suppl.), S760 (2002).
[79] S. Ghosh, T.F. Rosenbaum and G. Aeppli, *Phys. Rev. Lett.*, **101**, 157205 (2008).
[80] M.T. Anderson, K.B. Greenwood, G.A. Taylor and K. Poeppelmeier, *Prog. Solid State Chem.*, **22**, 197 (1993).
[81] C.J. Howard, B.J. Kennedy and P.M. Woodward, *Acta. Cryst.*, **B59**, 463 (2003).
[82] A.K. Azad, S.A. Ivanov, S.-G. Ericksson, J. Eriksen, H. Rundlof, R. Mathieu and P. Svedlindh, *Mater. Res. Bull.*, **36**, 2215 (2001).
[83] G. Blasse, *Philips Res. Rep.*, **20**, 327 (1965).
[84] A. Munoz, J.A. Alonso, M.T. Casais, M.J. Martinez-Lope and M.T. Fernandez-Diaz, *J. Phys.: Condens. Matter.*, **14**, 8817 (2002).
[85] P.D. Battle and C.W. Jones, *J. Solid State Chem.*, **78**, 108 (1989).
[86] P.D. Battle, C.P. Grey, M. Hervieu, C. Martin, C.A. Moore and Y. Paik, *J. Solid State Chem.*, **175**, 20 (2003).
[87] Y. Sasaki, Y. Doi and Y. Hinatsu, *J. Mater. Chem.*, **12**, 2361 (2002).
[88] C.R. Wiebe, J.E. Greedan, G.M. Luke and J.S. Gardner, *Phys. Rev. B*, **65**, 14413 (2002).
[89] C.R. Wiebe, J.E. Greedan, P.P. Kyriakou, G.M. Luke, J.S. Gardner, A. Fukaya, I.M. Gat-Malureaunu, P.L. Russo, A.T. Savici and Y.J. Uemura, *Phys. Rev. B*, **68**, 134410 (2003).
[90] K.E. Stitzer, M.D. Smith and H.-C. zur Loye, *Solid State Sci.*, **4**, 311 (2002).
[91] A.S. Erickson, S. Misra, G.J. Miller, R.R. Gupta, Z. Schlesinger, W.A. Harrison, J.M. Kim and I.R. Fisher, *Phys. Rev. Lett.*, **99**, 016404 (2007).
[92] H.J. Xiang and M.H. Whangbo, *Phys. Rev. B*, **75**, 052407 (2007).
[93] K.-W. Lee and W.E. Pickett, *Eur. Phys. Lett.*, **80**, 37008 (2007).
[94] E.J. Cussen, D.R. Lynham and J. Rogers, *Chem. Mater.*, **18**, 2855 (2006).
[95] M.A. Subramanian, G. Aravamudan and G.V. SubbaRao, *Prog. Solid State Chem.*, **15**, 55 (1983).
[96] M.A. Subramanian and A.W. Sleight, *Handbook on the Physics and Chemistry of Rare Earths, Vol. 16, Rare Earth Pyrochlores*, K.A. Gschneidner and L. Eyring (Eds), Elsevier, 1993.
[97] N.P. Raju, M. Dion, M.J.P. Gingras, T. Mason and J.E. Greedan, *Phys. Rev. B*, **59**, 14489 (1999).
[98] A.P. Ramirez, B.S. Shastry, A. Hayashi, J.J. Krajewski, D.A. Huse and R.J. Cava, *Phys. Rev. Lett.*, **89**, 067202 (2002).

[99] J.D.M. Champion, A.S. Wills, T. Fennell, S.T. Bramwell, J.S. Gardner and M.A. Green, *Phys. Rev. B*, **64**, 140407 (2001).

[100] J.R. Stewart, G. Ehlers, A.S. Wills, S.T. Bramwell and J.S. Gardner, *J. Phys.: Condens. Matter*, **16**, L321 (2004).

[101] S.E. Palmer and J.T. Chalker, *Phys. Rev. B*, **62**, 488 (2000).

[102] O.A. Petrenko, M.R. Lees, G. Balakrishnan and D. McK. Paul, *Phys. Rev. B*, **70**, 012402 (2004).

[103] J.S. Gardner, S.R. Dunsiger, B.D. Gaulin, M.J.P. Gingras, J.E. Greedan, R.F. Kiefl, M.D. Lumsden, W.A. McFarlane, N.P. Raju, J.E. Sonier, I. Swainson and Z. Tun, *Phys. Rev. Lett.*, **82**, 1012 (1999).

[104] J.S. Gardner, B.D. Gaulin, A.J. Berlinsky, P. Waldron, S.R. Dunsiger, N.P. Raju and J.E. Greedan, *Phys. Rev. B*, **64**, 224416 (2001).

[105] J.P.C. Ruff, B.D. Gaulin, J.P. Castellan, K.C. Rule, J. Clancy, J. Rodriguez and H.A. Dabkowska, *Phys. Rev. Lett.*, **99**, 237202 (2007).

[106] H.R. Molavian, M.J.P. Gingras and B. Canals, *Phys. Rev. Lett.*, **98**, 157204 (2007).

[107] I. Mirabeau, I.N. Goncharenko, P. Cadavez-Peres, S.T. Bramwell, M.J.P. Gingras and J.S. Gardner, *Nature*, **420**, 54 (2002).

[108] B.C. den Hertog and M.J.P. Gingras, *Phys. Rev. Lett.*, **84**, 3430 (2000).

[109] S.V. Isakov, R. Moessner and S.L. Sondhi, *Phys. Rev. Lett.*, **95**, 217201 (2005).

[110] C. Castelnovo, R. Moessner and S.L. Sondhi, *Nature*, **451**, 42 (2008).

[111] T. Fennell, P.P. Deen, A.R. Wildes, K. Schmalzl, D. Prabhakaran, A.T. Boothroyd, R.J.Aldus, D.F. McMorrow and S.T. Bramwell, *Science*, **326**, 415 (2009).

[112] D.J.P. Morris, D.A. Tennant, S.A. Grigera, B. Klemke, C. Castelnovo, R. Moessner, K.C. Rule, J.-C. Hoffmann, K. Kiefer, S. Gerischer, D. Slobinsky and R.S. Perry, *Science*, **326**, 411 (2009).

[113] D.J. Flood, *J. Appl. Phys.*, **45**, 4041(1974).

[114] S. Rosenkranz, A.P. Ramirez, A. Hayashi, R.J. Cava, R. Siddharthan and B.S. Shastry, *J. Appl. Phys.*, **87**, 5914 (2000).

[115] I. Mirabeau and I.N. Goncharenko, *Physica B*, **350**, 250 (2004).

[116] G.C. Lau, B.D. Muegge, T.M. McQueen, E.L. Duncan and R.J. Cava, *J. Solid State Chem.*, **179**, 3126 (2006).

[117] G.C. Lau, R.S. Freitas, B.G. Ueland, B.D. Muegge, E.L. Duncan, P. Schiffer and R.J. Cava, *Nat. Phys.*, **2**, 249 (2006).

[118] H.D. Zhou, C.R. Wiebe, Y.J. Jo, L. Balicas, Y. Qui, J.R.D. Copley, G. Ehlers, P. Fouquet and J.S. Gardner, *J. Phys.: Condens. Matter*, **19**, 342201 (2007).

[119] H.W.J. Blöte, R.F. Wielinga and W.J. Huiskamp, *Physica*, **43**, 549 (1969).

[120] J.D.M. Champion, M.J. Harris, P.C.W. Holdsworth, A.S. Wills, G. Balakrishnan, S.T. Bramwell, E. Cizmar, T. Fennell, J.S. Gardner, J. Lago, D.F. McMorrow, M. Orendac, A. Orendacova, D. McK. Paul, R.I. Smith, M.T.F. Telling and A. Wildes, *Phys. Rev. B*, **68**, 020401 (2003).

[121] A. Poole, A.S. Wills and E. Lelièvre-Berna, *J. Phys.: Condens. Matter*, **19**, 452201 (2007).

[122] J.A. Hodges, P. Bonville, A. Forget, A. Yaouanc, P. Dalmas de Reotier, G. Andre, M. Rams, K. Krolas, C. Ritter, P.C.M. Gubbens, C.T. Kaiser, P.J.C. King and C. Baines, *Phys. Rev. Lett.*, **88**, 077204 (2002).

[123] Y. Yasui, S. Soda, S. Iikubo, M. Ito, M. Sato, H. Hamaguchi, T. Matsushita, N. Wada, T. Takeuchi, N. Aso and K. Kakurai, *J. Phys. Soc. Jpn.*, **72**, 3014 (2003).

[124] J.S. Gardner, G. Ehlers, N. Rosov, R.W. Erwin and C. Petrovic, *Phys Rev. B*, **70**, 180404 (2003).

[125] V. Bondha-Jagalu and S.T. Bramwell, *Can. J. Phys.*, **79**, 1381 (2001).

[126] K. Matsuhira, K. Sekine, C. Paulsen and Y. Hinatsu, *J. Magn. Magn. Mater.*, **272–6**, e981 (2004).

[127] S.R. Giblin, J.D.M. Champion, H.D. Zhou, C.R. Wiebe, J.S. Gardner, I. Terry, S. Calder, T. Fennell and S.T. Bramwell, *Phys. Rev. Lett.*, **101**, 237201 (2008).

[128] K.Matsushira, Y. Hinatsu, K. Tenya, H. Amitsuka and T. Sakakibara, *J. Phys. Soc. Jpn.*, **71**, 1576 (2002).

[129] G. Luo, S.T. Hess and L.R. Corruccini, *Phys. Lett. A*, **291**, 306 (2001).

[130] A.S. Wills, M.E. Zhitomirsky, B. Canals, J.P. Sanchez, P. Bonville, P. Dalmas de Reotier and A. Yaouanc, *J. Phys.: Condens. Matter*, **18**, L37 (2006).

[131] I. Mirabeau, A. Apetrei, J. Rodriguez-Carvajal, P. Bonville, A. Forget, D. Colson, V. Glazkov, J.P. Sanchez, O. Insard and E. Suard, *Phys. Rev. Lett.*, **94**, 246402 (2005).

[132] P. Dalmas de Réotier, A. Yaouanc, L. Keller, A. Cervellino, B. Roessli, A. Forget, C. Vaju, P.C.M. Gubbens, A. Amato and P.J.C. King, *Phys. Rev. Lett.*, **96**, 127202 (2006).

[133] F. Bert, P. Mendels, A. Olariu, N. Blanchard, G. Collin, A. Amato, C. Baines and A.D. Hillier, *Phys. Rev. Lett.*, **97**, 117203 (2006).

[134] Y. Chapuis, A. Yaouanc, P. Dalmas deReotier, S. Pouget, P. Fouquet, A. Cervellino and A. Forget, *J. Phys.: Condens. Matter*, **19**, 446206 (2007).

[135] K. Matsuhira, Y. Hinatsu, K. Tenya and T. Sakakibara, *J. Phys.: Condens. Matter*, **12**, L649 (2000).

[136] H. Kadowaki, Y. Ishii, K. Matsuhira and Y. Hinatsu, *Phys. Rev. B*, **65**, 144421 (2002).

[137] J. Lago, T. Lancaster, S.J. Blundell, S.T. Bramwell, F.L. Pratt, M. Shirai and C. Baines, *J. Phys.: Condens. Matter*, **17**, 979 (2005).

[138] K.W.H Stevens, *Proc. Phys. Soc., London*, **A65**, 209 (1952).

[139] J.E. Greedan, M. Sato, X. Yan and F. Razavi, *Solid State Commun.*, **59**, 895 (1986).

[140] P.H. Hubert, *Bull. Chim. Soc. Fr.*, **11**, 2385 (1974); P.H. Hubert, *Bull. Chim. Soc. Fr.*, **11-1**, 2463 (1975).

[141] K. Miyoshi, Y. Nishimura, K. Honda, K. Fujiwara and J. Takeuchi, *J. Phys. Soc. Jpn.*, **69**, 3517 (2000).

[142] N.P. Raju, E. Gmelin and R. Kremer, *Phys. Rev. B*, **46**, 5405 (1992).

[143] M.J.P. Gingras, C.V. Stager, N.P Raju, B.D. Gaulin and J.E. Greedan, *Phys. Rev. Lett.*, **78**, 947 (1997).

[144] J.N. Reimers, J.E. Greedan and M. Sato, *J. Solid State Chem.*, **72**, 390 (1988).

[145] C.H. Booth, J.S. Gardner, G.H. Kwei, R.H. Heffner, F. Bridges and M.A. Subramanian, *Phys. Rev. B*, **62**, R755 (2000).

[146] A. Keren and J.S. Gardner, *Phys. Rev. Lett.*, **87**, 177201 (2001).

[147] J.E. Greedan, D. Gout, A.D. Lozano-Gorrin, S. Derahkshan, H.-J. Kim, Th. Proffen, E. Bozin and S.J.L. Billinge, *Phys. Rev. B*, **79**, 014427 (2009).

[148] M. Ito, Y. Yasui, M. Kanda, H. Harashina, S. Yoshii, K. Murata, M. Sato, G. Okumura and K. Kakurai, *J. Phys. Chem. Solids*, **62**, 337 (2001).

[149] J. vanDuijn, N. Hur, J.W. Taylor, Y. Qui, Q.Z. Huang, S.-W. Cheong, C. Broholm and T.G. Perring, *Phys. Rev. B*, **77**, 020405 (2008).

[150] B. Canals and C. Lacroix, *Phys. Rev. Lett.*, **80**, 2933 (1998).

[151] D. Yanagishima and Y. Maeno, *J. Phys. Soc. Jpn.*, **70**, 2880 (2001).

[152] N. Taira, M. Wakeshima and Y. Hinatsu, *J. Phys.: Condens. Matter*, **13**, 5527 (2001).

[153] K. Matsuhira, M. Wakeshima, R. Nakanishi, T. Yamada, A. Nakamura, W. Kawano, S. Takagi and Y. Hinatsu, *J. Phys. Soc. Jpn.*, **76**, 043706 (2007).

[154] G.V. Bazuev, O.V. Marakova, V.Z. Oboldin and G.P. Shveikin, *Dokl. Akad. Nauk*, **230**, 869 (1976); L. Soderholm and J.E. Greedan, *Mater. Res. Bull.*, **17**, 707 (1982).

[155] H. Ichikawa, L. Kano, M. Saitoh, S. Miyahara, N. Furukawa, J. Akimitsu, T. Yokoo, T. Matsumura, M. Takeda and K. Hirota, *J. Phys. Soc. Jpn.*, **74**, 1020 (2005).

[156] H.D. Zhou, E.S. Choi, J.A. Souza, Y. Xin, L.L. Lumata, B.S. Conner, L. Balicas, J.S. Brooks, J.J. Neumeier and C.R. Wiebe, *Phys. Rev. B*, **77**, 020411 (2008).

[157] J. N. Reimers, J.E. Greedan, R.K. Kremer, E. Gmelin and M.A. Subramanian, *Phys. Rev. B*, **43**, 3387 (1991).

[158] Y. Shimakawa, Y. Kubo and T. Manako, *Nature*, **379**, 53 (1996).

[159] M. Sato, Xu Yan and J.E. Greedan, *Z. Anorg. Allg. Chem.*, **540/541**, 177 (1986).

[160] J.E. Greedan, M. Sato, Naushad Ali and W.R. Datars, *J. Solid State Chem.*, **68**, 300 (1987).

[161] I. Solovyev, *Phys. Rev. B*, **67**, 174406 (2003).

[162] N. Cao, T. Timusk, N.P. Raju, J.E. Greedan and P. Gougeon, *J. Phys.: Condens. Matter*, **7** 2489 (1995).

[163] I. Kezsmarki, N. Hanasaki, D. Hashimoto, S. Iguchi, Y. Taguchi, S. Miyasaka, and Y. Tokura, *Phys. Rev. Lett.*, **93**, 266401 (2004).

[164] Y. Yasui, Y. Kondo, M. Kanada, M. Ito, H. Harashina, M. Sato and K. Kakurai, *J. Phys. Soc. Jpn.*, **70**, 284 (2001).

[165] T. Katsufuji, H.Y. Hwang and S.-W. Cheong, *Phys. Rev. Lett.*, **84**, 1998 (2000).

[166] Y. Taguchi, Y. Oohara, H. Yoshizawa, N. Nagaosa and Y. Tokura, *Science*, **291**, 2573 (2001).

[167] Y. Yasui, M. Soda, S. Iikubo, M. Ito, M. Sato, N. Hamaguchi, T. Matsushita, N. Wada, T. Takeuchi, N. Aso and K. Kakurai, *J. Phys. Soc. Jpn.*, **72**, 3014 (2003).

[168] Y. Yasui, T. Kageyama, T. Moyoshi, H. Harashina, M. Soda, M. Sato and K. Kakurai, *J. Phys. Soc. Jpn.*, **75**, 084711 (2006).

[169] Y. Taguchi and Y. Tokura, *Phys. Rev. B*, **60**, 10280 (1999); Y. Taguchi and Y. Tokura, *Physica B*, **284**, 1448 (2000).

[170] A. S. Singh, R. Suryanarayanan, R. Tackett, G. Lawes, A.K. Sood, P. Berthet and A. Reveolevschi, *Phys. Rev. B*, **77**, 020406 (2008).

[171] I. Mirabeau, A. Apetrei, I. Goncharenko, D. Andreica, P. Bonville, J.P. Sanchez, A. Amato, E. Suard, W.A. Crichton, A. Forget and D. Colson, *Phys. Rev. B*, **74**, 174414 (2006).

[172] J.E. Greedan, J.N. Reimers, S.L. Penney and C.V. Stager, *J. Appl. Phys.*, **67**, 5967 (1990).

[173] B.D. Gaulin, J.N. Reimers, T.E. Mason, J.E. Greedan and Z. Tun, *Phys. Rev. Lett.*, **69**, 3244 (1992).

[174] S.R. Dunsiger, R.F. Keifl, K.H. Chow, B.D. Gaulin M.J.P. Gingras, J.E. Greedan, A. Keren, K. Kojima, G.M. Luke, W.A. McFarlane, N.P. Raju, J.E. Sonier, Y. Uemura and W.D. Wu, *Phys. Rev. B*, **54**, 9019 (1996).

[175] N. Taira, W. Makoto, H. Yukio, T. Aya and O. Kenji, *J. Solid State Chem.*, **176**, 5527 (2001).

[176] C. Bansal, H. Kawanaka, H. Bando and Y. Nishihara, *Phys. Rev. B*, **66**, 052406 (2002).

[177] C.R. Wiebe, J.S. Gardner, S.-J. Kim, G.M. Luke, A.S. Wills, B.D. Gaulin, J.E. Greedan, I. Swainson, Y. Qui and C.Y. Jones, *Phys. Rev. Lett.*, **93**, 076403 (2004).

[178] J.S. Gardner, A.L. Cornelius, L.J. Chang, M. Prager, Th. Brukel and G. Ehlers, *J. Phys.: Condens. Matter*, **17**, 7089 (2005).

[179] Y. Machida, S. Nakatsuji, H. Tonomura, T. Tayama, T. Sakakibara, J. van Duijn, C. Broholm and Y. Maeno, *J. Phys. Chem. Solids*, **66**, 1435 (2005).

[180] S. Nakatsuji, Y. Machida, Y. Maeno, T. Tayama, T. Sakakibara, J. van Duijn, L. Balicas, J.N. Millican, R.T. Macaluso and J.Y. Chan, *Phys. Rev. Lett.*, **96**, 087204 (2006).

[181] U. Konig, E.F. Bertaut, Y. Gros, M. Mitrikov and G. Chol, *Solid State Commun.*, **8**, 759 (1970).

[182] W. Schiessl, W. Potzel, H. Karzel, M. Steiner, G.M. Kalvius, A. Martin, M.K. Krause, I. Halgvy, J. Gal, W. Schaefer, G. Will, M. Hillberg and R. Wappling, *Phys. Rev. B*, **53**, 9143 (1996).

[183] E.F. Bertaut, V.V. Qui, R. Pauthenet and A. Murasik, *J. Phys. (France)*, **25**, 516 (1964).

[184] R. Plumier, *C.R. Acad. Sci.*, **264**, 278 (1967).

[185] S. Diaz, S. deBrion, M. Holzapfel, G. Chouteau and P. Strobel, *Physica B*, **346–7**, 146 (2004).

[186] J.C. Lashley, R. Stevens, M.K. Crawford, J. Boerio-Goates, B.F. Woodfield, Y. Qui, J.W. Lynn, P.A. Goddard and R.A. Fisher, *Phys. Rev. B*, **78**, 104406 (2008).

[187] S. Diaz, S. deBrion, G. Chouteau, B. Canals, V. Simonet and P. Strobel, *Phys. Rev. B*, **74**, 092404 (2006).

[188] Y. Kino and B. Luthi, *Solid State Commun.*, **9**, 805 (1971).

[189] H. Martinho, N.O. Moreno, J.A. Sanjurjo, C. Rettori, A. Garcia-Adeva, D.L. Huber, S.B. Oseroff, W. Ratcliff II, S.-W. Cheong, P.G. Pagliuso, J.L. Sarro and G.B. Martins, *J. Appl. Phys.*, **89**, 7050 (2001).

[190] S.-H. Lee, C. Broholm, T.H. Kim, W. Ratcliff II and S.-W. Cheong, *Phys. Rev. Lett.*, **84**, 3718 (2000).

[191] S.-H. Lee, G. Gasparovic, C. Broholm, M. Matsuda, J.-H. Chung, Y.J. Kim, H. Ueda, G. Xu, P. Zschack, K. Kakurai, H. Takagi, W. Ratcliff II, T.H. Kim and S.-W. Cheong, *J. Phys.: Condens. Matter*, **19**, 145259 (2007).

[192] Y. Tsunoda, H. Suzuki, S. Katano, K. Siratori, E. Kita and K. Kohn, *J. Phys. Soc. Jpn.*, **75**, 064710 (2006).

[193] S.-H. Lee, C. Broholm, G. Gasparovic, W. Ratcliff II, Q. Huang, T.H. Kim and S.-W. Cheong, *Nature*, **418**, 856 (2002).

[194] W. Ratcliff II, S.-H. Lee, C. Broholm, S.-W. Cheong and Q. Huang, *Phys. Rev. B*, **65**, 220406R (2002).

[195] V.N. Glazkov, A.M. Farutin, V. Tsurkan, H.-A. Krug von Nidda and A. Loidl, *Phys, Rev. B*, **79**, 024431 (2009).

[196] J. Rodriguez-Carvajal, G. Rousse, C. Masquelier and M. Hervieu, *Phys. Rev. Lett.*, **81**, 4660 (1999).

[197] J. Akimoto, Y. Takahashi, N. Kijima and Y. Gotoh, *Solid State Ionics*, **172**, 491 (2004).

[198] A.S. Wills, N.P. Raju and J.E. Greedan, *Chem. Mater.*, **11**, 1510 (1999).

[199] Y. Oohara, J. Sugiyama and M. Kontani, *J. Phys. Soc. Jpn.*, **68**, 242, (1999).

[200] I. Tomeno, Y. Kasuya and Y. Tsunoda, *Phys. Rev. B*, **64**, 094422 (2001).

[201] J.E. Greedan, C.R. Wiebe, A.S. Wills and J.R. Stewart, *Phys Rev. B*, **65**, 184424 (2002).

[202] J.E. Greedan, N.P. Raju, A.S. Wills, C. Morin, S.M. Shaw and J.N. Reimers, *Chem. Mater.*, **10**, 3058 (1998).

[203] H. Berg, H. Rundlov and J.O. Thomas, *Solid State Ionics*, **144**, 65 (2001).

[204] D. Morgan, B. Wang, G. Ceder and A. van de Walle, *Phys. Rev. B*, **67**, 134404 (2003).

[205] A.S. Wills, N.P. Raju, C. Morin and J.E. Greedan, *Chem. Mater.*, **11**, 1936 (1999).

[206] C.R. Wiebe, P.L. Russo, A.T. Savici, Y.J. Uemura, G.J. MacDougall, G.M. Luke S. Kuchta and J.E. Greedan, *J. Phys.: Condens. Matter*, **17**, 6469 (2005).

[207] W. L. Roth, *J. Phys.*, **25**, 507 (1964).

[208] J. Soubeyroux, D. Fiorani, E. Agostinelli, S.C. Bhargava and J.L. Dormann, *J. Phys. Coll.*, **C8**, 1117 (1988).

[209] N. Tristan, J. Hemberger, A. Krimmel, H.-A. von Nidda, V. Tsurkan and A. Loidl, *Phys. Rev. B*, **72**, 174404 (2005).

[210] A. Krimmel, V. Tsurkan, D. Sheptyakov and A. Loidl, *Physica B*, **378–80**, 583 (2006).

[211] T. Suzuki, H. Nagai, M. Nohara and H. Takagi, *J. Phys.: Condens. Matter*, **19**, 145265 (2007).

[212] S.-B. Lee and L. Balents, *Phys. Rev. B*, **78**, 144417 (2008).

[213] G. Liu and J.E. Greedan, *J. Solid State Chem.*, **122**, 416 (1996).

[214] C.A. Bridges and J.E. Greedan, *J. Solid State Chem.*, **177**, 4516 (2004).

[215] C.A. Bridges, T. Hansen, A.S. Wills, G.M. Luke and J.E. Greedan, *Phys. Rev. B*, **74**, 024426 (2006).

[216] S. Derakhshan, J.E. Greedan, T. Katsumata and L.M.D. Cranswick, *Chem. Mater.*, **20**, 5714 (2008).

3

Lithium Ion Conduction in Oxides

Edmund Cussen

Department of Pure and Applied Chemistry, University of Strathclyde, Glasgow, Scotland

3.1 INTRODUCTION

Definitions of a crystalline material make use of a number of phrases to indicate the immutable, static, and ordered arrangement of atoms typically encountered in a crystal. This is elegantly encapsulated by the origins of the word; the ancient Greeks believed that crystals of quartz were formed from water that had been cooled so dramatically as to undergo an irreversible freezing process. This idea is recorded in the shared roots of the words crystal and cryogen. Whilst our understanding of crystals has developed in the intervening millennia it is interesting to note that the idea that the Greeks had formed remains one the strongest theme of a crystal. The atoms are frozen.

Of course, the main feature of a crystal, and what distinguishes it from other solids, is that it contains order. Conceptually, we describe crystals as being formed from a perfectly ordered array of atoms or molecules. This order has a helpful consequence; by describing a small portion of the structure and the symmetry of crystal we can map out the atomic positions of an infinite lattice. The power of this approach is beguiling. It allows us to map all of the atoms in a crystal, which may be metres in

Functional Oxides   Edited by Duncan W. Bruce, Dermot O'Hare and Richard I. Walton size,[1] by using a handful of parameters. Time is not one of these parameters; the structure of crystals is considered to be permanent and in the vast majority of crystals this is an acceptable approximation.

In this review we will be looking at an array of fascinating materials that provide extreme examples of both disorder and dynamic behaviour whilst retaining a crystalline arrangement of some or all of the atoms in the structure. The degree of movement that can occur within a crystal is astonishing; at room temperature sodium β-alumina shows the same ionic conductivity as an aqueous solution of 0.1 M sodium chloride. Such fast ion mobility can be achieved in materials that have mechanical properties that often resemble those of conventional ceramics. The polycrystalline powders can be sintered to give mechanically robust pellets that will not only retain shape but will also withstand compressive forces, high temperatures and in many cases be chemically unreactive. Whilst these properties are, of themselves, unremarkable in ceramics they certainly provide a strong contrast to those of electrolytes which are typically liquids. Consequently there is considerable interest in the technological development of these solid state electrolytes for a number of applications in which mechanical robustness, possibly combined with thermal or chemical stability, is important.

The principle technological motivation for studying lithium ion conductors comes from the potential advantages conferred by a thermally stable, chemically inert and mechanically robust electrolyte on the performance and safety of lithium batteries. The widespread usage of mobile electronic devices over the last two decades has, in part, been facilitated by developments in lithium battery technology. These have substantially arisen from the development of new electrode materials that have given substantial increases in energy density,[2] although improvements in volume and gravimetric density are both sought in order to facilitate the development of new applications, most obviously in electrically powered vehicles. Hand in hand with these improvements in energy density comes an increase in danger associated with the failure of a fully charged battery. Any short circuit across the electrolyte can lead to rapid discharge of the cell and associated heating and the potential for ignition if sufficient fuel sources are nearby. Consequently it seems probable that existing safety concerns will become more pressing as the energy densities of batteries are increased further.

Polymeric materials containing dissolved lithium salts are currently used as electrolytes and a range of polymer/solvent systems have been studied.[3] These materials perform effectively and have some considerable advantages in terms of processing, but many of the systems that provide the highest conductivity are highly reactive. $LiPF_6$ provides excellent conductivity of *ca* 10^{-3} S cm^{-1} and is compatible with the electrode materials,[4] but is subject to reaction with water leading to the formation of the extremely toxic product hydrofluoric acid.[5] This tendency to undergo hydrolysis is a problem that can be tolerated by using appropriate quality control to ensure that water is absent when the devices are assembled and employing suitable containment to prevent subsequent exposure to moisture during the operational lifetime of the battery. However, if these reactive electrolytes could be replaced by electrically insulating, inert, mechanically robust ceramics that show fast lithium ion conductivity in order to give improved temperature stability and eliminate the problems of chemical reactivity there is scope for considerable improvements in volumetric density and safety. Despite a growing number of families of fast lithium ion conducting ceramics the search for a material that meets these criteria and is chemically compatible with the electrode materials under the conditions of operation of a cell is ongoing.

It should be borne in mind that the lithium conductivity in the majority of crystalline solid state electrolytes occurs in compounds where all species, including the lithium cations, appear to be fixed on certain positions in the structure. Of course, this statement requires some justification and illustrates a potential difficulty in applying the language of a static model (a crystal) to a dynamic process (ionic mobility). In order to clarify this point and because of the unique utility of crystallographic models in the description of structure, it is useful to remember what is being described in such a model and how the data are generated. Any diffraction experiment relies on the interaction between an atomically ordered solid and an incident wave with a wavelength similar to the interatomic spacing. This wave, whether an X-ray, electron or neutron, is propagated through the material and scattered by the atoms. We note here that whilst the scattering of X-rays and electrons occurs *via* interaction with the electron cloud of an atom the scattering of neutrons occurs *via* interactions with the nucleus. Hence, whilst the scattering of X-rays and electrons increases rapidly with atomic number the scattering of neutrons shows no such dependency. In fact, the scattering of neutrons shows no regular dependence on atomic number and so neutrons may be scattered by a light element more efficiently than heavier elements. In our case it is of particular interest that the scattering length of ^7Li, the most abundant isotope of lithium, is similar in magnitude to other nuclei and so lithium can be readily detected. By comparison the intensity of an X-ray beam scattered by an atom will vary with the square of the number of electrons and so the intensity arising from a lithium atom in $Li_{0.5}La_{0.5}TiO_3$ will be *ca* 1/360th that scattered by a lanthanum atom. Consequently the detection of lithium *via* X-ray diffraction can be extremely imprecise and under some circumstances, such as in the presence of dominant scatterers in the example here, may provide no useful accuracy whatsoever.

The vast majority of scattered waves will be out of phase with other scattered waves. As these are propagated through the lattice this will tend towards cancellation and so no diffracted beam is observed. However, under the Bragg condition constructive interference will occur between the waves and a diffracted beam of radiation will be formed. The key point to be remembered from this is that the formation of a diffracted beam relies on the summation of multiple scattered waves. These waves are scattered from different regions of the material and in a typical diffraction experiment are collected over a time period measured in seconds. These two features mean that any model which is derived from diffraction experiments will necessarily present a picture of the structure that has been integrated over both time and space. Of course in the case of a regular array of immobile atoms in a crystalline lattice such an averaging process presents a structural picture which may conform very closely to reality, assuming we neglect the thermal vibration of atoms. As we shall see, in the case of fast lithium ion conductors the results of diffraction experiments can produce structural pictures which most definitely do not conform to any chemical sensible representation of a structure. Instead it may be necessary to interpret what the averaged structure derived from the diffraction data can tell us about what situations exist locally within the material and then consider how these chemically sensible arrangements can be assembled to produce the ensemble represented by the crystallographic model.

Structural models derived in such a manner are useful in understanding why exceptional ionic mobility occurs in some materials. However diffraction gives us no direct information on mobility; it is unable to observe an atom moving between positions. The reason for this is that the most common mechanism for ion movement through a solid material involves hopping of the mobile species onto a nearby position. The ions hop from an occupied position to a nearby, vacant position at a speed which is similar to those that arise from thermal vibration. Consequently the time taken for an atom to move from an occupied site to a vacant position, *ca* 10^{-12} s, is only an order of magnitude larger than the period of an atomic vibration and occurs considerably quicker than the propagation of a diffracted wave through a portion of crystalline material. This means that a diffraction experiment will detect an atom at the start or end of a hop, or an average of the two, but will contain no information about the path that the atom takes whilst travelling between these two points.

In order to achieve rapid ion mobility this hopping mechanism suggests two key points that are widely useful in our understanding of fast lithium ion conduction in crystalline solids. First, in order to obtain lithium mobility it is necessary to have vacant sites suitable for lithium coordination. These must be accessible to a lithium cation on an occupied site and this means that they must be a short distance away and also have a similar energy. Clearly a site which is located 1.5 Å away from a lithium ion may be accessible by hopping a short distance, but if this vacant site lies within the van der Waal's radius of another atom then electrostatic repulsion will prevent both sites from being occupied for any significant period of time. Similarly a lithium cation will be unable to hop from one site to another energetically similar vacant position if the distance between them is too large, typically >2–3 Å. An important consequence of this is that there in order to optimise the probability of an ion hopping event occurring there is a trade-off. The greater the number of lithium ions in a structure the greater probability of one of these ions having sufficient energy to leave a site. But in order for an ion to hop it needs to have the greatest number of vacant sites available within an accessible volume and energy. Consequently it is anticipated that conductivity occurring via this simple mechanism should be greatest when the number of lithium cations and lithium site vacancies are equal.

A second consideration that arises from this model is that hopping is most likely to occur if there is minimal energetic barrier between the two sites. Lithium cations in oxides are most commonly found in tetrahedral or octahedral coordination. For a cation occupying a position at the centre of either of these polyhedra it will be necessary for the cation to move closer to one or more oxide anions. This will lead to electrostatic repulsion between the electrons of the ions and so introduces an activation barrier for ion migration out of the site. The activation energy will be lowest if the path for ion hopping maintains the largest separation possible between the cation and the anions and so it follows that the cation is most likely to exit a regular tetrahedron or octahedron by passing through one of the triangular faces. Such a consideration leads to the idea of ions passing through an aperture, or window, during a hopping event. One approach to increasing the probability of ion hopping is to chemically modify a structure in order to increase the size of the aperture and so reduce the activation energy. This is most commonly undertaken by trying to increase the separation between the atoms that constitute the limiting dimensions of the window. An alternative approach is apparent if it is remembered that the atomic positions represent the average position about which the oxide ion is vibrating. The degree of atomic displacement due to thermal vibration will clearly vary with temperature but it is useful to remember that studies of the passage of various guest molecules through the apertures of zeolites[6, 7] have shown that the window sizes determined crystallographically can underestimate the size of guests that can be admitted by ca 0.4 Å. It is likely that a similar degree of mismatch between effective aperture size and that determined crystallographically by subtracting the relevant ionic radii from the interatomic distances exists in many of the fast ion conducting phases discussed here. It has been shown in zeolites that the size of this disparity can depend on the bonding arrangements and the polarisabilities of ions involved and it is probably that such chemical adjustments also play a role in the lithium ion conduction.

This idea of hopping is a useful one, but it represents only part of the story of ionic conductivity. Once an ion has hopped from a site to another site of similar energy then one of two things can happen. The most likely event is that the ion will reside for a short period of time on the second site before hopping back to the first site. At the end of this the arrangement will be as if nothing had happened. If the second site has a high energy then such a reversal is more likely. Indeed at an extreme illustration of this situation, it could be considered that atomic vibrations represent hopping events between sites with negligible residence time. Of course such an event cannot be described as ion conduction because no charge has been transported. For a material to be described as an ionic conductor it is necessary for the ion to reside on the second site for some period before then hopping again to a third site, then fourth site and so on through the material. It is thus necessary for a continuous network of vacant sites to exist in order for ionic conductivity to occur. Of course atoms may move in this manner in any crystalline substance and, at least for simple ionic systems, the time taken for an atom to hop is largely constant regardless of the material. What causes some materials to be classed as fast ion conductors, with an ionic mobility that may match a liquid salt, and other crystals to be ionic insulators is the frequency with which these hopping events occur. In a fast ion conductor the ion may only be resident on a site for a few nanoseconds before hopping again. This does mean that an ion spends only ca 1/1000 of the time hopping, but this mobility seems rather brisk when compared with the geological residence times that occur in conventional, ionically insulating compounds.

The discussion here is a brief summary of some of the considerations in a simple model for lithium conduction in an ionic solid. In practice, the movement of a lithium ion may be strongly correlated the vibrational modes of the lattice and the movement of other lithium ions and is likely to involve some degree of local relaxation of the lattice around sites which switch from containing vacancies to accommodating a lithium ion. In some remarkable cases, such as lithium sulfate, the lithium ion conductivity is associated with partial melting of the anion sublattice within the crystalline material. Despite the complexities of individual systems the basic hopping model provides a useful framework to discuss ion conductivity in the solid state and it is supported by considerable experimental evidence.[8]

Fast ion conduction is most likely to occur if the occupied and vacant sites provide the same environment for lithium cations and so occupation of either site makes the same contribution to the lattice energy.[9] If such a situation arises then it is likely that the lithium cation will be distributed across multiple sites in the material with partial occupancy, *i.e.* occupational disorder exists in the material. The identification of small quantities of either lithium cations or vacancies is a challenge for diffraction experiments due to the observation of an average structure. In compounds where the lithium ions are mobile it necessarily follows that there are multiple coordination sites for lithium which provide similar lattice energies and so an additional complication arises if sample preparation is not rigorously reproduced. Due to the dynamic properties of lithium at relatively low temperatures, and the plethora of low lying excited states for the crystal structure, it would not be surprising to stabilise structures with varying lithium distributions by using different synthetic protocols. As we shall see, there are several classes of compounds where relatively subtle changes in synthesis temperature or cooling rate can have a profound effect on both the crystallographic structure and the lithium conduction observed in the resultant product.

Despite the occasional difficulties of interpreting the average structure obtained from diffraction experiments a crystallographic description of a compound is the single most useful piece of information in understanding ionic conductivity in a material. However, a range of other techniques have been brought to bear on these problems and in many cases it is only by combining the information from a range of local probes with structural information and transport properties that it is possible to build a picture of the mechanism for ion transport in a certain system. A description of the various experimental techniques and the role of calculations in understanding ion mobility would be out of place here and have been reviewed elsewhere.[10] Instead we shall focus on how ionic conductivity in crystalline phases can be understood with reference to the structure of these inorganic materials.

3.2 SODIUM AND LITHIUM β-ALUMINA

The field of fast ion transport in solids was revolutionised by the observation of fast Na^+ conduction in a series of compounds known, somewhat confusingly, as the β-aluminas. These compounds were extensively studied[11–13] through the 1970s and 1980s and, although more advanced materials have since been developed, a study of the lithium conduction in these phases can serve us as both a useful overview of mechanisms for fast ion conduction and also illustrate some of the challenges and pitfalls that are encountered in trying to understand and optimise ionic mobility.

The compounds that were to become known as β-alumina have been known, and studied, since 1916.[14] The interest in these materials was initially stimulated by practical considerations arising in the glass making industry; a new phase was observed forming in the aluminium oxide bricks that line the refractory furnace used in glass manufacture. Understanding this phase became a concern for both furnace engineers and glass manufacturers as well as an interesting problem for structural chemists to investigate. Preliminary analysis suggested that a new polymorph of Al_2O_3 had been found and so the name β-alumina was attached to this material.[14] However, as early as 1926, analyses were identifying the presence of a relatively small and variable quantity of sodium in the material. Moreover the presence of sodium was identified as a crucial factor in the formation of this phase.[15] Early crystallographic studies suggested a composition $Na_2Al_{22}O_{34}$ that, whilst at odds with initial compositional studies, has since been proved to be an accurate description of the prototypical structure which in the most straightforward compound can be represented as being composed of intergrowths of sodium oxide and aluminium oxide.[16] This has lead to the widely employed simplification of the composition; $Na_2O \cdot 11Al_2O_3$.

Interest in these compounds was revived by the report of exceptional ionic mobility;[17] the Na^+ cations in this crystalline solid give a conductivity at room temperature, 0.2 S cm^{-1}, that is similar to that of an aqueous solution of 0.1 mol dm^{-3} of NaCl. Our interest in these compounds arises from the facile cation exchanges that can be performed in order to replace Na^+ and so give a range of new compounds, many of which demonstrate fast ion mobility. Treatment of β-alumina with a molten lithium salt gives a range of compounds in which some, or almost all, of the sodium cations are replaced with lithium.[12, 17] In order to understand how the exchange and the ion mobility can occur it is necessary to have an appreciation of the structure of β-aluminas. As we shall see, these compounds contain a number of structural features that are a common ingredient in the recipe for ion mobility in crystalline materials.

The structure of these materials can be considered as composites between blocks of spinel structure, predominantly composed of Al_2O_3, separated by sheets that contain the majority of the alkali metal cations. The idealised structure of $Na_2O \cdot 11Al_2O_3$ is represented in Figure 3.1. The structure contains a close-packed arrangement of oxide anions that are stacked in an *ABCA* arrangement to gives blocks of material that resemble the structure of spinel, $MgAl_2O_4$. In the idealised structure these blocks are uniquely occupied by Al^{3+} though, as we shall see shortly, there is scope for considerable positional and occupational disorder in these phases. These spinel blocks are separated by a layer of oxide anions *B* that contains oxide vacancies on up to 75% of the positions and this space is instead occupied by the sodium cations. Due to the large number

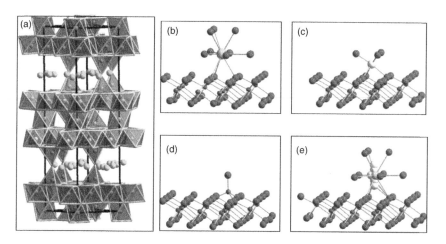

Figure 3.1   (a) A simplified representation of the structure of sodium β-alumina. The spinel block is composed of AlO_4 and AlO_6 units represented by tetrahedra and octahedra respectively. The disordered arrangement of sodium cations in the interlayer region is indicated by grey spheres. The partially occupied oxide positions in the interlayer region have been omitted for clarity. Chemically reasonable coordination environments are shown for interlayer (b) Na^+, (c) Li^+ and (d) Al^{3+} cations. The complexity of the observed Li^+ and Na^+ distribution in the same large interstice illustrated in (a) is shown in (e)[22]

of vacancies within this layer there are a number of cation sites available to sodium as shown in Figure 3.1. For the ideal stoichiometry $Na_2O \cdot 11Al_2O_3$ exactly 50% of these sites would be occupied by a sodium ion. In the β-alumina phases this disordered alkali metal layer lies on a mirror plane and so leads to an inversion of the stacking sequence *(ABCA)B(ACBA)B(ABCA)...* and so on. The β″-alumina phases are similar but lack the mirror symmetry in the alkali metal layer and so are formed by a continuous *(ABCA)B(ABCA)* stacking of oxide anions throughout the structure.[18]

This picture of the β-alumina structure masks a large degree of compositional variation that can be accommodated by disorder within the structure. Although often referred to using the $Na_2O \cdot 11Al_2O_3$ formalism, in practice most samples contain a considerable excess of sodium. The sodium cation is too large to fit into either the tetrahedrally or octahedrally coordinated sites of the spinel blocks and so the excess sodium is incorporated into the cation-rich layers. Charge neutrality is preserved by simultaneous incorporation of half of an equivalent oxide anion into this region and so the general composition of sodium β-alumina can be written $[Na_2O]_{1+x} \cdot 11Al_2O_3$. It is this ability to continually vary the Na:Al ratio within the structure that leads to variation in the reported composition, structure and most importantly a change in the properties of these materials. Indeed this extended tour of the structure of a fast Na^+ conductor serves as a both a guide to the structure and a cautionary tale that shall be encountered repeatedly when examining fast ion conducting phases; ion mobility in a crystal almost always goes hand-in-hand with some continuously variable disorder parameter. A consequence of this is that samples prepared using subtly different synthetic conditions or apparently the same conditions, but as practised in different laboratories, can lead to samples with dramatically different properties.

Examination of the Li^+ mobility in β-alumina requires exchange of Na^+ from the sodium based precursors. However, the equilibrium for the Na^+/Li^+ exchange strongly favours retention of Na^+ in the structure and so in order to achieve anything near a complete cation exchange it is necessary to use an extremely pure lithium salt. If mixed Na/Li phases are required then exchange is carried out using ≤1% sodium nitrate in a lithium nitrate melt.[19] Even at this low sodium concentration it is possible for an equilibrated sample to have a composition as lithium-poor as 20:80 Li:Na. In every case it is vital to determine experimentally the compositions of the samples after the cation exchange.

The ionic conductivity in the pure sodium β-alumina is the highest of any alkali metal cation in the β-alumina structure. Depending on the synthesis conditions the room temperature conductivity varies slightly in the range 0.014–0.030 S cm^{-1}.[17, 19, 20] Studies on single crystals show that, as could be anticipated from the structure of the material, the ion transport is highly anisotropic; Na$^+$ can move readily within the Na-O layers, but cations are blocked from migrating through the spinel blocks. The ion transport in sodium β-alumina follows an Arrhenius dependence over a remarkably wide temperature range with an activation energy of 0.16 eV being observed between −150 and 820 °C.[17] Identifying the lithium mobility in the analogous lithium material proved to be a much greater challenge. In part this arose from the difficulties encountered in preparing a sample that contained purely lithium. However, the significance of this can only be appreciated after the successful study of mixed Lithium-sodium β-alumina phases. These showed that the not only is Li$^+$ an order of magnitude less mobile that Na$^+$ in these phases, but the presence of a mixture of alkali cations gives conductivities that show a strong nonlinear dependence with composition as shown in Figure 3.2.[12] This nonlinearity combined with the tendency of the structure to selectively scavenge sodium cations from impure lithium salts made the determination of the transport properties of a lithium β-alumina a considerable challenge for solid state scientists. It was over a decade after the report of fast ion mobility in the β-alumina structure that a definitive picture of the mobility of lithium ions in this structure was built up. The transport in pure lithium β-alumina is thermally activated with an activation energy of 0.24 eV over the temperature range −80 to 400 °C and a room temperature conductivity of 3 × 10^{-3} S cm^{-1}.[19]

A naive approach to the question of alkali metal cation mobility in this, or any, crystalline structure would observe that as lithium is both lighter and smaller than sodium it should be the more mobile

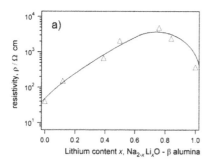
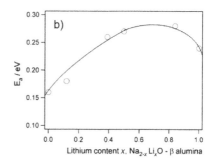

Figure 3.2   (a) The electrical resistivity and (b) the activation energy for conduction of various β-alumina at 25 °C as a function of Li/Na composition[19]

species. Observations in the β-alumina structure show that such an approach has considerable pitfalls in failing to take into account either the constraints imposed by the host lattice or the possibilities of cooperative mechanisms for ionic motion. The identification of lithium from X-ray diffraction data can be fraught with potential ambiguities. This is especially true when disordered lithium shares crystallographic sites with other cations. The absence of unambiguous, precise structural information has led to the development of innovative experiments designed to probe the cation environments in this structure. An ingenious experiment studied the conductivity of a series of samples of various alkali metal β-aluminas as a function of applied hydrostatic pressure.[21] The resulting data showed the highly informative pressure dependencies shown in Figure 3.3.

These data show that the passage of lithium through the structure is eased by the application of pressure. The crude picture of ion hopping through a narrow aperture would always lead to a reduction in mobility with increasing applied pressure as the compression of the crystal structure leads to a narrowing of the window and an increase in resistivity, as observed for K^+. The opposite observation for Li^+ mobility in this case suggested that lithium cations are in some way too small for optimal movement through the structure. The passage of Na^+, through some happy combination of circumstances, is independent of pressure suggesting that this cation is optimally matched to the structure.

This curious observation can be rationalised by looking at the coordination environments of Li^+ and Na^+ in the interlayer region.[22] These are most starkly illustrated in the mixed phase $Li_{1.1}Na_{0.9}O \cdot 11Al_2O_3$ that

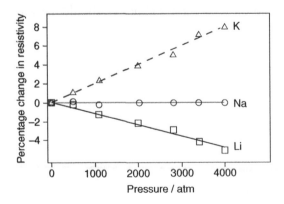

Figure 3.3 The change in resistivity of potassium, sodium, and lithium β-alumina as a function of applied pressure[21]

shows a displacement of Li^+ from the mirror plane towards the spinel sheets.[23, 24] This is driven by the bonding requirements of the Li^+, the lithium cation is simply too small to reside on a position equidistant from the two spinel blocks. Instead it undergoes a displacement that, whilst lengthening the $Li \cdots O$ distance to one of the layers reduces the distance to the other and so permits the formation of a lithium oxide bond. It was speculated that the increased polarising effect of Li^+ compared with Na^+ gave a stronger, more covalent bond than for Na^+. This has been supported by XANES studies of the local coordination environment in these materials that show a much stronger interaction between Li^+ cations and the spinel block than for any other alkali metal cations.[24]

The displacement of lithium from the plane has an interesting effect on the dielectric properties of β-aluminas.[25, 26] The transport of Li^+ though the two-dimensional sheets between the spinel layers involves a motional component perpendicular to the sheets as the lithium cations migrate from sites below this conduction plane to the neighbouring sites that lie above it. This motion has been detected by dielectric measurements on single crystals that show Li^+ hopping perpendicular to the conduction plane. This enhanced interaction with one of the pair of spinel blocks is the origin of the increase in activation energy for cation motion of Li^+ compared with Na^+. It also provides a clue as to why these structures tend to form with sodium at high temperature, but require low temperature modification to yield phases containing other cations. The sodium phase is thermodynamically stable due to the excellent match in size between the interlayer void and the radius of Na^+. The mismatch in radius for all other alkali cations means that these materials can only be formed by kinetically controlled manipulation of the β-alumina structure to yield materials that are only metastable at high temperature.

The mechanism for ion motion in the β-aluminas involves a complex interstitialcy mechanism.[20] There are three distinct sites within the conducting layers that can accommodate alkali metal cations. The stoichiometric phase $A_2O \cdot 11Al_2O_3$ would contain all of the A^+ cations on the filled site (1) but as these compounds always contain an excess of A^+ cations, additional interstitial sites (3) are also partially occupied. Initial calculations on ion hopping between sites showed that the energy to hop from (1) to a vacant site (2) is an order of magnitude greater than the reported value of 0.26 eV. Subsequent work has revealed the crucial importance of site (3). The activation energy for a cation to move from one (3) position to another via an intermediate (2) site is reduced to 0.26 eV in excellent agreement with the experimentally determined values.

The closely related β''-alumina structures exhibit an even wider range of ion exchange and mobility. Whilst the movement of divalent cations into the β-alumina structure is slow, it is possible to readily exchange a wide variety of divalent cations into Na^+ β''-alumina.[27] Remarkably the mobility seems to be a strong function of the structure rather than the cation and so the ionic conductivity in Pb^{2+} β''-alumina is almost equal to that of Na^+ β''-alumina, 10^{-3} S cm^{-1} around room temperature.[13]

The mobility of Na^+ in the β-alumina structure is similar to that in a molten salt and whilst the lithium cation is somewhat less mobile in the lithium exchanged analogue it may be surprising that these materials are not widely used and that the search for new conductors is still so vigorously prosecuted. Whilst the β-alumina and β''-alumina structures provide a general route to ion mobility in crystalline oxide they are critically limited in application due to moisture sensitivity.[19, 28] The ease of exchange is a pitfall that is revealed by the ready replacement of the desired cation with H^+ or H_3O^+ cations. Such exchange clearly has a huge impact on the mobility of the desired cation and has prevented these phases from realising the potential that could have been expected.

3.3   AKALI METAL SULFATES AND THE EFFECT OF ANION DISORDER ON CONDUCTIVITY

As befits an simple metal salt, the structural chemistry of lithium sulfate was studied relatively early in the history of crystallography.[29] This compound contains a face-centred array of sulfate anions that accommodate lithium cations in the tetrahedral interstices. With cubic symmetry this would be an example of an anti-fluorite phase, but at room temperature the compound possesses monoclinic symmetry and a small metric distortion from a regular cubic unit cell.[30] The structure of β-Li_2SO_4 is shown in Figure 3.4 and contrasted with the α-Li_2SO_4 phase that results in an increase in space group symmetry from $P2_1/a$ to the cubic group $Fm\bar{3}m$.[31] The room temperature phase β-Li_2SO_4 shows a fully ordered arrangement of lithium and sulfate ions and the latter shows no rotational disorder; the tetrahedral shape of the anion can be uniquely located by diffraction experiments indicating that the oxide anions are static. Under atmospheric pressure, this phase is stable on heating up to 575 °C at which temperature it undergoes a phase transition to the α polymorph. As can be seen by the strong similarity between these two crystal

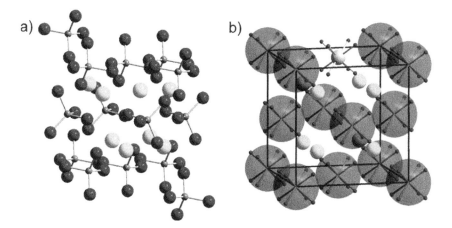

Figure 3.4 The structure of (a) the crystallographically ordered phase β-Li$_2$SO$_4$ at room temperature.[30] The high temperature polymorph, α-Li$_2$SO$_4$, has a disordered arrangement of sulfate groups as shown in (b).[31, 32] The [SO$_4$]$^{2-}$ anions can be modelled using a centrosymmetric distribution of eight oxide anions around the sulfate group or alternatively as a sphere of scattering density as represented by the large transparent spheres. Lithium cations are shown as light grey spheres

structures, this martensitic transition does not involve any substantive changes in the crystal structure, the sulfate group remains centred on the points of a face-centred cubic (f.c.c.) lattice and the lithium cations occupy the tetrahedral interstices.

The high temperature polymorph has been extensively studied since the report in 1965 of exceptionally high ionic conductivity in this phase that reaches a maximum value of 3 S cm^{-1} just below the melting temperature of 850 °C. Despite the simple structure implied by the crystallographic picture of both of these lithium sulfate structures the subtleties of local ion arrangements and the mechanism for lithium ion migration have been a subject of an occasionally fierce debate that lasted three decades. Even now, over 40 years after the initial report of fast ion conductivity in α-Li$_2$SO$_4$ there are features of the mechanism for ion transport that would still benefit from further illumination.

The structure of the fast ion conducting phase can be most clearly understood with reference to the straightforward, fully ordered and insulating α phase observed at room temperature. The monoclinic cell of this structure can be related to a face-centred pseudo-cubic array as illustrated in Figure 3.4. The departure from cubic symmetry leads to the presence of two crystallographically inequivalent lithium cations in the structure. However, the chemical environments for these two cations are similar; they are both tetrahedrally coordinated and show four inequivalent lithium oxide distances in the range 1.93 Å $\leq$ L $\cdots$ O $\leq$ 1.98 Å. There are four crystallographically distinct oxide anions in this structure and diffraction experiments unambiguously show these to be precisely located with no positional disorder beyond that expected for the usual thermal libration effects.

On heating above the transition temperature the structure adopts the higher symmetry structure and, somewhat counter-intuitively, this simplification in the structural description leads to a considerably more complex picture of the chemical bonding in the structure. In the f.c.c. α structure the lithium cations occupy a single crystallographic site, as do the oxide anions.[31] The latter are arranged around a single, centrosymmetric sulfur position. Clearly the point group symmetry of this position is incompatible with the regular tetrahedral shape of the $[SO_4]^{2-}$ anion and so instead of observing four oxide anions around the sulfur, the neutron diffraction experiments on a powdered sample show a perfect cubic arrangement of eight oxide anions, i.e. the arrangement that arises from the placement of a tetrahedron and the inverted mirror image on the same site. It must be remembered that a diffraction experiment produces an average picture of the unit cell by integrating the contents of hundreds of unit cells. Whilst diffraction experiments can provide an unambiguous picture of the time and spatially averaged structure, the interpretation of these data, particularly when related to dynamic processes involving ion motion, is often more subjective. The interpretation of the structure of α-Li_2SO_4 is an example of a particularly challenging problem and is at the heart of the debate concerning the mechanism for ion motion in the structure.

In order to resolve dynamic and static contributions to crystallographically observed disorder it is often useful to perform scattering experiments as a function of temperature; dynamically disordered processes will show variation whilst static displacements will be independent of temperature. Due to the high space group symmetry and the large degree of disorder in the structure, a typical powder diffraction profile shown in Figure 3.5 illustrates the limited information that can be derived from powder samples.[31] Such a problem would usually be overcome by the preparation and crystallographic characterisation of a single crystal of the material. In the present case this presents a considerable challenge; the β to α phase transition involves a large volume change of 3.2% that introduces catastrophic strain into the material and so it is not possible to study the β phase by heating a single crystal from room temperature.

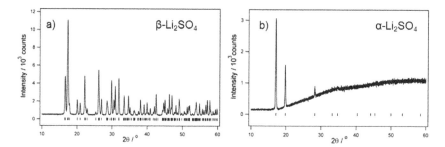

Figure 3.5 Representative neutron diffraction patterns generated from the published structures of β- and α-Li$_2$SO$_4$ using $\lambda = 1.22$ Å. (a) Data collected from the low temperature form β-Li$_2$SO$_4$ contain a large number of peaks allowing a routine determination of the crystal structure of the ordered, monoclinic phase.[30] Data collected at temperatures above the structural transition to the fast ion conducting phase α-Li$_2$SO$_4$ (b) contain only a few peaks with significant intensity. The unusual background intensity in (b) arises from liquid-like scattering due to disorder in the α-Li$_2$SO$_4$ structure.[31] Vertical markers indicate the positions of allowed Bragg reflections

This challenge was overcome in a remarkable experiment which grew a suitable single crystal from the melt, protected it under vacuum and mounted it in a neutron beamline.[32] Throughout the process the temperature of the sample was maintained above the structural transition temperature of 575 °C in order to prevent fracture of the crystal. This careful study showed the sulfate anion can be best described using a largely spherical picture of the anion; a single [SO$_4$]$^{2-}$ anion scatters as a unit centred on a sulfur atom with a sphere of oxide scattering centred on a radius of 1.49 Å corresponding to the sulfur-oxide interatomic distance. The scattering from this spherical assembly can be modelled using a single set of displacement parameters to describe the whole anion that reveal anisotropy in the oxide anion distribution in the spherical shell. Anisotropy is also found in the lithium distribution. This advanced model showed that 90% of the lithium ions are found on the tetrahedral interstices occupied in the classic antifluorite structure, but they show a strong anisotropic displacement in the vibration towards the four neighbouring sulfate anions giving a lithium-sulfur distance of around 2.8 Å. The remaining scattering from Li$^+$ is distributed in a spherical manner around the sulfate anions at a radius, 2.8 Å, that matches the distance derived from the anisotropic displacement of the lithium cations observed on the interstitial positions.

The derivation of this picture represented a challenge of what can be achieved using diffraction techniques but some of the findings are unambiguous. Foremost amongst these is the observation that the spherical picture of the sulfate anion leads to a separation of only 1.57 Å between the lithium cation on the interstice and the nearest surface of the oxide scattering shell. This is significantly shorter that the sum of the ionic radii, 1.90 Å, and so must indicate the presence of either a time or spatial correlation between the occupation of these two positions that avoids the local existence of such an energetically unfavourably short lithium-oxide interatomic distance.

From the above discussion it is clear that the transition from β- to α-Li_2SO_4 involves the introduction of disorder in the sulfate anion. This necessarily involves movement of the oxide anions around the sulfur centre and it is now generally agreed that in the high temperature phase the sulfate anions are capable of rotation. Additional evidence for this has come from both experimental calorimetry measurements[33] and computer simulation.[34] Calorimetric measurements from a number of groups have shown a consistently high value for the heat of transition of *ca* 25 kJ mol^{-1} compared with the enthalpy of melting of 9 kJ mol^{-1}. Large enthalpies of transition are a common feature of these fast-ion conducting sulfates and imply that in this case partial melting of the structure occurs on conversion from the β- to the α-Li_2SO_4 phase. Molecular dynamic calculations have been performed in order to study both the room temperature and high temperature phases and these have also shed light on the curious behaviour of the sulfate groups in the fast-ion conducting regime.[34] These show a broad distribution in oxide-oxide distances between ions in neighbouring sulfate groups indicating a rotationally disordered distribution of the anions. Most importantly, such a distribution was not observed in the lithium-oxide distances. Instead the lithium-oxide distribution functions showed a relatively sharp feature centred on 1.9 Å; a distance that provides good agreement with a typical Li-O distance observed in the ordered low temperature polymorph. This ordering in the Li-O distance in the presence of disorder in the oxide position indicates a strong correlation between the Li^+ cation positions and the local orientation of the $[SO_4]^{2-}$ tetrahedra.

The observations summarised here show that the high temperature polymorph shows fast Li^+ conductivity and dynamic orientational disorder of the sulfate groups. The coexistence of these two highly unusual properties in a crystalline salt leads to the attractive interpretation that these phenomena are linked. Considerable effort has gone into the examination of a model for correlated motion in this structure and this has led to the description of ionic mobility in the α-Li_2SO_4 structure proceeding *via* what has become known as the paddle-wheel mechanism. This mechanism describes the sulfate as acting as a rotor which provides a

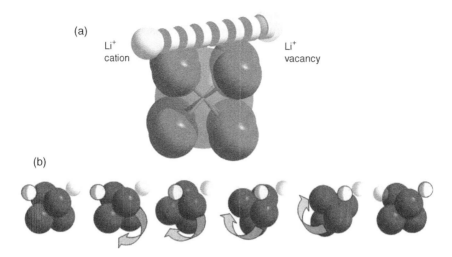

Figure 3.6 A schematic representation of the paddle-wheel mechanism for fast Li^+ conduction in α-Li_2SO_4. The direct path between a Li^+ cation, shown as a solid light grey sphere, and a vacant site, shown as a transparent sphere, is indicated by the broken cylinder. The oxide ions of the sulfate group are shown as large spheres with a radius equal to ½ of the van der Vaals radius of this ion. Even at this greatly reduced radius, the oxide ions clearly block the direct passage of Li^+ between adjacent sites. In (b) the Li^+ cations and oxide ions are shown in a space filling representation and it can be seen how progressive rotation of an individual tetrahedral sulfate anion can be coupled with a relatively facile passage of Li^+ cations between sites. The arrows indicate the direction of rotation of the sulfate groups in this conceptual model

variable barrier to Li^+ ion migration as illustrated in Figure 3.6. When the rotors are in an unfavourable arrangement the window for ion migration is small, leading to a large energy barrier for Li^+ hopping and minimal conductivity. However, the dynamic freedom of the sulfate groups permits rotation of the groups such that the window for Li^+ hopping can be temporarily opened to give facile passage of the cation and fast-ion conductivity. This argument has an attractive elegance and is compatible with a wide range of experimental observations but an unambiguous proof of the mechanism has proved elusive. As the proposed mechanism does not involve the Li^+ ion having any significant residence lifetime on any intermediate sites any clear evidence of paddle-wheel motion must come from directly observing an ion-hopping event. These events are exceedingly rare and so are undetectable by any technique that produces time-averaged data.

Some of the strongest support for the paddle-wheel model has come from simulations of ion mobility. These have shown that the β to α transition is initiated by disordering in the orientation of the sulfate anion and that the Li^+ motion is a secondary event. There is some experimental support for such a statement from Raman spectroscopy measurements on the high temperature phase. Unfortunately the rarity of Li^+ hopping events in the simulations of the high temperature phase meant that it was not possible to unambiguously ascertain whether the movement of Li^+ cations is coupled with the rotation of the sulfate groups.[34]

Whilst a definitive answer on the efficacy of rotor motion in enhancing Li^+ hopping proves elusive, molecular dynamics calculations have shown a path for ion migration through the structure. A variety of ions show spectacular mobility in a number of systems that adopt fluorite-related phases. This is due to the presence of unoccupied octahedral interstices in the prototype structure that lend themselves to acting as intermediate sites in a pathway for ion motion. The schematic diagram of the path for Li^+ motion in β-Li_2SO_4 shows that these vacant sites indeed play an important role in Li^+ hopping. Lithium cations migrate by hopping from a tetrahedral site to an adjacent unoccupied octahedral interstice where it resides for typically a picosecond before jumping to an adjacent tetrahedrally coordinated site.[34] Diffraction experiments show that there is no lithium on this site, but of course this time-averaged picture cannot provide information on positions that are only occupied for a transitory period of time. Precise examination of the calculated path for lithium migration shows that the cations move near the octahedral interstice and there is some correlation between the curved path derived from these calculations and the route proposed on the basis of the neutron diffraction study as shown in Figure 3.7.[32]

Of course this mechanism could operate independently of the motion of the sulfate groups and it has been argued that it is not necessary to invoke a cooperative paddle-wheel effect in order to rationalise the high Li^+ mobility in this phase.[35] Alternative explanations have argued that a consideration of the number of vacant sites can give a probability of successful hopping events that agrees with the experimentally observed Li^+ mobility.[36] Such a percolation model clearly has a part to play in the description of the conductivity of this phase, even if only crudely. After all, the probability of a successful Li^+ hop will clearly be influenced by the number of vacant sites and this must have an impact on the conductivity of the phase. It can be argued that concrete experimental support for the paddle-wheel mechanism is lacking.[37] Whilst it must be accepted that at some level a complete picture of Li^+ motion in this structure necessarily lies beyond the scope of current experimental techniques the arguments in favour of anion mobility providing some enhancement to Li^+ mobility in the structure are irresistible. Quasi inelastic neutron scattering experiments

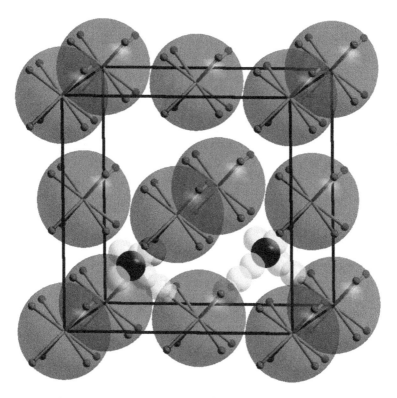

Figure 3.7 The conduction pathway for Li^+ conduction in α-Li_2SO_4 suggested by experimental results. Rotating sulfate anions are drawn as large spheres. Two of the lithium sites in the unit cell are shown as black spheres. Analysis of diffuse scattering and calculations suggests that lithium migration between these two sites involves movement along the path indicated by small transparent spheres and rotation of the sulfate group[32]

show that the Li^+ hopping and sulfate rotational motion operate on similar timescales suggesting a dynamic correlation[38] to complement the spatial correlation derived from the painstaking crystallographic study.[32] A quantitative computational simulation of the ion dynamics in β-Li_2SO_4 has suggested that rotation of the sulfate group gives an enhancement of *ca* 50% to the diffusion coefficient of Li^+ in the structure.[39]

Given the difficulty of proving the positive correlation between anion rotation and Li^+ migration a number of studies have examined the counter-argument and looked at how ion mobility is diminished by replacing the sulfate anion. Such an introduction of other tetrahedral anions into the structure should disrupt the cooperative anion-rotation. To this end a series of Li_2SO_4 compounds containing up to 4% $[WO_4]^{2-}$ were studied to see if the slower rotation of the heavier tungstate tetrahedron impeded the passage of Li^+.[40] These samples showed a progressive reduction in Li^+ mobility with replacement of sulfate that provide strong circumstantial evidence in favour of the paddle-wheel mechanism. Further support comes from a study of Li_2SO_4 that identified two new polymorphs of this compound that can be stabilised at elevated pressure.[41] Analysis of the higher pressure ϵ-Li_2SO_4 phase indicates that the reduction in the size of the interstitial sites due to pressure leads to the occupation of the octahedrally coordinated sites by Li^+. From a mechanistic point of view a more interesting observation is that the application of pressure introduces intersulfate group oxide-oxide interactions that are absent in β-Li_2SO_4. These interactions substantially reduce the degree of anion rotational motion and molecular dynamics calculations show that this is accompanied by a significant reduction in the Li^+ mobility.

Many of the features of fast Li^+ conduction observed in lithium sulfate have been reproduced in mixed metal sulfates.[42, 43] In common with Li_2SO_4 these phases undergo a phase transition on heating to yield a highly disordered material that shows facile Li^+ migration. The conductivity in the high temperature phases are of the same order as that seen in β-Li_2SO_4 but in some cases it has been possible to reduce the transition temperature and so access fast lithium conduction at a lower temperature than in the lithium sulfate parent.

Lithium sodium sulfate has been studied more extensively than any other mixed metal sulfate and shows ionic conductivity of $1\ S\ cm^{-1}$ at $550\ ^\circ C$. At room temperature $LiNaSO_4$ adopts a trigonally distorted variant of a body-centred cubic (b.c.c.) structure.[44, 45] As in lithium sulfate, the lithium cations are located in tetrahedrally coordinated interstices in the sulfate lattice. The larger sodium cation is too large to fit into this position and instead is found to occupy the highly irregular coordination environments positions shown in Figure 3.8. This fully ordered structure shows minimal conductivity[46] but on heating to $515\ ^\circ C$ β-$LiNaSO_4$ undergoes a structural transformation to form a crystalline material that has proved resistant to detailed structural characterisation. As in the case of α-Li_2SO_4 the powder diffraction profiles contain a scarcity of Bragg peaks; a structural study carried out at $575\ ^\circ C$ using a combination of neutron and X-ray diffraction only detected a total of nine Bragg peaks with significant intensity and a background that contained considerable scattering indicative of short range order.[42] This β to α transition involves an increase in unit cell volume of 6% and this large expansion and the increase in symmetry implies a spherical disordering of the sulfate groups. Hence the chemistry

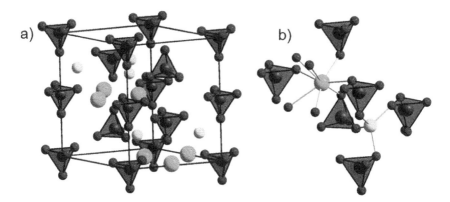

Figure 3.8 The structure of β-LiNaSO₄ is observed at low temperature and displays minimal ionic conductivity. Larger and smaller grey spheres represent Li^+ and Na^+ respectively. The coordination environments of Na^+ in Li^+ are shown in (b)[44]

of $LiNaSO_4$ strongly mirrors that of Li_2SO_4 despite the difference in anion arrangement and a more complex cation distribution. A tentative analysis of the neutron diffraction data collected from the high temperature phase concluded that the sodium cations occupy the regular octahedrally coordinated sites in the structure with lithium found exclusively in the tetrahedra. This assignment has found support from a reverse Monte Carlo analysis of neutron diffraction profile that included the diffuse scattering in the analysis.[47] Molecular dynamics simulations suggest a much more complex distribution of sodium cations over a range of environments. However, it should be noted that the substantial disorder identified in the total profile analysis of the neutron diffraction data led the authors to conclude that a temporally and spatially averaged structure derived from this analysis may correspond to a structure which exists only rarely within any local configuration of atomic coordinates. Despite these difficulties a great deal of information has been derived on the conductivity mechanism in α-LiNaSO₄ using a range of novel experiments.

By using isotopic enrichment, the contributions of Li^+ and Na^+ mobility to the total conductivity have been decoupled.[43] Despite the difference in cation size and typical coordination environments lithium and sodium show similar activation energies of *ca* 0.64 eV and make similar contributions to the overall conductivity of the phase. This observation is incompatible with the simple view of a positive cation migrating through a static window and indeed it has been shown that the sulfate anions in $LiNaSO_4$ are dynamic in the β phase. There is an energy barrier to anion rotation, 0.86 eV, that means that there is a complex temperature dependent contribution to the activation energy for cation hopping but it is accepted that an Arrhenius temperature dependence of conductivity may represent a bearable approximation in these phases.[43]

The presence of vacancies on both the tetrahedrally and octahedrally coordinated interstices in α-LiNaSO$_4$ provide a sharp contrast to the filled tetrahedral sites of α Li$_2$SO$_4$ and suggest that cation hopping could occur quite readily in α-LiNaSO$_4$ without invoking a paddle-wheel mechanism involving anion mobility. However, it seems that in the mixed cation phase the anion motion makes a larger contribution to conductivity than in lithium sulfate. A study of the pressure dependence of ion mobility in LiNaSO$_4$ found that the activation volume for conductivity is approximately zero.[48] This parameter is associated with the change in lattice volume caused by the migration of an ion. This observation of negligible expansion implies that the passage of cations through the structure does not involve any local translation of the other atoms in the lattice and so strongly suggests that anion rotation is a key factor governing ionic conductivity in α-LiNaSO$_4$. A study of the phonon spectra of LiNaSO$_4$ has provided a local probe of the dynamic processes occurring in the insulating regime, i.e. well below the structural transition temperature.[49] These showed dramatic changes in the Li$^+$ and Na$^+$ vibrational modes around 200 °C and the onset of sulfate rotational disorder at temperatures as low as 350 °C, over 150 °C below the structural transition and the onset of fast ion conductivity. The bands due to Li$^+$ vibration are largely unaffected by the onset of sulfate disorder at 350 °C and the authors argued that this suggests that there is only a weak correlation between anion rotation and Li$^+$ mobility. However, it must be remembered that these observations are being made in the insulating β-LiNaSO$_4$ phase. Consequently it may be that these phonon modes are indicative of behaviour within isolated portions of the sample and that only when these higher energy states become distributed throughout the lattice does a structural transition occur and so switch-on fast ion conduction. Of course, such an argument does admit that percolation is a key component in cation conduction in α-LiNaSO$_4$ and seems to operate in conjunction with the paddle-wheel effect. Indeed, examination of the density distributions of ions in LiNaSO$_4$ derived from neutron scattering shows that there is no clear distinction between the two mechanisms and that the two operate in tandem to give fast ion conduction.[47]

The effects of Ag$^+$ substitution in the lithium sulfate system have been widely studied and the findings have illuminated the mechanisms for fast ion mobility in these structures. The phase diagram[50] for Li_2SO_4–Ag_2SO_4 identifies three different structure types at high temperature and the properties of the two lithium-richer phases have shown interesting ion mobility. Ag^+ cations can be doped into the f.c.c. structure of Li_2SO_4 up a limiting composition of *ca* 21 mol% Ag^+. The presence of silver cations in the structure depresses the structural transition temperature and the limiting composition $Li_{1.58}Ag_{0.42}SO_4$ transforms to a stable face-centred phase around 510 °C. A structural study of this compound at 565 °C showed the same features in the diffraction pattern as the pure lithium composition, although the presence of the electronically dense silver cation greatly aided the identification of these cation positions from the X-ray diffraction pattern.[51] Neutron diffraction data contained only eight significant Bragg peaks and considerable diffuse scattering implying a large degree of disorder within the crystal structure.

The structural model of f.c.c. Li_2SO_4 can be used as a starting point for the description of the silver-doped phases. The sulfate anions show the same positional disorder, with a spherical shell of oxide anions around the sulfur core, and the lithium cations are exclusively found on the tetrahedral sites of the fluorite structure. Individual analyses of X-ray and neutron diffraction data failed to give a complete picture of the structure and led to over-simplifications in the description of cation arrangement. In order to produce a satisfactory description of the Ag^+ distribution it was necessary to simultaneously model both the X-ray and neutron diffraction data. The best fit was obtained by allowing 30% of silver to occupy the octahedrally coordinated sites that are vacant in Li_2SO_4, with the remaining cations sharing the tetrahedrally coordinated sites with the lithium cations. Due to the presence of Ag^+ on the octahedral sites this necessarily leads to the presence of vacancies on the tetrahedrally-coordinated sites that are filled in lithium sulfate and for the composition $Li_{1.6}Ag_{0.4}SO_4$ this site is only 94% occupied.

The presence of vacancies on this position could be anticipated to increase the cation mobility. Indeed a conventional picture of lithium conduction *via* cation hopping between sites would rely on such a partial occupation to permit conduction. Therefore it is particularly noteworthy that the ionic conduction of $Li_{1.6}Ag_{0.4}SO_4$ is, to a first approximation, broadly similar to that reported for Li_2SO_4. Despite the introduction of vacancies onto the tetrahedrally coordinated site that is favoured by lithium cations, the activation energy for conduction in f.c.c. $Li_{2-x}Ag_xSO_4$ is invariant with composition. Even more surprisingly, the conductivity in these phases increases with increasing silver content[52] and it is probable that this arises from a greater mobility of Ag^+ through the structure compared with Li^+.[50] This clearly indicates that the conventional indicators of cation mobility, such as ionic radius and mass, are not applicable in the face-centred sulfates and that the dynamic properties of the anion sublattice play a key part in facilitating cation motion.

Of course, the silver content of the system can be increased beyond the 21 mol% solubility limit found for the f.c.c. structure. This leads to a two-phase region where the f.c.c. structure coexists with a body-centred phase. The latter cannot be formed for silver contents less than *ca* 35 mol%. As for the face-centred phase, increasing the silver content depresses the structural transition temperature and so the body-centred lattice can be stabilised at temperatures as low as 420 °C. This phase has been widely studied at the composition $LiAgSO_4$ and shows a conductivity of 1 S cm^{-1} at 530 °C.[52] Structural studies of this phase show the diffuse scattering and limited number of Bragg peaks familiar from other compounds in these systems, but an evaluation of the total diffraction profile suggests that this compound is isostructural with $LiNaSO_4$, *i.e.* the tetrahedra are half filled by lithium cations whilst silver is found in the octahedrally coordinated sites. There is some evidence that there is partial disorder in this distribution with some indication of silver cations on the tetrahedrally coordinated sites. However, the lithium, as in all other fast ion conducting sulfates, is not found on the octahedrally coordinated positions.[42]

Some general conclusions can be drawn on cation mobility in different sulfate rotor phases and a study of diffusion of various mono- and divalent cations gives a useful insight into these.[43] The greatest monovalent cation mobility is seen in Li_2SO_4, with Li^+ and Na^+ showing the highest diffusivities. The mobility of these cations in either $LiNaSO_4$ or $LiAgSO_4$ is somewhat reduced at the same temperature. However, this reduction in performance is greatly offset by the stability of the fast ion conducting phases in the mixed-metal sulfates at lower temperature α-Li_2SO_4. An additional point is that the mixed-metal phases show a more generalised cation mobility. Whilst Li_2SO_4 shows minimal mobility of small divalent cations, the expansion in the lattice and stabilisation of the body-centred structures caused by the introduction of Na^+ or Ag^+ reduces the discrimination shown by the host lattice and substantially increases the diffusivity of Mg^{2+} and Ca^{2+} through the lattice. Interestingly, lithium sulfate showed a strong size-dependence in the ionic mobility and this leads to the mobility of Pb^{2+} being appreciably greater in Li_2SO_4 structure than in the body-centred phases $LiNaSO_4$ and $LiAgSO_4$. It is possible that the readily polarisable nature of the lead cation is a key component in easing the passage of this species through the anion sublattice.

3.4   LISICON AND RELATED PHASES

Following the discovery of fast ion conduction in the β-alumina structure a number of structural systems were studied in the search for other
fast ion conductors. One of the first new families to be discovered was a
series of compounds based on lithium silicate, lithium germanate, and
related compounds. Li_4GeO_4 has a relatively simple structure based on
the stacking of distorted close-packed layers of oxide anions. The germanium and lithium cations occupy tetrahedrally coordinated interstitial sites between the layers to form isolated GeO_4 tetrahedra that are
linked together by vertices shared with LiO_4 tetrahedra that are themselves linked *via* a mixture of corner- and edge-sharing.[53, 54] Edge-
sharing of metal oxide tetrahedra is often avoided in crystal structures
as the short separation between the metal cation and the anion associated with the relatively low fourfold coordination number, combined
with the tetrahedral angle results in a relatively short separation
between the metal cations in adjacent edge-sharing tetrahedra.
However, in the lithium germanate structure the separation between
lithium cations, *ca* 2.45 Å, is significantly larger than the sum of the
radii. The orthorhombic crystal structure contains two inequivalent
lithium positions and the tetrahedra show significant differences; a
crystallographic study of Li_4GeO_4 shows that one LiO_4 unit is smaller
and more regular with a variation of only 0.01 Å from the mean Li-O
distance of 1.93 Å whilst the other lithium position is coordinated to
four oxide anions at distances that range from 1.94 to 2.14 Å about a
mean value of 2.03 Å. The origin of these irregularities is unclear as the
structure is otherwise quite simple, with no evidence of positional or
occupational disorder. What is noteworthy about the structure is that it
contains a considerable amount of unoccupied space. Consideration of
the ionic radii of the component ions suggests that the crystallographic
volume is only 45% filled and as shown in Figure 3.9 there are channels
long the z-direction of the unit cell.

The oxide lattice and the SiO_4 tetrahedra of Li_4SiO_4 are similar to those
observed in lithium germanate but the lithium distribution is considerably
more complex in the silicate.[54, 55] Lithium is disordered over a number of
four-, five- and six- coordinate positions in Li_4SiO_4 including a number of
face-sharing tetrahedral sites that are separated by distances as short as
0.92 Å.[55] Clearly such short distances are to be avoided due to the high
electrostatic repulsion between the two cations, and the occupancy of these
two positions is sufficiently low that local ordering of the lithium

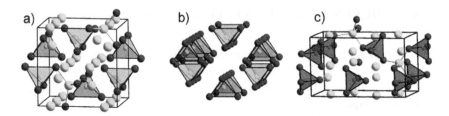

Figure 3.9 The crystal structure of Li_4GeO_4 contains an ordered arrangement of GeO_4 units, shown as tetrahedra, linked together by Li^+ cations.[53] The tetrahedral coordination of Li^+ is illustrated in (a). The $[GeO_4]^{4-}$ sublattice contains channels running along [001] as shown in (b). The structure can adapt to changes in lithium content as illustrated in (c) for the low temperature structure of Li_3PO_4[64]

and vacancies is capable of avoiding simultaneous occupation of adjacent sites. The lithium distribution and the structure of lithium silicate are exceptionally complex.[56]

Despite the large structural differences between lithium silicate and lithium germanate the conductivities of these phases are similar. Both materials show minimal conductivity at room temperature, $\sigma_{RT} \approx 10^{-9}$ $S\,cm^{-1}$.[57] Lithium germanate shows a single activation energy of 0.21 eV between room temperature and 600 °C and this low value means that the conductivity improves slowly as a function of temperature to reach a value of only $10^{-3}\,S\,cm^{-1}$ at 600 °C. Despite the presence of occupational disorder in the lithium silicate structure, the conductivity of this phase is largely the same as that of the fully ordered germanium analogue indicating that the presence of vacancies on the lithium sublattice of Li_4SiO_4 has minimal impact on the mechanism for charge transport. The lithium conductivity of compounds in the solid solution $Li_4Si_{1-x}Ge_xO_4$ can be up to an order of magnitude larger than the end members and shows a similar activation energy.[57]

The interest in these materials was stimulated by the realisation of fast lithium ion conductivity by doping lithium germanate or silicate with aliovalent cations. By changing the charge of the cations in the structure it is necessary to change the lithium content of the material in order to maintain charge balance. If the charge of the nonlithium cations is increased it is necessary to reduce the number of lithium cations in the material and this leads to the introduction of vacancies on the lithium sites. An alternative strategy is to reduce the charge of the nonlithium cations and so introduce an excess of lithium into the structure by occupying interstitial sites in order to provide a neutrally charged lattice.

The introduction of vacancies into lithium germanate or silicate can be illustrated by two substitutions. The replacements:

$$\text{Si}^{4+} + \text{Li}^+ \rightarrow \text{P}^{5+} + \square_{\text{Li}} \quad \text{and} \quad 2\text{Li}^+ \rightarrow \text{Zn}^{2+} + \square_{\text{Li}}$$

both introduce vacancies (designated $\square$) onto the lithium cation positions of lithium silicate and this approach has been widely applied to optimise the transport and stability properties of these phases.

Initial studies focused on the effect of introducing pentavalent cations into the structure and the phase behaviour of systems such as $\text{Li}_{3+x}\text{X}_{1-x}\text{Si}_x\text{O}_4$ ($X = \text{P}$, As, V).[58–60] The behaviour of these materials can be illustrated with a detailed examination of the properties of lithium arsenate. Lithium arsenate has a fully ordered structure at low temperatures. As for lithium germanate and lithium silicate, this is based on a hexagonal stacking of buckled, close-packed layers of oxide anions. The Li^+ and As^{5+} cations fill tetrahedral interstices in a fully ordered arrangement in the low temperature β phase.[61] On heating this sample above a transition temperature of ca 745 °C this transforms to γ-Li_3AsO_4.[61, 62] This transformation is associated with disordering of the lithium positions and this disorder increases further at higher temperatures.[63] This complex lithium distribution involves a range of four- and five-coordinate sites in the structure. The high temperature of the β to γ transition precludes the use of the disordered γ polymorph as a solid electrolyte and the conductivity of the β polymorph is poor; even at 280 °C it reaches a value of only 10^{-8} S cm^{-1}.[58] The structural chemistry of Li_3PO_4 is similar, showing a fully ordered low-temperature phase[64] and a high-temperature γ phase. The structure of the latter shares the same arrangement of buckled hexagonally close packed layers as the low temperature phase, but the distribution of the lithium cations has not been conclusively established and there are conflicting reports of order and disorder across multiple interstitial sites in the structure.

The substitution of Si^{4+} for As^{5+} in the compounds $\text{Li}_{3+x}\text{As}_{1-x}\text{Si}_x\text{O}_4$ achieves two effects: (i) the variable quantity of lithium in the structure favours a disordered cation arrangement and so lowers the temperature at which the γ structure is stabilised; and (ii) the increased quantity of lithium cations leads to population and partial occupancy of additional sites in the structure. Both of these are key ingredients in the generation of fast lithium conducting phases based on these γ tetrahedral structures. Stabilising the disordered high-temperature phase generates a large number of equivalent sites throughout the structure and introducing the excess of lithium cations increases the number of mobile lithium cations that can act as charge carriers. The highest conductivity, ca 10^{-6} S cm^{-1}, is achieved for approximately equal concentrations of Si^{4+} and As^{5+}.[58]

It is a noteworthy observation that the activation energy for these compounds is largely the same as the poorly conducting phase γ-Li$_3$AsO$_4$ indicating that it is the change in carrier concentration, rather than a change in mechanism for carrier mobility, that is the key factor in the fourfold increase in conductivity observed on increasing the lithium content from Li$_3$AsO$_4$ to Li$_{3.5}$Si$_{0.5}$As$_{0.5}$O$_4$.[58] The introduction of other pentavalent cations has similar effects on the Li$_4$SiO$_4$ and Li$_4$GeO$_4$ phases. The highest lithium ion conductivity is observed for compounds that use vanadium doping to stabilise the γ tetrahedral phase; room temperature conductivities reach maxima of ca 10^{-5} S cm^{-1} for Li$_{3.3}$Si$_{0.3}$V$_{0.7}$O$_4$ and 4×10^{-5} S cm^{-1} for Li$_{3.6}$Ge$_{0.6}$V$_{0.4}$O$_4$.[59, 65]

The conductivity in the γ tetrahedral structure undergoes a most dramatic change on doping Li$_4$GeO$_4$ with the Zn^{2+} cation to give a series of compounds Li$_{4-2x}$Zn$_x$GeO$_4$.[66] The conductivity of this series reaches a maximum around $x = 0.25$ (this composition is often described as Li$_{14}$Zn(GeO$_4$)$_4$) and the name LISICON was coined to describe the high lithium conductivity of this LIthium SuperIonic CONducting material. A similar acronym has led to the naming of another family of compounds NASICON and it should be noted that the similarity in the names reflects only a single property. We shall encounter the NASICON family of materials and see that there is no chemical or structural link between LISICON and NASICON.

The Zn^{2+} cations enter the structure in two distorted, tetrahedrally coordinated interstitial sites and the partial occupation of these positions causes a displacement of the lithium from the sites occupied in the fully ordered Li$_4$GeO$_4$ parent phase. The lithium is distributed across three crystallographically distinct sites in the $x = 0.5$ compound Li$_{3.0}$Zn$_{0.5}$GeO$_4$; two are octahedrally coordinated and the third contains an uncommon trigonal bipyramidal arrangement of oxide anions around the lithium cation as shown in Figure 3.10. All three environments are highly irregular and considerably more open than the tetrahedrally coordinated positions found in Li$_4$GeO$_4$.

It should be remembered that the structure is derived from neutron diffraction experiment and, as for all diffraction experiments, it represents a time-averaged summation through space of the contents of multiple unit cells. In the case of a fully ordered structure, such as that of room temperature phase of lithium phosphate, the averaging process is not apparent in the resulting structural model. However, in the case of

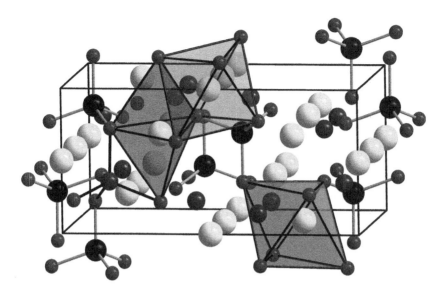

Figure 3.10 The simplified representation of the crystal structure of $Li_3Zn_{0.5}GeO_4$. Black and dark grey spheres represent Ge^{4+} and Zn^{2+} cations respectively. The coordination environments around the three crystallographically distinct Li^+ cations are shown as grey polyhedra centred on the Li^+ cations. The actual cation distribution is more complex than indicated; there is substantial mixing of the Zn^{2+} and Li^+ cations across the sites[67]

$Li_{3.0}Zn_{0.5}GeO_4$, there are five interstitial sites that have a minority occupancy of lithium or zinc cations and many of these positions lie sufficiently close together that simultaneous occupancy of both is extremely unlikely due to electrostatic repulsion at distances as short as the 0.97 Å separation between the trigonal-bipyramidal lithium site and a neighbouring tetrahedrally coordinated zinc position. Consequently the structural model must be interpreted and the different disordered cations separated into discreet components that provide chemically plausible regions of the crystal. These fragments can then be integrated to give an average structure that matches the results of diffraction experiments. In the case of $Li_{3.0}Zn_{0.5}GeO_4$ such an analysis gives rise to a model of the disorder that eliminates Li^+-Li^+ separations of less than 1.98 Å.[67] Thus a chemically sensible arrangement is achieved by invoking local cation ordering of cations into distinct clusters that are disordered over the length scales, typically hundreds of angstroms, sampled by diffraction experiments.

The first of these arrangements has the same structure as the stoichiometric material Li_2ZnGeO_4, a fully ordered material that is isostructural with Li_3PO_4. The replacement of a single Zn^{2+} cation with two Li^+ cations clearly requires the insertion of cations onto positions that are unoccupied in the Li_2ZnGeO_4 parent. These are accommodated in octahedral interstitial sites between the distorted close packed oxide layers. The occupation of these sites leads to a local re-arrangement of the cation occupancies of the adjacent tetrahedrally coordinated positions. There are two octahedrally coordinated lithium environments in the crystallographic description of the structure. The first of these shares faces with two of the Zn/Li tetrahedral sites and occupation of this octahedron leads to short interatomic distances between the cation positions in face-sharing polyhedra. The position of only one of the tetrahedrally coordinated cation sites is displaced away from the central octahedron. This suggests that both of these positions are simultaneously occupied in portions of the material and this arrangement is denoted a Type-I cluster. In order to minimise electrostatic repulsion it is likely that the tetrahedral position is exclusively occupied by Li^+ in order to avoid the charge build up associated with the presence of the Zn^{2+} cation.

If two Type-I clusters exist in adjacent portions of the lattice then further vacancy ordering can occur around another octahedrally coordinated site that shares four faces with tetrahedral positions that are fully occupied in Li_2ZnGeO_4. Two of these tetrahedra contain vacancies and the cations are displaced in a manner that suggests that they are not occupied simultaneously with the central octahedron. This arrangement of a central occupied octahedron linked *via* shared faces with two full and two empty tetrahedral interstices is a Type-II cluster.

This assessment of the cation ordering over multiple crystallographic sites in a structure provides a useful picture of the structural changes in response to increasing cation content in the LISICON related phases. The presence of three cations per germanate, phosphate or arsenate unit can be readily accommodated using a fully occupied set of interstitial sites that, in an undistorted hexagonal close packed anion array, correspond to half of the tetrahedrally coordinated interstitial sites in the structure. As the cation content is increased lithium cations occupy octahedral interstices that share faces with some of the tetrahedral sites, leading to depopulation of some of the tetrahedra and a local portion of Type-I defect embedded within the Li_2ZnGeO_4 matrix. As the cation content is increased pairs of Type-I defects are more likely to coexist and coalesce to form Type-II clusters. Increasing the cation content further will lead to the elimination the Li_2ZnGeO_4 matrix and individual Type-I defects until for the Li_4GeO_4 compounds only the Type-II structure exists. These different arrangements of cations are illustrated in Figure 3.11.

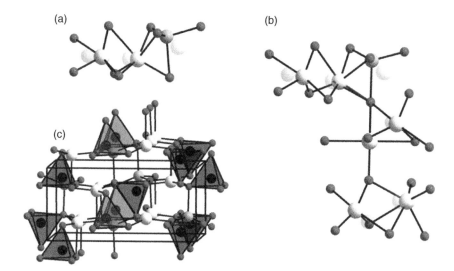

Figure 3.11 The types of aggregated defects in $Li_3Zn_{0.5}GeO_4$. Transparent cations indicate sites that are partially occupied in the structure but are unoccupied in the local cluster. A small degree of local ordering leads to the formation of (a) Type-I clusters formed by face-sharing between a LiO_5 trigonal prism, a LiO_6 octahedron and a ZnO_4 tetrahedron. Pairing of these clusters leads to the formation of (b) the larger Type-II cluster. Both clusters are embedded in a fully ordered matrix of Li_2ZnGeO_4 that has the structure shown in (c)[67]

The presence of local ordering in this manner has large implications for the ionic conductivity of the material. Quantitative discussions of ion mobility become, at best, approximations based on estimates of the charge carrying unit. Clearly if ionic conduction occurs by the formation and migration of clusters then the mass of the mobile species cannot be equated with the atomic mass of lithium cations. This situation is complicated further by the possibility that multiple mechanisms are in operation involving either or both Type-I or Type-II clusters; mobility may arise from the wholesale movement of an individual cluster or the formation or destruction of the Type-II cluster from the Type-I components.

An interesting investigation of mobility in the LISICON structure has employed ion exchange to examine which lithium cations can be replaced with protons. Treatment of a number of compositions found that no protons could be introduced into the Li_2ZnGeO_4 parent, but that lithium cations beyond this composition could be replaced with H^+ by treatment in dilute acetic acid.[68] It is tempting to conclude that this indicates that only the octahedrally coordinated lithium cations are mobile and that the lithium cations occupying the tetrahedra are immobilised and so unable to be replaced by H^+ and migrate out of the lattice. Such a conclusion is at odds with the cooperative model of lithium migration that is implied by the presence of local cation ordering in the cluster model. An alternative conclusion from the observations of the proton/lithium exchange process is that instead of proceeding under kinetic control, the removal of lithium from the lattice is limited by the thermodynamic stability of protons in the tetrahedrally coordinated sites. An NMR study of $Li_{3.5}Zn_{0.25}GeO_4$ showed that this compound contains a mixture of mobile and relatively immobile lithium cations in the highly conducting temperature regime.[69] However a definitive assignment of the origin of this observation has not been made and it is unclear how these observations can be integrated with the clustering mechanism inferred from neutron diffraction experiments. A more rigorous approach to this problem could use the availability of two isotopes of lithium, ^6Li and ^7Li, to investigate the mobility of lithium cations through the structure, although no such study has yet been attempted.

The conductivity in this system reaches a maximum[66] for the compound $Li_{3.5}Zn_{0.25}GeO_4$ with a value of $0.12 \, S \, cm^{-1}$ at 300 °C although on cooling to room temperature phase segregation occurs[70] that appears to be driven by the formation of lithium germanate and results in a reduction of conductivity to ca $10^{-7} \, S \, cm^{-1}$. Slightly lithium-poor compositions show increased stability but characterisation at room temperature reveals aging effects that are manifested in a gradual reduction in conductivity.[70] Consequently these zinc-based compounds have failed to find widespread usage despite the excellent Li^+ mobility in $Li_{3.5}Zn_{0.25}GeO_4$.

In order to overcome these stability problems a wide-range of alternative dopants have been studied in an effort to reproduce the transport properties of LISICON whilst eliminating the problems of aging. The introduction of trivalent cations into the basic lithium germanate structure can cause an adjustment in the lithium concentration, but in this case it is possible to introduce additional, interstitial, lithium cations or vacancies on the lithium position. Which of these types of doping occurs depends on the nature of the substitution that occurs. This can be most clearly illustrated by looking at the introduction of aluminium cations. These can enter the structure in place of Si^{4+} and so introduce an interstitial lithium cation:

$$Si^{4+} + \square_{Li} \rightarrow Al^{3+} + Li^+$$

or aluminium can occupy a lithium cation position and introduce lithium vacancies:

$$3Li^+ \rightarrow Al^{3+} + 2\square_{Li}$$

Which one of these two situations arises depends on the nature of the dopant. Many cations, including Cr^{3+}, Fe^{3+}, B^{3+}, Ni^{2+} and Co^{2+},[71, 72] tend to replace Si^{4+} to give interstitial Li^+ whilst others, such as In^{3+},[73] replace Li^+ and so are associated with the formation of vacancies. In other cases the site preference of the dopant mechanism is unknown and for some cations, most notably Al^{3+} and Ga^{3+}, either situation may arise[74–76] and so compositions can be found in two difference series, e.g. $Li_{4+x}Si_{1-x}Al_xO_4$[74, 77] and $Li_{4-3x}Al_xSiO_4$[74] are both observed. It should be noted that there is a scarcity of structural studies in many of these systems and in some studies the low dopant levels meant that differentiation between the two possible compositions, e.g. $Li_{4.20}Fe_{0.067}Si_{0.933}O_4$ and $Li_{3.78}Fe_{0.072}SiO_4$ is not straightforward and mis-assignments are possible. Nevertheless, both interstitial and vacancy mechanisms have been unambiguously identified in some systems.[73, 74] It is an unexpected observation that similar conductivity can result for systems that show different dopant mechanisms and this can be illustrated by looking at the effect of doping on the properties of lithium germanate.

Studies of the effect of Ga^{3+} doping of lithium germanate have identified a solid solution that exists across the composition range $Li_{4-3x}Ga_xGeO_4$ $0.08 \leq x \leq 0.35$ although the upper limit for Ga^{3+} doping can be extended considerably above 1100 °C.[76] The dependence of conductivity on lithium content in this system differs considerably from that of LISICON. The maximum conductivity for the $Li_{4-3x}Ga_xGeO_4$ system, is reached close to the lithium-rich dopant limit ca $Li_{3.7}Ga_{0.1}GeO_4$. The conductivity of this compound at 300 °C is ca 0.02 S cm^{-1}, i.e. an order of magnitude less than LISICON as shown in Figure 3.12.

The behaviour of these compounds seems to be largely independent of the trivalent cation used to dope the lithium germanate; Fe^{3+} and Al^{3+} reach similar conductivities to the $Li_{4-3x}Ga_xGeO_4$ system and an activation energy of ca 0.5 eV is consistently observed for all compositions across the solid solution ranges,[78] albeit with a small dependence on the thermal history of the sample.[76]

A closely related family of lithium conducting compounds with the antifluorite structure have been studied in some detail. Instead of the distorted hexagonally close packed array of oxide anion observed in the Li_3PO_4, Li_4GeO_4 and LISICON-related phases and cubic close packed array is found for compounds of general composition Li_5MO_4, where M is a trivalent cation. Cubic close packing avoids the face-sharing linkages between interstitial tetrahedral sites that cause short cation-cation distances and displacement-driven distortions in the lithium-rich compounds discussed above. Consequently all of the tetrahedral sites can be

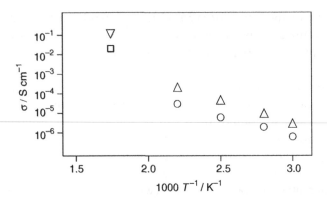

Figure 3.12 The intragrain and intergrain conductivities of $Li_{3.1}Zn_{0.45}GeO_4$ are shown by circles and triangles respectively.[189] The maximum conductivity in this system is found for the composition $Li_{3.5}Zn_{0.25}GeO_4$ and is shown as a square.[66] The maximum conductivity in the Ga^{3+} doped series is indicated by an inverted triangle[76]

occupied and such a situation is found in the fluorite structure; fluoride occupies all of the tetrahedral interstices in a face-centred lattice of calcium cations in CaF_2. In the case of the Li_5MO_4 compounds the cubic close packed array is composed of oxide anions. The 2:1 ratio observed in fluorite is more obvious if the formula $Li_5M\square_2O_4$ is used to indicate the presence of vacancies on 25% of the cation sites.

Li_5GaO_4 Li_5AlO_4 and Li_5FeO_4 are isostructural and at lower temperatures display a fully ordered arrangement of cations across 75% of the interstitial sites.[79, 80] On heating they undergo a transition to give a disordered arrangement of lithium cations across partially occupied sites.[81] Unfortunately the conductivity of Li_5GeO_4, and the aluminium- and iron-containing analogues, is strongly dependent on the environment. Reaction with atmospheric moisture occurs readily to give lithium hydroxide and $LiGaO_2$.[82] However, with careful control of composition the properties of the high-temperature polymorph β-Li_5GaO_4 have been established. This phase shows significantly higher conductivity than the α polymorph and reaches a value of *ca* 2×10^{-5} S cm^{-1} at 300 °C. Attempts have been made to increase this value further by doping with divalent cations Zn^{2+} and Mg^{2+}. These have established that the introduction of additional interstitial lithium cations, in series such as $Li_{5+x}Ga_{1-x}Zn_xO_4$, can give a modest increase in conductivity, whilst the introduction of additional lithium vacancies, in compounds from the series $Li_{5-x}Zn_{1-x}Ga_xO_4$, reduces the conductivity.[83] Similar effects are seen[84] in doping Li_5AlO_4 and Li_5FeO_4, but the problems of reactivity and particularly of moisture sensitivity and associated phase segregation mean that this family of compounds have not been developed further as possible practical electrolytes.

3.5 LITHIUM CONDUCTION IN NASICON-RELATED PHASES

Fast ion conductivity is a feature of a large number of mixed metal phosphates related to the parent compound $NaZr_2(PO_4)_3$.[85] This contains a framework composed of PO_4 tetrahedra and ZrO_6 octahedra linked in a manner that generates a large number of possible coordination sites for sodium cations. This availability of multiple, partially occupied sites allows facile cation hopping and the resulting sodium mobility led to this, and related compounds, being known as NASICON; a potentially confusing abbreviation designed to convey the presence of sodium ion superionic conductivity. Confusion can arise because a number of cations, including our species of interest; lithium, are also mobile in NASICON materials. This confusion is only enhanced by the adoption of a similar abbreviation (LISICON) to describe a series of fast lithium ion conductors that are chemically unrelated to the NASICONs. This unfortunate situation serves as an introduction to, and an explanation of, the elaborate naming of, what can best be described as lithium ion conduction in NASICON materials. Although the mobile cation is variable, and a number of structural modifications and structural transitions occur in these materials, the negatively charged metal phosphate framework is a relative constant in these materials and so we will begin our description here; in the prototype phase sodium zirconium phosphate.

The framework is built from a corner-sharing arrangement of alternating ZrO_6 and PO_4 tetrahedra. As all oxide anions are bridging between a pair of polyhedra there are no terminal oxide anions in the framework. The structure has rhombohedral symmetry and two crystallographically distinct oxide anions and this indicates that the single negative charge on the $[Zr_2(PO_4)_3]^-$ framework must be highly distributed over the framework. It is likely that this is a factor in the ease of passage of cations through the structure as the presence of lower symmetry or terminal oxide anions would lead to localisation of charge and the formation of a strong local potential that could trap cations and prevent significant ionic conduction.

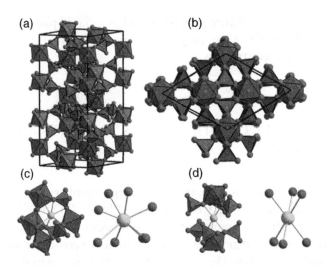

Figure 3.13 The anionic $[Zr_2(PO_4)_3]^-$ framework of NASICON viewed in projections close to (a) the [110] and (b) the [001] directions. The different coordination sites for the extra-framework Na^+ cations are shown in (c) and (d)[85]

The anionic zirconium phosphate framework is shown in Figure 3.13. Despite the high connectivity of this metal oxide framework it can clearly be seen that a large amount of nonframework volume exists in the unit cell. Consideration of the standard radii of ions suggests that the unit cell is 50% occupied by the $[Zr_2(PO_4)_3]^-$ framework. The presence of a single sodium cation per formula unit contributes another 2% to the filled volume of the unit cell. It should be remembered that the most efficient filling of three-dimensional space by hard, spherical atoms can give a maximum of 74% filled volume. Nevertheless it is clear that there is considerable extra-framework volume and that the introduction of sodium cations into this space leaves considerable space in the lattice. Examination of this space identifies a number of likely bonding environments for the extra-framework cations; there is one centrosymmetrically elongated octahedrally coordinated site and three irregular eight-coordinate sites per formula unit as shown in Figure 3.13. In the proto-type $NaZr_2(PO_4)_3$ compound the sodium cations fill the octahedrally coordinated site.[85]

Much work has been undertaken in order to modify the sodium content of related phases. As the octahedral site is filled in $NaZr_2(PO_4)_3$ it is clear that any increase in the extra-framework cation concentration requires population of additional sites. Considerable effort has been made to understand the distribution of these extra sodium cations.

There is considerable complexity in the local arrangements, but a simplified description would identify the additional sodium as occupying the eight-coordinate sites. The lithium cation is smaller than sodium and so is unable to achieve a satisfactory bonding arrangement in the eight-coordinate position. Whilst the lithium-analogue of NASICON can contain the same basic zirconium phosphate anion framework as NASICON, the smaller lithium cation introduces an additional degree of complexity to the system and the formation of four different polymorphs of $LiZr_2(PO_4)_3$.

Lithium zirconium phosphate is commonly prepared using solid state reactions around 1000 °C. The compounds produced in this manner have structures that show an unusual history dependence.[86] Preparation at 1200 °C gives a rhombohedral structure that is directly analogous to the sodium parent phase described above. This phase, α-$LiZr_2(PO_4)_3$, is stable on cooling to 60 °C and undergoes a transition at this temperature to a triclinically distorted α' phase.[86] This transition is fully reversible and it is not possible to form a kinetically stable sample of the high temperature α phase by rapid quenching.

Surprisingly, preparation of the material at 900 °C does not yield the α phase as would be expected from the observations made on cooling a sample from 1200 °C. Instead a new orthorhombic polymorph is formed with the $Fe_2(SO_4)_3$ structure-type.[87] This β-$LiZr_2(PO_4)_3$ structure contains the same corner linkages between alternating FeO_6 and SO_4 polyhedra as found in NASICON. On cooling, this β polymorph undergoes a structural transition at 300 °C driven by further distortion and leads to the formation of a monoclinically disorted phase, β'-$LiZr_2(PO_4)_3$.[88] Again, this is a fully reversible transition and attempts to kinetically trap the higher temperature phase by quenching have been unsuccessful.

The structural differences between these four polymorphs are subtle but are strongly coupled with the ionic conductivity of this compound and so merit detailed discussion. First we shall examine the polymorph with the highest symmetry and hence the simplest description; the α-$LiZr_2(PO_4)_3$ polymorph that is formed at 1200 °C and stable at temperatures above ca 30−60 °C. This structure of this phase has been the subject of some dispute due to the difficulties of locating the lithium cations in the structure. Some preliminary studies concluded that lithium would be found[89] in the same environment that sodium occupied in the NASICON structure of $NaZr_2(PO_4)_3$.[90] However, this elongated octahedral coordination would require the stabilisation of Li^+ by coordination to six oxide anions at a distance, 2.53 Å,[85] that is too large for the relatively small lithium cation.

Neutron diffraction has been used to study the structure at a range of temperatures above the structural transition and show a complex lithium distribution.[91] Whilst the anionic $[Zr_2(PO_4)_3]^-$ lattice is fully ordered with a regularly alternating arrangement of octahedra and tetrahedra, the majority of lithium is in a disordered arrangement in the elongated octahedral site identified in NASICON. The displacement of Li^+ from the centre of this octahedron results in the cation being found on six crystallographically equivalent sites that are displaced 1.4 Å from the inversion centre at the middle of the octahedron as shown in Figure 3.14. The distortion changes the lithium coordination environment and leads to a pseudo-tetrahedral configuration of oxide anions around the cation with a mean lithium oxide distance of 2.3 Å. This is an exceptionally long lithium oxide bond and suggests that the cation is considerably under-bonded. An *in situ* study of the structure in the temperature range $150 \leq T\ (°C) \leq 600$ shows that this site is no more than 15% occupied and so accounts for, at most, 90% of the lithium in stoichiometry $LiZr_2(PO_4)_3$.

As the temperature is increased the population of this site is decreased further, and a study of the difference Fourier map showed that there is significant concentration of lithium cations on another distorted four-coordinate position within the structure. This second lithium position is more strongly enclosed by the oxide anions and the mean lithium oxide

(a)   (b)   (c)

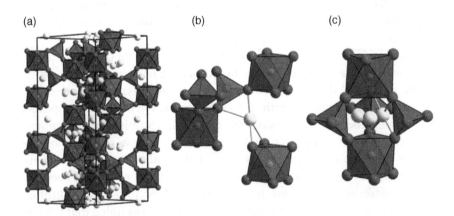

Figure 3.14 (a) The structure of the α-$LiZr_2(PO_4)_3$. Octahedra and tetrahedra represent ZrO_6 and PO_4 units respectively and lithium is shown as spheres. The large majority of lithium cations are found in (b) a heavily distorted tetrahedral coordination environment at room temperature.[91] As the temperature is increased the population of this site is reduced as the population of the unusual square planar coordinated position shown in (c) is increased

distance, 1.93 Å, is more representative of the interatomic separations typically seen for lithium in four coordination. This distorted environment has been described as tetrahedral, but the O-Li-O angle approaches 180° and so it can be better regarded as a distorted square planar arrangement of anions around the lithium cations. As can be seen in Figure 3.14, this site lies between two of the ZrO_6 octahedra that are arranged along the z direction of the unit cell and so the Li1 and Li2 sites fill in the coordination environments between these octahedra. The orientation of the ZrO_6 units is such that the interstices between them alternate between elongated octahedra, which are fully occupied in $NaZr_2(PO_4)_3$, and trigonal prisms. The latter are too large to be filled by lithium and so the introduction of lithium into these interstices, as observed in α-$LiZr_2(PO_4)_3$, requires displacement from the central axis towards the pseudo-square faces of the prism to give the unusual square planar coordination for Li^+.

On cooling this compound below the transition temperature the structure undergoes a distortion to triclinic symmetry.[92] It seems that this distortion is largely driven by changes in coordination of the lithium cations as the departure from pseudo-rhombohedral symmetry is minimal for the zirconium phosphate framework. The reduction in symmetry arises from an ordering of the lithium cations over half of the available distorted octahedral interstices. As observed in the high temperature phase, the lithium cation is too small to adopt a six-coordinate arrangement and so is displaced from the centre of the octahedra to give two, crystallographically distinct lithium positions which are both tetrahedrally coordinated by oxide anions at distances ranging from 1.90 to 2.43 Å with mean distances of 2.1 and 2.2 Å that are intermediate between those observed in the high temperature α polymorph.[93] These two lithium positions are both partially occupied. However, the two sites are only separated by 1.34 Å and such a short $Li^+ \cdots Li^+$ distance cannot exist locally in the solid. The fractional occupancies of these two lithium positions can be summed to give a value of unity and this implies that there are no vacant voids in the structure and that each void contains lithium in only one of the two positions.

Similar structural features are found in the two polymorphs of $LiZr_2(PO_4)_3$ that are accessed if the sample is prepared by heating at 900 °C. Such a preparation leads to the formation of β-$LiZr_2(PO_4)_3$ with an orthorhombic structure containing a complex distribution of lithium. As in the α- and α'-$LiZr_2(PO_4)_3$ polymorphs the zirconium phosphate framework is fully ordered. However, in β-$LiZr_2(PO_4)_3$ the lithium cations partially occupy two crystallographically distinct sites that are

(a)   (b)

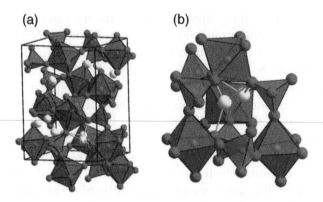

Figure 3.15   (a) The structure of β-LiZr$_2$(PO$_4$)$_3$ contains a network of ZrO$_6$ and PO$_4$ units shown as octahedra and tetrahedra respectively. (b) The lithium cations, shown as light grey spheres, occupy four-coordinate sites within a large cavity in the structure[94]

separated by only 1.56 Å as shown in Figure 3.15. Both of these sites have a minority occupation and so this unacceptably short Li$^+$ $\cdots$ Li$^+$ separation can be avoided by local ordering of lithium such that only no two adjacent sites are simultaneously occupied.[94] Both of these two sites, Li(1) and Li(2), are found within a single large cavity within the structure that is bounded by nine oxide anions within 3.1 Å of either of the two lithium sites. This extra-framework space is clearly too large to provide a bonding environment for a lithium cation located in the centre of the void and so both Li(1) and Li(2) sites are displaced from the centre and achieve distorted tetrahedrally coordinated environments. The bond lengths show considerable variation of around 0.15 Å about mean values of 2.17 and 2.16 Å. Both the substantial departure from regular tetrahedral geometry and the relatively large mean bond lengths suggest that the lithium cation is relatively poorly constrained by this bonding environment and this is reflected empirically in bond valence sums for the two lithium sites, of 0.63 and 0.69, that fall considerably short of the value of unity anticipated for a monovalent cation.

The β to β′ structural transition that occurs at 300 °C contains the same key features seen for the analogous α to α′ transition when the sample is prepared at 1200 °C. The β′ phase has lower symmetry than its higher temperature parent phase; the transition β to β′ involves a change in space group symmetry from orthorhombic *Pbna* to monoclinic *P2$_1$/n*.[94] These two groups are closely related and the difference between the two phases is crucially dependent on the removal of a twofold symmetry axis from *Pbna* to give *P2$_1$/n*. The absence of this symmetry operation in the lower temperature crystal structure requires a doubling of the number of atoms used to describe the structure, *i.e.* a doubling of the crystallographic asymmetric unit occurs. However, whilst the number of symmetry-inequivalent framework atoms is doubled it is clear that strong pseudo-symmetry relationships exist between pairs of framework atoms. There are small, but significant, displacements between these paired atoms but it is clear that examination of the framework atoms alone may lead to an overestimation of the symmetry of this phase and the erroneous conclusion that the twofold axis of *Pbna* is retained. It is probable that twinning would be a considerable problem for single crystal samples of β'.

As was observed for the α to α' transition, the driver for the formation of a lower symmetry structure is a redistribution of the lithium cations. In the β'-LiZr$_2$(PO$_4$)$_3$ phase the lithium cations are fully ordered onto a single, filled position. This site is within the same large cavity that houses the lithium cations in the high temperature β polymorph but the ordering of lithium permits a more regular tetrahedral coordination with the lithium oxide bond lengths varying by around 0.07 Å from the mean value of 2.02 Å. This contraction in bond lengths is much larger than could be anticipated to arise from thermal contraction and the bond lengths are approaching those for a lithium cation in typical, tightly bound, tetrahedral coordination environment.

There are thus similarities in the transitions observed between the high temperature phases and the corresponding low temperature polymorphs. What is still unclear is why the α and β phases cannot be interconverted. It is noteworthy that the α' polymorph is 3.6% less dense than the β' phase and this suggests that the latter may be the thermodynamically more stable phase at room temperature. It may be possible to drive the α' to β' conversion by the application of pressure but, to date, a successful prosecution of this approach has not been reported. Studies of related systems have found that pressure-stabilised phases cannot be quenched but instead readily convert to the ambient pressure phase.[95] Given that the differences between the four polymorphs of LiZr$_2$(PO$_4$)$_3$ are substantially manifested only in the lithium distribution and that the lithium cations show considerable mobility; it is likely that any attempts to kinetically trap the highly conducting phases by quenching from high pressure face a very considerable challenge.

All four polymorphs conduct lithium ions but the level of mobility in the different structures varies considerably. Some data from these polymorphs are collected in Figure 3.16.

The β' polymorph is stable below 300 °C and shows a conductivity of only 10^{-10} S cm^{-1} at room temperature.[86, 96] However, there is a large

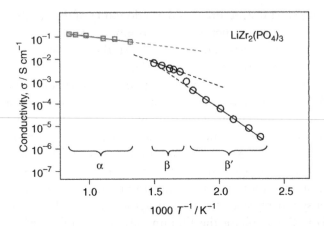

Figure 3.16 Conductivity of different polymorphs of $LiZr_2(PO_4)_3$ as a function of temperature. Data shown as circles were collected from a sample of β'-$LiZr_2(PO_4)_3$ that transformed to β-$LiZr_2(PO_4)_3$ on heating through a transition temperature around 320 °C.[88] The gradients used to determine the activation energy for conduction in these two phases is illustrated by the solid lines. Data collected from α-$LiZr_2(PO_4)_3$ are shown as squares.[98] The dotted lines indicate where the data for one polymorph have been extrapolated into the stability range of a different polymorph

activation energy associated with this conductivity process and so the conductivity increases rapidly as a function of temperature. Impedance measurements have shown a significant conductivity in this material over the temperature range $150 \leq T / (^{\circ}C) \leq 300$. This conductivity shows Arrhenius behaviour with an activation energies reported in the range $0.79^{[88]}$ to 0.91 eV[86] and a maximum conductivity of *ca* 7×10^{-4} S cm^{-1} at 300 °C. On increasing the temperature further the conductivity passes through a discontinuity around 320 °C; the conductivity increases by an order of magnitude and the activation energy for conduction above this temperature is substantially reduced. Exact estimates of the activation energy in the higher temperature regime vary but it should be noted that measurements from a single laboratory have suggested that the change in activation energy on passing through the β' to β phase transition may be as large as 0.91 to 0.28 eV.[86] The β' to β transition has been studied using calorimetry and these have identified a small endothermic process associated with the transition.[88] However, the size of this process, *ca* 2–3 kJ mol^{-1}, is over an order of magnitude smaller than the pseudo-melting transitions seen in the rotor phases related to Li_2SO_4 and discussed in Section 3.3 and so is indicative of only modest changes to the lattice energy.

The conductivity of phases prepared at 1200 °C show similar dramatic temperature sensitivities around the α' to α structural transition at 50 °C. At room temperature the conductivity of α'-LiZr$_2$(PO$_4$)$_3$ is ca 5×10^{-8} S cm^{-1}.[97] On heating the sample there is a large discontinuous increase in the conductivity at 50 to 60 °C and thereafter the conductivity follows an Arrhenius relationship with an activation energy of 0.39 eV.[98] The structural transition is endothermic and, as for the β' to β transition, the relatively small enthalpy of 4.1 kJ mol^{-1} suggests that the transition is associated with only modest increase in disorder in the crystal structure.[97] Although the activation energy of the α and β phases are similar the conductivities differ considerable. At 300 °C α-LiZr$_2$(PO$_4$)$_3$ has a conductivity of ca 10^{-2} S cm whilst that of β-LiZr$_2$(PO$_4$)$_3$ is over an order of magnitude smaller.[86]

By taking the observations of the conductivities of these phases in conjunction with the structural findings of the neutron diffraction experiments it is possible to draw some conclusions concerning the lithium mobility in these lithium zirconium phosphates. Both of the phases stable at room temperature (α' and β') have lower symmetry structures that contain an ordered arrangement of lithium. In the case of β' the lithium fills a single site whereas in α' there is a positional displacement that leads to the presence of partial occupancy in the crystallographic model but nevertheless each of the voids in the structure contains exactly one lithium cation. On heating these lower temperature polymorphs above key temperatures of ca 30 to 60 °C for α' to α and 300 °C for β' to β the lithium cations become disordered across multiple positions and show a considerable extension in the lithium oxide bonds. The presence of partial occupancy of lithium cations on multiple positions implies that these sites have similar energies for lithium occupation and so provide an energetically accessible pathway for ion migration. Moreover the distortion and expansion of the lithium coordination environment imply an increased ease of lithium hopping from the temporally and spatially averaged sites identified by the diffraction experiments. Both of these effects are manifested in the conductivity data. The activation energy undergoes a substantial decrease, from 0.91 to 0.28 eV for the β' to β transition indicating that lithium cations have a lower energy pathway for migration in the higher temperature phase and both α' to α and β' to β transitions show a stepwise increase in the conductivity that would be anticipated to arise from a sudden increase in the number of energetically accessible vacant sites for ion transport.

The NASICON structure is capable of accommodating considerable compositional variation and a large number of related compounds have been studied in order to try an improve the lithium ion conducting properties. There have been two distinct approaches to this. One approach has been to try and reduce the temperature of the structural transition and reduce the barrier to ion mobility and so access a compound that shows fast lithium conductivity under ambient conditions. An alternative strategy is to adjust the number of mobile cations and vacant sites in order to increase the conductivity of the cations in the higher temperature, disordered phase. Both approaches have had considerable success in both illuminating the mechanism for ion motion in the structure and in changing the physical properties towards those of a useful fast ion conductor.

3.6 DOPED ANALOGUES OF $LiZr_2(PO_4)_3$

The replacement of zirconium with other tetravalent cations gives a series of closely related phases with the NASICON structure that reproduce many of the features of α- and α'-$LiZr_2(PO_4)_3$. Crystallographic structures have been reported for compounds based on tetravalent hafnium, titanium, germanium and tin and in every case a high temperature phase has been reported with rhombohedral symmetry that suggests the formation of phases isostructural with α-$LiZr_2(PO_4)_3$.[99–102] Whilst this is broadly correct there remain some differences in the crystallographic models used to describe these structures. For example a characterisation of a sample described as '$LiHf_2(PO_4)_3$' identified the presence of disordered lithium cations in the elongated octahedral site occupied in the α $LiZr_2(PO_4)_3$ structure, but could not identify any additional lithium in the structure and concluded that the negative charge of the anionic framework was balanced by a small quantity of extra-framework hafnium. This lithium deficit was confirmed by elemental analysis and a composition for this sample of $Li_{0.87(3)}Hf_{2.032(7)}(PO_4)_3$ was proposed.[100] This structure also undergoes a structural transition to a triclinic phase, although the transition takes place over the subambient temperature range $221 \leq T / (K) \leq 256$.

Another sample described as $LiHf_2(PO_4)_3$ shows a structural transition at 0 °C, although the absence of compositional information on this phase suggests that the variation in transition temperature may arise from compositional variations between the two samples.[103]

The reduction in symmetry is driven by an ordering of lithium cations and, as for the zirconium analogue, the tetrahedrally coordinated lithium site undergoes a considerable contraction with the mean lithium oxide distance reduced from 2.27 Å in the rhombohedral structure to around 2.0 Å in the triclinic phase. There is some evidence from NMR studies[104] of LiHf$_2$(PO$_4$)$_3$, and also LiSn$_2$(PO$_4$)$_3$, that the low temperature distorted phases contain lithium in the larger cavities occupied in the β- and β'-LiZr$_2$(PO$_4$)$_3$ phases although such a lithium distribution is not supported by observations from neutron diffraction studies.

The low temperature polymorph of LiHf$_2$(PO$_4$)$_3$ shows negligible ion mobility, but the rhombohedral phase shows a total conductivity of *ca* 10^{-7} S cm^{-1} at room temperature. This value incorporates both inter- and intra-grain conductivities and so represents an estimate of the lower boundary for lithium conductivity in this phase. Other reports suggest that the room temperature conductivity of this compound may be as high as 5×10^{-6} S cm^{-1}.[96] The conductivity shows an Arrhenius dependence at temperatures up to at least 600 °C with an activation energy consistently reported to be around 0.45 eV,[96, 100] that is similar to the value of 0.39 eV reported for the rhombohedral analogue, α-LiZr$_2$(PO$_4$)$_3$.

The analogous compound LiTi$_2$(PO$_4$)$_3$ adopts the rhombohedral NASICON structure and, surprisingly, the lithium is fully localised on the centre of the elongated octahedral position that is occupied by sodium cations in the NaZr$_2$(PO$_4$)$_3$ parent. This robust assignment is based on analysis of high resolution neutron diffraction data and clearly shows that in this compound the lithium is coordinated to six oxide anions at a distance of 2.027 Å.[99] Despite the absence of positional and occupational disorder in LiTi$_2$(PO$_4$)$_3$ the conductivity of this sample is comparable with that of the hafnium analogue; $\sigma \approx 10^{-7}$ S cm^{-1} at room temperature[96] with an activation energy of *ca* 0.48 eV.[105]

In addition to replacing Zr^{4+} with elements in the same group it is possible to stabilise the rhombohedral structure using the d^{10} cations Ge^{4+} and Sn^{4+}. LiSn$_2$(PO$_4$)$_3$ contains lithium in a filled six-coordinate site[101] that makes it a direct analogue of LiTi$_2$(PO$_4$)$_3$. However, LiGe$_2$(PO$_4$)$_3$ contains lithium in the octahedron, but displaced from the central position onto a partially occupied site[102] in a similar manner to that reported for α-LiZr$_2$(PO$_4$)$_3$. Despite the strong structural similarities between these two compounds the room temperature conductivity, $\approx 5 \times 10^{-9}$ S cm^{-1} of the germanium compound[105] is exceptionally low for a NASICON related phase. Moreover the activation energy of 0.60 eV suggests that the mechanism for ion mobility is significantly different from α-LiZr$_2$(PO$_4$)$_3$ and many of the related compounds.

This substantial variation in the activation energy of LiM$_2$(PO$_4$)$_3$ compounds was noted and has been investigated in considerable detail by a number of research groups. By preparing compounds containing a mixture of tetravalent cations it is possible to adjust the structure continuously between the end members and this adjustment of the nature M^{4+} cation can be used to tune the activation energy. Replacement of the Ti^{4+} cation with Ge^{4+}, Sn^{4+}, Hf^{4+} or Zr^{4+} showed that the activation energy could be reduced to a minimum value of ca 0.30 eV and that this value was associated with a cell volume of 1310 Å^3.[106] This is considerably smaller than α-LiZr$_2$(PO$_4$)$_3$ (vol. = 1504 Å^3)[91] and larger than LiGe$_2$(PO$_4$)$_3$ (vol. = 1214 Å^3).[102] Initial discussions of the origin of this enhancement were hamstrung by an absence of crystallographic information on the nature of the fast ion conducting phases and particularly the presence of both cation-ordered and cation-disordered phases at room temperature in some compositional systems. Even when these subtleties could be appreciated the identification of the origin of this optimum conductivity remained a considerable challenge. Ion size is a well quantified concept in structural chemistry and because of this it can be tempting to regard the tetravalent cations in the anion framework as hard spheres that merely contribute a steric consideration to the stabilisation of the crystal structure. However, a more nuanced analysis of these different compositions would suggest that changing the nature of the tetravalent cation can have an impact on the size of the cavity that accommodates lithium and indeed whether lithium is found within one or two of the possible cavities. Moreover the position and occupancy of the lithium position along with the strength of the lithium oxide bonds may be adjusted by the polarisability of the [M$_2$(PO$_4$)$_3$]$^-$ framework. Finally, changes in the size of the MO$_6$ octahedra may have additional subtle effects on the size of the window for Li$^+$ migration than is formed by oxide anions.

The crucial aperture, or bottleneck, for lithium migration out of the elongated octahedron that is consistently favoured in these compounds is defined by three oxide ions that compose the faces of this cavity. This cavity is illustrated in Figure 3.14. It should be remembered that although this is an irregular octahedron it retains a single axis of threefold symmetry and a centre of inversion. Consequently the upper and lower faces of octahedron are identical equilateral triangles and the remaining four faces are symmetry related. An elegant study of the literature structures for a number of phases has allowed an interpretation of the

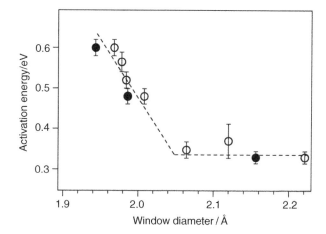

Figure 3.17 The variation in activation energy for conductivity in various LiM$_2$(PO$_4$)$_3$ compounds as a function of window size. Filled symbols represent data extracted from crystallographic studies and empty circles indicate that data were derived from calculations[105]

transport properties of a wide range of compounds from the systems LiM$_{2-x}$M'$_x$(PO$_4$)$_3$ (M, M' = Ge, Ti, Sn, Hf) by examining the radius of the void between the oxide anions that compose the four, irregular triangular faces.[105] The results of this analysis are shown in Figure 3.17 and clearly show that for apertures that have a radius of less than *ca* 2.1 Å the activation energy increases linearly with the contraction of the aperture. For compounds, such as LiHf$_2$(PO$_4$)$_3$, where the octahedral faces are larger than this critical value the activation energy takes a constant value of *ca* 0.33 eV. Presumably this energy reflects the barrier for Li$^+$ migration in another part of the ion pathway and for the larger structures it is this other effect that limits ion mobility.

There are a number of doping strategies that have been employed to modify the structure and increase the conductivity of lithium cations in the NASICON structure. These have focused on increasing the concentration of lithium in the compound and this approach has largely been successful; the conductivity at room temperature has been increased by orders of magnitude beyond that of the simple, stoichiometric quaternary phases discussed in the previous section. The general approach to increasing the room temperature conductivity can be illustrated by the system Li$_{1+x}$Ti$_{2-x}$Al$_x$(PO$_4$)$_3$. These compositions use the substitution of trivalent aluminium cation for tetravalent titanium to increase the negative charge on the anionic [M$_2$(PO$_4$)$_3$] framework and rely on the incorporation of additional lithium cations into the extra-framework volume to preserve overall charge neutrality. The aluminium and titanium occupy the oxide octahedra in a disordered fashion with no change in the rhombohedral symmetry of the structure.[107–109]

Similar approaches have been used in sodium-based NASICON phases and have shown that the concentration of extra-framework cations can be doubled in a fairly straightforward manner to give compositions such as $Na_2TiAl(PO_4)_3$.[110] However, in the analogous lithium-based system the compositional range of this system is bounded by a limiting composition that contains considerably less of the alkali metal. A number of reports exist[109, 111] of the properties of materials that contain a higher level of aluminium and lithium doping up to ca $Li_{1.7}Ti_{1.3}Al_{0.7}(PO_4)_3$, but there is an absence of structural analysis of these phases and considerable variation in the reported lattice parameters. For example the c-lattice parameter of the rhombohedral phases of the series $Li_{1+x}Ti_{2-x}Al_x(PO_4)_3$ has been reported to either undergo a significant decrease[112, 113] or increase[114] with increasing aluminium content. Moreover an aluminium phosphate impurity has been identified for compositions $x \geq 0.3$ and it is probable that the solubility limit for aluminium in the $Li_{1+x}Ti_{2-x}Al_x(PO_4)_3$ system lies close to this value.[111, 113] Samples have been prepared using an excess of lithium carbonate or ammonium hydrogen carbonate and it has been shown that glassy phases, that are largely undetectable using a typical X-ray diffraction experiment, can readily form in these systems.[112] These findings suggest that the reported compositions may depart quite significantly from the actual contents of the NASICON phases.

Given the potential utility of these phases it is curious that there is a paucity of the detailed structural characterisation that is necessary to explain the disparity between the different reports of maximum lithium content and the variation in lattice parameters. In many cases the composition of the sample has been assumed to be homogeneous and to be constrained to exactly match that of the starting reagents. These assumptions, combined with the potential for phase separation signified by the presence of aluminium phosphate impurity, have made it difficult to rationalise the lithium conductivity of compounds in this system. Indeed there is surprisingly limited information available on the lithium distribution in these compounds. We have seen that in the $LiTi_2(PO_4)_3$ end member the lithium fully occupies a site in the centre of the elongated octahedron. Remarkably no comparable structural studies have been carried out on the lithium-stuffed phases. The most detailed analysis of the lithium cations in this system is based on a ^7Li NMR study of the composition $Li_{1.2}Ti_{1.8}Al_{0.2}(PO_4)_3$ that clearly falls inside the solubility range for aluminium in the structure. This identified lithium signals arising from two different sites with a temperature dependent population of these two environments. However, given that in $LiM_2(PO_4)_3$ ($M = Zr$, Ti, Hf, Sn) the lithium cation is found in square planar, octahedral and a range of tetrahedral coordination environments in a number of different regions of the structure it is unlikely that it is possible to identify the lithium distribution from the NMR signals alone. In order to obtain an insight into likely lithium arrangement in the $Li_{1+x}Ti_{2-x}Al_x(PO_4)_3$ compounds the most illuminating information comes from a neutron diffraction study of the compound $Li_{1.15}Ti_{1.85}In_{0.15}(PO_4)_3$.[107] This identified a complex lithium distribution across three crystallographically distinct positions in the rhombohedral unit cell. The majority of the lithium is on the octahedrally coordinated position observed in $LiTi_2(PO_4)_3$, and this site is 96.9(6)% occupied. The remaining lithium cations are heavily disordered over two crystallographically distinct positions within the large cavities occupied in the β-$LiZr_2(PO_4)_3$ phase.

Whilst there is uncertainty about the location of the lithium in the structure, there is no doubt that some of these cations are highly mobile. Studies of the room temperature conductivities of a range of compositions in the $Li_{1+x}Ti_{2-x}Al_x(PO_4)_3$ system show a dramatic effect of aluminium doping and increasing lithium content. The conductivity rises sharply and reaches a maximum value for the composition $Li_{1.3}Ti_{1.7}Al_{0.3}(PO_4)_3$. The measured value of conductivity can vary substantially with porosity and the addition of a 20% excess of lithium oxide can increase the conductivity by up to two orders of magnitude. A separate study of recrystallisation from glasses of composition $Li_{1.64}Ti_{1.36}Al_{0.64}(PO_4)_3$ showed that even above the recrystallisation temperature of the glass the sintering treatment could give samples with conductivities that vary over three orders of magnitude.[111] These large effects are ascribed to alterations in the microstructure of the sample, and particularly a near elimination of the porosity in the sintered pellet used for the measurements.[112]

The total conductivity of a crystalline sample of the optimum composition, $Li_{1.3}Ti_{1.7}Al_{0.3}(PO_4)_3$, of 7×10^{-4} S cm^{-1} at room temperature shows a similar contribution from the intra-grain conductivity and the lithium conductivity through the grain boundaries. Interestingly, some authors have reported non-Arrhenius thermal dependence of the conductivity behaviour for this composition. Measurements between room temperature and 400 °C show a smooth evolution in apparent activation energy that reduces from ca 0.35 eV from 25 °C to 200 °C to ca 0.20 eV between 200 °C and 400 °C.[108, 112] A study over a smaller temperature range found a variation of an order of magnitude in the conductivity between samples prepared by grinding and heating reagents compared to a sol gel route, but the activation energies for conduction in both samples was similar at *ca* 0.40 eV.[114] Given the importance of sample density and microstructure in this system it could be surmised that the apparent change in activation energy is a result of evolution in the grain boundary structure during the conductivity measurements. However, ^{7}Li NMR has been used to study the local lithium motion and the temperature dependence of these signals provide a good match to the activation energies derived from the conductivity measurements carried out on the bulk sample.

It is argued that this temperature dependence of the activation energy arises from local correlation effects and is intimately involved with the presence of cation vacancies in the elongated octahedron site. Whilst such vacancies must remain speculative in the absence of an accurate structure determination of this composition, the presence of such vacancies in the closely related phase $Li_{1.15}Ti_{1.85}In_{0.15}(PO_4)_3$ suggests that this is the most likely origin for non-Arrhenius conductivity in the $Li_{1+x}Ti_{2-x}Al_x(PO_4)_3$ system.

A wide range of other trivalent elements can be substituted into the NASICON structure. There is limited information available on the structure of these compounds, and in some cases it has not been established the reported properties arise from a single phase sample. However, for low doping levels the lattice parameters of the NASICON structure are adjusted in a linear manner indicative of incorporation of the dopant atoms into the lattice and so the properties of these samples can be usefully examined.

The trivalent scandium ion can be substituted for titanium in the octahedra. Samples have been prepared with bulk compositions up to $Li_2TiSc(PO_4)_3$. However, these authors recognised that the lattice parameters of the rhombohedral unit cell evolved in a linear manner for compositions up to $Li_{1.4}Ti_{1.6}Sc_{0.4}(PO_4)_3$ but complex diffraction profiles that could not be indexed were obtained for the compositions that are richer in scandium.[96] A separate report has identified a similar solubility limit in the rhombohedral structure and states that a distortion to a monoclinic structure occurs beyond this scandium content,[112] although no unit cell dimensions have been reported for monoclinic structures of this compound, or any other of the doped NASICON phases in these compositional ranges.

We note here that an extensive body of literature exists on compounds based on trivalent metals, *e.g.* $Li_3Sc_2(PO_4)_3$, but as the transport behaviour and lithium distribution of these compounds is quite different to the Li M_2^{4+}(PO$_4$)$_3$ phases these will be discussed in subsequent sections. The room temperature conductivity of the Li$_{1+x}$Ti$_{2-x}$Sc$_x$(PO$_4$)$_3$ system rises to a maximum value of ca 6×10^{-4} S cm^{-1} for the composition[96, 112, 115] Li$_{1.3}$Ti$_{1.7}$Sc$_{0.3}$(PO$_4$)$_3$ and shows two distinct regions on heating to 400 °C. The conductivity occurs via a thermally activated mechanism with a clear, reversible increase in the activation energy on heating the sample through a transition around 100 °C.[96] Above this temperature the activation energy $E_a \approx 0.38$ eV is similar to that observed in the aluminium doped phases.

A similar enhancement of the conductivity of LiTi$_2$(PO$_4$)$_3$ has been achieved by the substitution of the titanium cations with a wide range of other trivalent cations including Fe^{3+}, In^{3+} and Ga^{3+}. In all cases the conductivity reaches a maximum or ca 4×10^{-4} S cm^{-1} around the 15–20 mol% doping level, but fails to surpass the conductivities achieved by aluminium or scandium doping.

By contrast the introduction of trivalent cations, such as indium, into lithium hafnium phosphate gives only a modest increase in the conductivity of the sample. The room temperature conductivity of Li$_{1.2}$Hf$_{1.8}$In$_{0.2}$(PO$_4$)$_3$ is 10^{-5} S cm^{-1} and this material is only marginally more conducting that the indium-free composition LiHf$_2$(PO$_4$)$_3$. Both materials shows a similar temperature-independent activation energy for conduction of ca 0.44 eV.[96] This modest enhancement in room temperature conductivity is a general observation for a number of other systems and it should be noted that even attempts to optimise the transport properties by using a mixture of Ge^{4+} and Ti^{4+} to adjust the structure have failed to significantly increase the lithium conductivity. A detailed study of Li$_{1+x}$Al$_x$Ge$_y$Ti$_{2-x-y}$(PO$_4$)$_3$ system produced a sample of composition Li$_{1.6}$Ge$_{0.8}$Ti$_{1.2}$(PO$_4$)$_3$ that demonstrated an intra-grain conductivity of 7×10^{-4} S cm^{-1} and an activation energy of 0.35 eV that both match that of the best compositions in the Li$_{1+x}$Ti$_{2-x}$Al$_x$(PO$_4$)$_3$ system, but fail to surpass them.[116]

Whilst it is clear that introducing trivalent cations into the octahedral site can modify the transport properties of the material it must be stressed that a change in lithium ion conductivity cannot be used as a proof of incorporation of a dopant cation into the structure. A detailed comparative study of the effect of Sc^{3+} and Y^{3+} on the properties of LiTi$_2$(PO$_4$)$_3$ showed that both Li$_{1+x}$Sc$_x$Ti$_{2-x}$(PO$_4$)$_3$ and Li$_{1+x}$Y$_x$Ti$_{2-x}$(PO$_4$)$_3$ showed maxima in the room temperature conductivity for dopant levels ca $x = 0.3$.[117] In the case of the optimum composition in the yttrium system, Li$_{1.3}$Y$_{0.3}$Ti$_{1.7}$(PO$_4$)$_3$, the conductivity was an order of magnitude greater than $LiTi_2(PO_4)_3$. However, examination of the lattice parameters for both systems showed that whilst the introduction of Sc^{3+} caused a linear expansion of the unit cell, the lattice parameters of the NASICON phase were constant and showed no dependence on the yttrium composition of the sample. Given the large increase in cation size on replacing Ti^{4+} with Y^{3+} it is impossible for the NASICON structure to incorporate this dopant without considerable expansion of the lattice. Consequently it was argued that the composition of the NASICON phase remains fixed at $LiTi_2(PO_4)_3$ and that the presence of dopant causes the formation of YPO_4. This impurity modifies the microstructure of the sample and so causes small, but significant, changes in the conductivity.[118] Similar effects result from introducing other impurity phases into the sample as a consequence of doping with La^{3+} cations.[119]

An alternative strategy to increase the lithium disorder in the NASICON structure has involved reducing the lithium content by doping with pentavalent cations to give compositions such as $Li_{1-x}Zr_{2-x}Nb_x(PO_4)_3$. The introduction of the Nb^{5+} cation onto the octahedral sites in the framework reduces the negative charge $[Zr_{2-x}Nb_x(PO_4)_3]^{(1-x)+}$ and the reduced quantity of extra-framework lithium necessary to provide charge neutrality is unable to fill the octahedral sites occupied in the undoped material. This approach stabilises the conducting rhombohedral phase at lower temperatures, but the reduced concentration of charge carriers gives a smaller conductivity than the lithium-rich phases produced by doping with trivalent species.[120, 121]

It should be noted that whilst the sodium conducting NASICON phases can be modified by partial replacement of the phosphorous cations with aliovalent species[122–124] that readily form MO_4 units, such as Si^{4+} or Ge^{4+}, this approach has not been successful for the lithium analogues that are of interest here. All reported crystal structures of lithium NASICON compounds contain framework tetrahedra that are fully and uniquely occupied by phosphorous.

As well as increasing the lithium content of these systems by introducing small quantities of trivalent cations in place of Zr^{4+} or Ti^{4+} it is possible to produce compounds based on trivalent metals where the anion framework carries three times the charge of the titanium or zirconium-based materials. An example is provided by the Fe^{3+}-based framework $[Fe_2(PO_4)_3]^{3-}$ that can be stabilised by the incorporation of three lithium cations into the extra-framework space to give a composition $Li_3Fe_2(PO_4)_3$.[125] Similar compounds can be prepared for a variety of trivalent transition metal cations and also the d^0 cation Sc^{3+}. The scandium-based material is of particular interest due to the resistance to reduction during any charge/discharge processes that occur during battery operation.

Li$_3$Sc$_2$(PO$_4$)$_3$ can be prepared by heating appropriate metal salts in air at temperatures up to 1100 °C. This compound contains an ordered arrangement of oxide tetrahedra and octahedra occupied by phosphorous and scandium respectively. This compound adopts three different polymorphs, the α polymorph is stable at temperatures up to 187 °C when it converts to a the β phase. Both of these crystallise in the monoclinic space group $P2_1/n$ with only small adjustments to the framework structure occurring at the transition. On heating the compound further a transition to an orthorhombic γ phase occurs at 245 °C. All of these are reversible transitions and single crystals suitable for crystallographic characterisation can be cycled through the transitions without excessive degradation.[126] This robustness is probably a result of the modest structural adjustments that occur at both transitions and these are largely limited to changes in the arrangement of the lithium cations, with only relatively small adjustments in the positions of the framework atoms. The room temperature structure contains a fully ordered arrangement of lithium across three, fully occupied sites within the large cavity. Interestingly there is no lithium in the elongated octahedral site favoured for the LiM$_2$(PO$_4$)$_3$ compounds. The transformation to the β polymorph appears to be driven by the disordering of the lithium cations as they are displaced from the fully occupied sites in the room temperature structure and instead distributed over three crystallographically distinct partially occupied sites tending to show highly irregular four and five coordination.[126] This disordering increases further on heating and drives the change in symmetry at 245 °C.[127]

All three phases show appreciable lithium ion conductivity. Figure 3.18 illustrates the conductivity of Li$_3$Sc$_2$(PO$_4$)$_3$ as a function of temperature and shows that the γ polymorph can be classified as a fast ion conductor with a conductivity of 10^{-2} S cm^{-1} in the high temperature regime that favours this structure. The Arrhenius plot of the conductivity data shows an obvious and large change in the activation energy associated with the β to γ transition, but also a more subtle change of gradient arising from the α to β phase change. The activation energies for lithium conduction in the α and β phases are relatively large and both of these phases show poor lithium ion conductivity.

Measurements on the transport properties of a single crystal of the high temperature γ polymorph show that the fast Li$^+$ conduction occurs preferentially along the z direction of the unit cell and is strongly anisotropic with a variation of over two orders of magnitude between

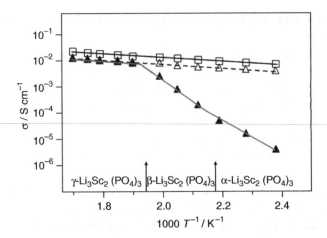

Figure 3.18 The conductivity of $Li_3Sc_2(PO_4)_3$ (shown as solid triangles) as a function of temperature. The data show three different activation energies for this compound as it undergoes phase changes at 187 and 245 °C (indicated by arrows). Data collected from the doped phases $Li_{2.8}(Sc_{0.9}Ti_{0.1})_2(PO_4)_3$ and $Li_{2.8}(Sc_{0.9}Zr_{0.1})_2(PO_4)_3$ are shown as empty triangles and squares respectively[127]

orthogonal directions. The activation energy also shows significant anisotropy and varies between 0.38 and 0.62 eV.[126]

The high ionic conductivity of γ-$Li_3Sc_2(PO_4)_3$ can be achieved at room temperature by doping this material with smaller tetravalent cations Ti^{4+} and Zr^{4+}. At room temperature $Li_{2.8}(Sc_{0.9}Ti_{0.1})_2(PO_4)_3$ and $Li_{2.8}(Sc_{0.9}Zr_{0.1})_2(PO_4)_3$ both crystallise with the orthorhombic structure that is stable in the pure scandium phase above 245 °C. The Sc^{3+} and Ti^{4+} or Zr^{4+} cations are fully disordered over the oxide octahedra of the framework and the lithium cations are disordered over three crystallographic positions in the cell with similar occupancies to γ-$Li_3Sc_2(PO_4)_3$. The cell volumes of the doped materials are significantly smaller than that of γ-$Li_3Sc_2(PO_4)_3$ and are similar to that of the poor conductor α-$Li_3Sc_2(PO_4)_3$ at room temperature.[127] This reduction in cell volume could be anticipated to favour an ordering of the lithium cations that could prevent the presence of short lithium–lithium separations in the material. The presence of disorder in the cation occupancy of the framework octahedra will lead to local variation in the oxide anions that compose the host environments for the lithium cations and so the mixture of trivalent and tetravalent cations provides an additional driver that favours occupational and positional disorder in the lithium positions. In stabilising the γ polymorph to lower temperatures the conductivity of these doped NASICONs is greatly increased compared with either the β- or α-$Li_3Sc_2(PO_4)_3$ phases observed below 245 °C with a conductivity of 3×10^{-3} S cm^{-1} at 150 °C for $Li_{2.8}(Sc_{0.9}Zr_{0.1})_2(PO_4)_3$ as shown in Figure 3.18.

The compounds $Li_3Cr_2(PO_4)_3$ and $Li_3Fe_2(PO_4)_3$ both show the same sequence of structural transitions[125, 126] as $Li_3Sc_2(PO_4)_3$. These compounds form orthorhombic fast lithium ion conducting phases above 265 °C and 312 °C respectively and show conductivities ca 10^{-2} S cm^{-1} at these temperatures.[125] These compounds contain trivalent chromium and iron and so are susceptible to reduction. Whilst this is associated with an increase in the electronic conductivity that would render these materials unsuitable for use as an electrolyte in lithium batteries it does offer potential for use as an electrode material for lithium storage. The related compound $Li_3V_2(PO_4)_3$ has been studied as a possible intercalation host for lithium cations based on the V^{3+}/V^{4+} redox couple and transformation between $Li_3V_2(PO_4)_3$ and $LiV_2(PO_4)_3$.[128, 129]

3.7 LITHIUM CONDUCTION IN THE PEROVSKITE STRUCTURE

The fastest lithium ion conductivity in oxides is found in the perovskite structure. This structure type is capable of accommodating a remarkable array of elements and oxidation states and the majority of the periodic table has been incorporated into this structure on one or more of a number of different coordination sites.[130] The general stoichiometry, ABO_3 contains a relatively small cation, B, and a larger cation, A, that approaches the size of the oxide anion. The structure can be described in a number of different ways but the coordination of the A and B cations can be most clearly visualised by considering the cubic unit cell of the prototypical structure as shown in Figure 3.19. It is noteworthy that the structure can be considered to be built from the close-packing of layers containing a mixture of oxide and the large A cations. These AO_3 layers are stacked in a cubic close-packed array and the B cations occupy the octahedral interstices between the layers.

By considering the relative sizes of the ions in the structure and the necessity of maintaining cation–anion contacts it can be seen that the ideal cubic structure will form if the ionic radii are such that the tolerance factor, t, is equal to unity:

$$t = \frac{(r_A + r_O)}{\sqrt{2}(r_B + r_O)}$$

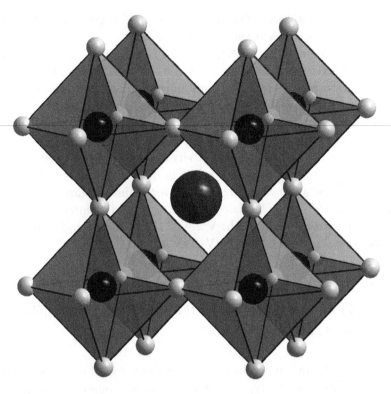

Figure 3.19   The structure of a prototypical perovskite ABO_3. A and B cations are shown as large grey and small black spheres respectively

The reason that perovskites are so widespread is that the structure is capable of accommodating deviations from ideal ion sizes. By undergoing one of an array of distortions the structure is capable of adjusting the coordination environment in order to provide the bond lengths necessary to stabilise cations that are other than the optimal size. These distortions can involve tilting of the BO_6 octahedra around various axes of rotation,[131–133] irregularities in the shape of the oxide octahedra and displacements of the A and/or B cations. In addition to these it is possible to introduce mixtures of cations onto either the A or B sites and so introduce the possibility of chemical ordering between the cations.[134, 135]

The structure is named after the mineral perovskite, calcium titanate, that contains Ti^{4+} in the octahedrally coordinated positions and Ca^{2+} in the large, 12-coordinate site in the centre of the unit cell. Lithium conductivity has been reported in closely related compounds that replace Ca^{2+} with various combinations of Li^+, La^{3+} and cation vacancies that maintain an overall divalent charge per formula unit.[136, 137] The substitution of a mixture of lithium and lanthanum for cation is a surprising one; the La^{3+} and Ca^{2+} have cations have similar ionic radii in 12-coordination, of 1.34 and 1.36 Å respectively, but 12-coordinate lithium is sufficiently unusual that there is not a standard definition of the effective radius of Li^+ in such an environment. The size difference between Li^+ and the other cations can be illustrated by considering radii of the various cations in eight-coordination: 1.16, 1.12 and 0.92 Å for La^{3+}, Ca^{2+} and Li^+ respectively. The relative size of Li^+ and the considerations involved in the tolerance factor implies the lithium should not be stable within the 12-coordinate site of the perovskite structure. Indeed a number of perovskites can be prepared that contain lithium in the smaller six-coordinate site instead.[138, 139]

The presence of lithium on the large interstitial site in lithium lanthanum titanate perovskites gives rise to exceptional ionic mobility. The lithium conductivity in this system can be as high as 10^{-3} S cm^{-1} at room temperature, *i.e.* several orders of magnitude higher than many other fast lithium ion conductors. However, it must be noted that this is the value of conductivity within a crystallite and the presence of grain boundaries reduces the total conductivity to 2×10^{-5} S cm^{-1} at room temperature. The compositions of the conducting phases in these materials have been the subject of some controversy due to the possible presence of vacancies on both the cation and the anion sites as well as the risk of lithium loss during the preparation of the sample. The highest conductivity is reached in compounds that contain La^{3+} on approximately half of the A site positions and reaches a maximum value for compositions of *ca* $Li_{0.34}La_{0.5}TiO_{2.94}$.[137]

These compounds have negligible electronic conductivity and very high intra-grain ionic conductivity up to at least 400 °C. This behaviour is reproducible over multiple heating and cooling cycles and whilst the conductivity increases smoothly with temperature it shows non-Arrhenius behaviour over this temperature range. It is unclear what the origin of this effect is, it has been argued that it may originate with a structural transition around 100 °C,[137] or that the rate-limiting window for Li^+ migration may show a thermal dependence due to rocking of the TiO_6 octahedra[140] or that disorder in the position of the lithium cations could cause an apparent temperature dependence in the activation energy for conduction.[141]

The data can be fitted using an activation energy *ca* 0.40 eV for temperatures between −50 °C and 110 °C and 0.26 eV at higher temperatures up to *ca* 300 °C.[140, 141] Whilst this material shows high Li^+ conductivity and good chemical and thermal stability it has failed to find widespread application. In part this is due to the large grain boundary resistance but a more serious shortcoming in these compounds is the ease of reduction of Ti^{4+} to Ti^{3+} that causes an increase in the electrical conductivity. The use of an electrolyte in a rechargeable lithium battery requires chemical stability in the presence of either elemental lithium or another highly reducing lithium source. Although the electronic conductivity of $Li_{0.34}La_{0.5}TiO_{2.94}$ is negligible, this family of compounds can be readily reduced by lithium and the associated reduction from tetravalent to trivalent titanium:

$$Li_{0.35}La_{0.55}Ti^{4+}O_{3.00} + x\,Li \rightarrow Li_{0.35+x}La_{0.55}Ti^{3+}_x Ti^{4+}_{1-x}O_{3.00}$$

Lithium can be inserted into the material up to at least $0.08\ Li^+$ per formula unit. This level of intercalation is insufficient for the number of lithium and lanthanum cations to exceed unity and so the A sites of the perovskite structure still contain some vacancies at this stoichiometry. Whilst this intercalation process is reversible, experiments using this electrolyte in conjunction with a graphite electrode show that an irreversible oxidation process occurs. The reduction of Ti^{4+} narrows the band gap and leads to electronic conductivity of $0.01\ S\ cm^{-1}$ at room temperature.[142] This reactivity and electronic conduction would lead to a rapid discharge *via* short circuit of a stored battery and so makes these materials unsuitable for use as an lithium electrolyte in these applications.

The perovskite structure can be adjusted by the introduction of different cations to either dope the system, by using aliovalent cations, or simply by introducing cations of different sizes or polarisabilities in order to adjust the structure. In the lithium lanthanum titanate system a large number of cations can be introduced onto the A site in order to replace lanthanum or lithium[143] and this survey has found that the combination of lithium and lanthanum is exceptional. The similarities in the chemistry of the lanthanides are commonly exploited in structural chemistry by the substitution of one lanthanide cation with another of slightly different size. In the case of lithium lanthanum titanates this approach has a spectacular impact on the transport properties.

Compounds, such as $Li_{0.34}Nd_{0.55}TiO_{3.01}$, that are analogous to the fast ion conducting phases show intra-grain conductivities that are between two and four orders of magnitude less than $Li_{0.34}La_{0.51}TiO_{2.94}$ with room temperature conductivities as low as $10^{-7}\ S\ cm^{-1}$ for $Li_{0.38}Sm_{0.52}TiO_{2.97}$.[144] These conductivities are associated with lithium mobility within the crystal structure and so are unaffected by microstructure, grain boundaries composition, sample history or variation in the details of the experimental arrangement. It follows that this large variation in performance must be linked with some substantive change in the crystallographic structure of the perovskite when the lanthanum cation is replaced with other lanthanides. Over a decade after the initial report of fast ion conduction in the $Li_{3x}La_{2/3-x}TiO_3$ system, the crystal structure of the fast ion conducting compounds remains a subject of active, and sometimes contentious, research.

A key stumbling block in the development of a plausible model for lithium conduction in these compounds arises from the pseudo close packed nature of the perovskite structure. If the lithium and lanthanum cations occupy the same 12-coordinate site in the centre of the prototypical perovskite cell shown in Figure 3.19, then the structure is considered to be composed of close packed layers of composition $Li_{1/2}La_{1/2}O_3$. This presents obvious impediments for lithium mobility through the structure on two fronts; first there are no vacant lithium sites for lithium cations to migrate through and secondly, the close packed structure fills a large amount of the volume enclosed by the material. Of course there are relatively small interstices between the layers of close packed atoms, but to make matters even less favourable for ion transport the octahedral interstitial sites are filled by the Ti^{4+} cations. Occupation of the remaining voids between the layers would lead to a tetrahedrally coordinated cation environment that is too small to be readily occupied.

The determination of the lithium environment in these phases has been a long-standing problem that has lagged behind the study of the transport properties. However, the structures of the poorly conducting analogues such as $Li_{0.35}Nd_{0.55}TiO_3$ and $Li_{0.38}Pr_{0.54}TiO_3$ were identified relatively early in the study of these systems.[145] These compounds both adopt a distorted variant of the perovskite structure that commonly arises due to a size mismatch between the size of the central, 12-coordinate interstice and the radii of the cations that occupy the site. The prototypical example of this distortion, in $GdFeO_3$, is commonly understood to arise from the small size of Gd^{3+} leading to a tilting distortion of the TiO_6 octahedra in order to reduce the gadolinium to oxide distance and so maintain a bonding interaction. The situation in $Li_{0.35}Nd_{0.55}TiO_3$ and $Li_{0.35}Pr_{0.55}TiO_3$ is complicated by the presence of a mixture of cations in this central interstice. In both cases the lanthanide cation is found in the centre of the large interstice. However, the lithium cation is displaced by a considerable distance of some 0.9 Å from the position occupied by the Ln^{3+} cation as shown in Figure 3.20. This gives rise to a wide range of Li–O bond lengths and large reduction in the coordination number;

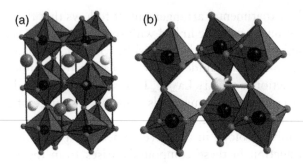

Figure 3.20   (a) The structure of $Li_{0.35}Nd_{0.55}TiO_3$. Titanium and oxygen atoms are contained in the octahedral units. The Li^+ (light grey) and Nd^{3+} (dark grey) cations are disordered in the central interstice of the structure. The unusual lithium tetrahedral coordination environment that results from the displacement of the lithium from the centre of the interstice is shown in (b)[145]

there are four oxide anions at a distances in the range 2.0 to 2.3 Å from lithium and these are arranged in the largely regular manner shown in Figure 3.20. There is no ordering between the Ln^{3+} and Li^+ cations and the close proximity of the lanthanide and lithium positions prevent simultaneous occupation of both sites within a simple, pseudo-cubic subcell of the structure. Consequently each large interstitial site contain either a lanthanide or lithium cation or, in the case of 10% of the interstices, a large vacant cavity capable of accommodating either Ln^{3+} or Li^+.

This structural model suggests that the limited lithium mobility that does occur in these compounds proceeds *via* a percolation pathway for lithium migration. This path is composed of portions of the material that contain either lithium cations or vacancies in the large interstitial void. The presence of a lanthanide cation in 55% of the subcells prevents the occupation of the immediately adjacent tetrahedrally coordinated lithium positions and so blocks the passage of Li^+. Thus the lithium conductivity in these structures will be limited first by the ability of lithium cations to exit the tetrahedrally coordinated site and secondly, by the availability of an empty interstitial site in an adjacent subcell. Given these limitations on ion movement through the structure, it would be surprising if these materials were fast lithium conductors.

It is clear that the fast ion conducting materials $Li_{3x}La_{2/3-x}TiO_{3-\delta}$ must contain some unique structural feature that permits the fast lithium mobility that is absent in the praseodymium and neodymium analogues, but the precise nature of this has been remarkably elusive. In part this difficulty arises from the synthetic challenge of obtaining reproducible samples. These compounds are typically prepared at temperatures of up to 1350 °C.[146] It is now appreciated that considerable lithium loss may occur during this process and charge balance can be maintained by the introduction of relatively small amounts of oxide anion vacancies that are difficult to detect reliably using the standard X-ray diffraction techniques that are routinely used. Moreover, the crystal structure of the material shows a strong dependence on the cooling conditions employed at the end of the synthesis.[146] A failure to appreciate these points can lead to the study of ion transport properties of a composition that is significantly different from that reported. It is likely that a wide variation in cooling rates has been employed in the preparation of samples for ion-transport measurements and, in the absence of careful reports of exactly how the samples were cooled, it is likely that this is responsible for some ambiguities in the early research literature on these compounds.

3.7.1   The Structures of $Li_{3x}La_{2/3-x}TiO_3$

In order to identify the lithium coordination site in these compounds a number of studies have focused on the most lithium-rich compositions. These studies identified a cubic perovskite structure that results from quenching the sample from high temperature.[147] However a detailed characterisation of the composition $Li_{0.5}La_{0.5}TiO_3$ showed that X-ray diffraction experiments were unable to detect a reduction from cubic to rhombohedral symmetry that largely arises from the tilting of the oxide octahedra and can be clearly detected by neutron diffraction.[148] The TiO_6 octahedra are largely regular and the lanthanide cations occupy 50% of the central interstices in the structure. However, the lithium cations are displaced towards the faces of the primitive pseudo-cubic cell to give a square planar environment for the lithium coordination as shown in Figure 3.21. The Li—O bond lengths are irregular with two short bonds, 1.81 Å, and two longer bonds, 2.07 Å, but this unusual arrangement provides a reasonable match for the bonding requirements of lithium as indicated by the valence of 1.18+ calculated for this set of interatomic distances.

This structural model has features which can be seen to favour fast lithium ion conductivity, especially when compared with the relatively poorly conducting Nd^{3+} and Pr^{3+} analogues that contain tetrahedrally coordinated lithium. The presence of Li^+ in pseudo square-planar coordination in $Li_{0.5}La_{0.5}TiO_3$ provides a large aperture for ion migration; movement of the Li^+ perpendicular to the plane defined by the four

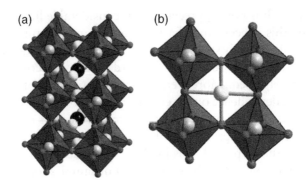

Figure 3.21 The structure of $Li_{0.5}La_{0.5}TiO_3$ contains a disordered arrangement of La^{3+} and Li^+ cations represented as dark and light spheres respectively. The Li^+ cations are displaced towards all six faces of the pseudo cubic cell to give a square-planar coordination environment[148]

oxide ions does not involve compression of any of the Li—O bonds and so will avoid the large energetic costs associated with repulsion between the electron shells of both species. This should make it relatively easy for the cations to hop out of an occupied position compared with the case of $Nd_{0.55}Li_{0.35}TiO_3$ where migration of the cation out of an LiO_4 unit will necessarily require the movement towards oxide anions.

A second feature of the structure of $Li_{0.5}La_{0.5}TiO_3$ that will facilitate fast ion conductivity is that the displacement of lithium towards the faces of the unit cell generates a large number of equivalent sites in the structure that are unoccupied. There are three times as many square-planar coordination sites as there are central interstices and so for a composition $Li_{0.5}La_{0.5}TiO_3$ the lanthanum fills 1/2 of the central interstices but lithium occupies only 1/6 of the faces of the subcell. This provides the large number of vacant sites necessary to ensure a continuous pathway for lithium hopping exists throughout the structure. Clearly, the presence of lanthanum in half of the subcells will prevent lithium migration through these portions of material, but a continuous three-dimensional pathway is proposed that involves lithium migration through the remaining unoccupied central interstices and the 5/6 of the square planar sites that are unoccupied in a random manner throughout the structure.

As the lithium content of the system is reduced the structure changes and the symmetry is reduced. There are a number of reports of samples exhibiting tetragonal symmetry[147, 149] and also a detailed electron microscopy study of similar compositions that show a definite orthorhombic distortion to the metric unit cell. Given the high resolution diffraction techniques employed it can be confidently asserted that these assignments are robust and that the difference in the crystal structures must arise from the different preparative conditions employed; quenching from 1300 °C gave a tetragonal phase $Li_{0.3}La_{0.567}TiO_3$,[149] whilst quenching from either 800 or 1000 °C affords an orthorhombically ordered phase $Li_{0.35}La_{0.55}TiO_3$.[150] A systematic study of the phase diagram has identified such a tendency for increased distortion at lower synthetic temperatures.[147] A high resolution neutron diffraction study of $Li_{0.3}La_{0.567}TiO_3$ identified a similar displacement to that observed in the $Li_{0.5}La_{0.5}TiO_3$ composition, with Li^+ close to the square windows that link the large central interstitial sites in three dimensions. A sample that had been quenched from 1300 °C shows a doubling of the unit cell along a single axis that is lost on annealing the material at 650 °C. The quenched sample and the structure observed at 650 °C both show a subtle distortion of the oxide sublattice, but the lanthanum and lithium cations remain disordered in the material, with no tendency for La^{3+} or Li^+ to avoid each other. Calculations on similar compositions show that the tilting of the TiO_6 units is dependent on the local mixing of the lithium and lanthanum cations.[151] These results also confirmed the assignment that the lithium cations that are more likely to migrate out of their sites when the adjacent interstices do not contain lanthanum cations.

When the lithium content is reduced further the structure becomes still more complex. A slow-cooled sample of $Li_{0.16}La_{0.62}TiO_3$ has been extensively studied and shows a distribution of lanthanum cations across two sites to give a layered structure shown in Figure 3.22. This contains sheets where the lanthanum site is almost filled which alternate with layers containing 70% vacancies on the lanthanum position.[152] These lanthanum-poor layers accommodate the lithium in square-planar coordination environments similar to the disordered lithium-rich compositions.

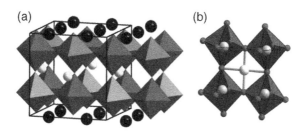

(a)   (b)

Figure 3.22 The structure of $Li_{0.16}La_{0.62}TiO_3$ contains a layered arrangement of La^{3+} and Li^+ cations represented as dark and light spheres respectively. The unusual square-planar lithium coordination environment is shown in (b)[152]

By applying maximum entropy methods to the diffraction data this analysis obtained a fuller picture of the lithium distribution than could be extracted from a conventional refinement of the structure.

The most striking observation to come out of this analysis is the determination of the pathway for lithium migration between the sites. This showed a tendency for lithium cations to migrate from a square-planar site to an adjacent, empty square-planar coordination site by following an arc that tracks around an oxide anion at an almost constant radius. This is at odds with the conventional picture of lithium movement in these perovskites that assumed that the large, central interstices that are unoccupied by lanthanum would be a key site in any lithium pathway. Instead it appears that lithium may shun this position simply because it is too large. Another key component of this structural model is that the lithium distribution is highly anisotropic, with a strongly two-dimensional aspect to the lithium distribution in this orthorhombic crystal structure. As well as introducing anisotropy into the properties of this material, this separation of lanthanum, lithium and vacancies into two distinct layers has a large impact on the distribution of carriers and carrier vacancies in the structure. Instead of considering this compound to be a perovskite of composition $\square_{0.22}Li_{0.16}La_{0.62}TiO_3$ (where $\square$ represents a vacant cation site) the layering can be indicated with a formulation such as $(La_{0.62}\square_{0.38})(Li_{0.160}\square_{.09})TiO_3$ that illustrates the quantity of lithium sites that are unoccupied. This understanding of the complex cation distribution is important when considering the percolation path and the mobility of carriers through the structure. Whilst some features of these experimental observations are supported by preliminary calculations,[153] there is a suggestion that the energy barrier for lithium migration through the lanthanum-rich layers may actually be smaller than the activation energy for migration within the lanthanum-poor sheets.[154]

Despite the exceptional intra-grain conductivity in these compounds, the conductivity that can be realised in a sample is severely limited by the grain boundary resistance; at room temperature the total conductivity, 10^{-5} S cm^{-1} is two orders of magnitude less than that indicated by the intra-grain lithium mobility.[137] The microstructure of these compounds shows considerable complexity and a strong dependence on the thermal history of the sample. The distortion of the oxide octahedra does not propagate readily through the lattice and introduces considerable orientational disorder leading to the formation of domains of variable sizes. In addition to this there are also local variations in both composition and lattice strain that are anticipated to have a large effect on the local lithium conductivity.[150] Indeed, a Raman spectroscopy study has shown that lithium ions remain mobile in the related compounds $Li_{1-x}Na_xLa_{0.5}TiO_3$ even when the lithium content is considerably below the percolation limit. In such a case, there is no continuous connected pathway for lithium ion conduction and the observed mobility arises from localised lithium movement within finite clusters.[155] Similar localised regions of cation ordering have been identified in the $Li_{3x}Nd_{1-x}TiO_3$ system where a superlattice ordering of Nd^{3+} and Li^+ cations occurs over length scales approaching 100 Å.[156] This ordering can be observed by bulk diffraction techniques, but the information provided by even high resolution neutron diffraction experiments is much too limited to control a traditional structural refinement. It is clear that a full understanding of the structure and lithium ion conductivity in this family of compounds requires detailed examination of data from a variety of sources and in some cases may lie beyond the current limits of powder diffraction techniques.

The scope for application of these materials as solid state electrolytes is severely limited by both the limited conductivity achieved in polycrystalline samples and the ease of reduction of Ti^{4+} and the associated increase in electronic conductivity. Attempts to overcome both of these limitations have resulted in a huge number of compounds being prepared that have either introduced new cations in order to modify the microstructure or the inter-grain conductivity or sought to replace titanium with more electropositive species such as Zr^{4+}, Hf^{4+}, Nb^{5+} or Ta^{5+} that will better withstand reducing conditions such as those in a rechargeable battery.

3.7.2   Doping Studies of Lithium Perovskites

Many attempts have been made to introduce larger cations in place of La^{3+} with a view to expanding the lattice and so easing the passage of lithium in the structure. Monovalent silver cations can be substituted for lithium to give the series $[Ag_yLi_{1-y}]_{3x}La_xTiO_3$. Whilst this does lead to in a reduction in the activation energy for lithium conductivity, assigned to an increase in size of the bottlenecks that limit ion transport, the overall conductivity is diminished by the reduction in lithium content and the increased disruption to the channels for ion mobility that results from introducing immobile Ag^+ into the structure.[157] Doping with Na^+ or K^+ has a similar effect.[158] The introduction of Sr^{2+} also causes an expansion in the unit cell in the series $(La_{0.5}Li_{0.5})_{1-x}Sr_xTiO_3$ and a modest increase in the conductivity is observed for $x = 0.05$.[142] However, for increased dopant levels the conductivity reduces substantially such that the room temperature conductivity, 10^{-4} S cm^{-1}, of $La_{0.375}Li_{0.375}Sr_{0.25}TiO_3$ is an order of magnitude less than the strontium-free end member. The arguments of lattice expansion would suggest that introduction of the larger Ba^{2+} cation would give the greatest increase in the lithium ion conductivity. Unfortunately the presence of barium in the system leads to a rapid reduction in the lithium ion mobility with barium content. This failure of barium to enhance the lithium ion conductivity is poorly understood but as the pathway for lithium ion mobility remains incompletely understood in the simplest $Li_{0.5}La_{0.5}TiO_3$ systems, especially when the complicating microstructural effects are also considered, it can be recognised that using larger cations to expand the lattice and ease the passage of lithium relies on a simplified picture of lithium mobility. It is likely that the contradictory successes and failures of the use of strontium and barium doping are a reflection of a range of complex secondary factors that will only be understood when these compounds receive the same degree of scrutiny as the $Li_{3x}La_xTiO_3$ system.

Replacement of Ti^{4+} with a range of cations in an attempt to improve inter-grain conductivity and stability has also led to mixed results. The introduction of a small amount of Al^{3+} gives a slight increase in the conductivity, such that the composition $La_{0.56}Li_{0.36}Ti_{0.97}Al_{0.03}O_3$ shows a room temperature conductivity of 2.95×10^{-3} S cm^{-1}. Unfortunately this trend does not persist to higher doping levels and the conductivity is reduced by increasing the aluminium content further.[159] It appears that the maximum in the conductivity arises from achieving an optimum number of vacancies on the A site of the perovskite structure. In the case of the $Li_{3x}La_xTiO_3$ system as well as the strontium and aluminium doped series the conductivity passes through a maximum when the total stoichiometry of the nominally A-site cations is ca 0.92 per formula unit. Unfortunately the inter-grain conductivity remains two orders of magnitude lower than the intra-grain value and the microstructure continues to play a crucial role in limiting the transport properties of samples in the $La_{2/3}Li_xTi_{1-x}Al_xO_3$ system.[160]

Alternative doping strategies have yielded a range of compounds containing Zr^{4+}, Ta^{5+} and Nb^{5+} on the octahedrally coordinated position occupied by Ti^{4+} in the parent material.[143] Several of these compounds show a total lithium ion conductivity that is equal to that of the $Li_{3-x}La_xTiO_3$ parent phases but are likely to show considerably greater electrochemical stability. Of particular interest is the report of conductivity of 1.3×10^{-5} S cm^{-1} at room temperature for the composition LiSr$_{1.65}$Zr$_{1.3}$Ta$_{1.7}$O$_9$.[161] The precise structural details of this compound are unknown but given the consistent observations of complicated cationic displacements combined with partial ordering and microstructural texture in the parent compounds it is possible that optimisation of the processing conditions could lead to an increase in this value. Although this composition does not provide an improvement in the lithium mobility compared with the titanium-based compounds the Zr^{4+} and Ta^{5+} cations contribute a greatly increased resistance to reduction compared with Ti^{4+}. Lithium conductivity has also been identified in a related layered tantalate phase Li$_2$La$_{2/3}$Ta$_2$O$_7$ although the intragrain conductivity of this phase is three orders of magnitude less than that observed in the Li$_{3-x}$La$_x$TiO$_3$ phases and a variable activation energy suggests a quite different mechanism for lithium mobility.[162]

Despite the problems of inter-grain resistance and reduction of Ti^{4+} it is possible that these shortcomings can be overcome by novel processing of the materials. Indeed it is possible to prepare samples of Li$_{0.3}$La$_{0.566}$TiO$_3$ by relatively low temperature processing to give uniform, highly crystalline particles of 100 nm in size.[163] If the microstructure of such particles differs significantly from the samples prepared using conventional high temperature routes then the total conductivity of these materials may be increased by orders of magnitude. Work continues to reproduce the conductivity of the titanates using other, more electrochemically stable cations, but it may be that application will be found where the ease of reduction of Ti^{4+} is not an issue such as pH sensing in a number of industrial applications, especially where stability at elevated temperatures is important.[164]

3.8 LITHIUM-CONTAINING GARNETS

One of the most exciting recent developments in fast lithium conducting oxides has been the identification of high lithium ion mobility in the garnet structure.[165] The garnet structure is a well studied, and largely well-understood, structural type that is capable of accommodating a wide range of cations due to the presence of square antiprismatic, octahedral and tetrahedral coordination environments in the structure. In the case of the prototypical material Ca$_3$Al$_2$Si$_3$O$_{12}$ these sites are filled by Ca^{2+}, Al^{3+} and Si^{4+} respectively as shown in Figure 3.23.[166]

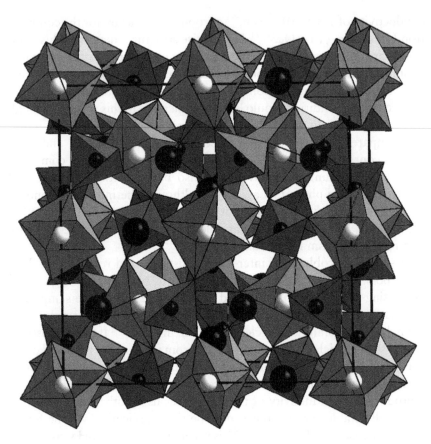

Figure 3.23   The complex structure of the garnet phase $Ca_3Al_2Si_3O_{12}$ contains AlO_6 and SiO_4 units shown as octahedra and tetrahedra respectively. The relatively large Ca^{2+} cations are shown as large black spheres[166]

The first report of lithium cations being introduced into this structure related to compounds $Li_3Ln_3Te_2O_{12}$ and $Li_3Ln_3W_2O_{12}$ which were identified as cubic garnet phases on the basis of the stoichiometries and lattice parameter considerations.[167] The first structural characterisations of lithium garnets relied on X-ray diffraction techniques and this led to a debate concerning both the stoichiometry and the lithium coordination environment of a number of cation-rich garnet phases with unusual stoichiometries such as $Li_5La_3M_2O_{12}$ (M = Nb, Ta, Sb).[168–170] As the prototype garnet structure fills all of the cation sites using a total of eight cations per formula unit it follows that in these lithium-rich garnets some additional sites must be occupied in order to accommodate ten cations per formula unit. Identification of these sites remained a considerable challenge and was given additional importance with the report in 2003 of fast lithium conduction in $Li_5La_3Ta_2O_{12}$ and $Li_5La_3Nb_2O_{12}$.[165] These compounds show similar transport properties with room temperature conductivity of the order 10^{-6} S cm^{-1} and activation energies of 0.43 eV and 0.56 eV for $Li_5La_3Nb_2O_{12}$ and $Li_5La_3Ta_2O_{12}$ respectively.

Whilst these values of conductivity fall some way short of the lithium mobility observed in other systems, most notably the lithium lanthanum titanate perovskites, there are a number of features of the garnets and particularly the tantalum-based materials, that make them potentially useful materials. Foremost amongst these is the observation that in $Li_5La_3Ta_2O_{12}$ the intra- and inter-grain conductivities are of the same order of magnitude. This means that although the intra-grain conductivity is considerably smaller than the observed in the best solid state lithium electrolytes, the total conductivity through a sample of the material compares well with many other materials as shown in Figure 3.24. It is of only marginally less importance that $Li_5La_3Ta_2O_{12}$ is particularly stable and shows fast lithium conductivity at temperatures up to at least 600 °C. The Ta^{5+} cation is also highly resistant to reduction[165] and so tantalum-based garnets can be used as an electrolyte in the presence of a metallic lithium or other common electrode materials with no reaction occurring at the electrolyte/electrode interface.[171] Hence, these tantalum-based garnets show none of the problems of reduction and associated increasing electronic conductivity, observed in the titanium-based perovskites.

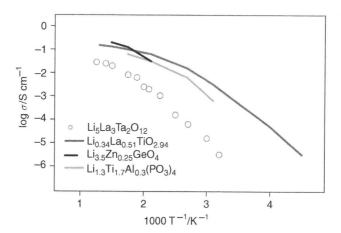

Figure 3.24 The total conductivity of $Li_5La_3Ta_2O_{12}$ as a function of temperature.[165] Data from other important lithium conductors are shown for comparison

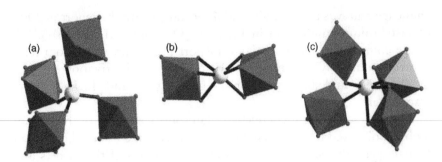

Figure 3.25   The possible coordination environments for Li^+ in $Li_5La_3Ta_2O_{12}$. In addition to (a) the three tetrahedral sites occupied in conventional garnets there are (b) two trigonal prismatic and (c) six large octahedral sites per formula unit that are capable of accommodating Li^{+}[165, 173]

Whilst the fast lithium mobility in these compounds was determined it was unclear what the mechanism for lithium conduction would be. Examination of the $[La_3Ta_2O_{12}]^{5-}$ framework shows that there is a considerable amount of extra-framework space and that three distinct interstices that could function as lithium coordination sites are available. As shown in Figure 3.25 the structure contains six distorted oxide octahedra and two square prisms in addition to the three tetrahedral interstices that are fully occupied in the conventional garnet stoichiometry. Some initial reports indicated that lithium was found exclusively on the distorted octahedral site,[172] but a series of neutron diffraction experiments identified a complex arrangement of lithium across both the oxide tetrahedra and octahedra.[173]

For both $Li_5La_3Nb_2O_{12}$ and $Li_5La_3Ta_2O_{12}$ the tetrahedral site is *ca* 80% occupied whilst each distorted octahedron contains lithium on three different positions; near the centre and displaced along the axis defined by the two triangular faces that are shared with adjacent tetrahedral sites. The displacement from the centre of the octahedron reduces four and increases two of the Li-O distances and gives what can be considered as a distorted tetrahedral coordination environment that is a consistent observation in these lithium-stuffed compositions.[173–177] By studying these structures over the temperature range 2 K to 873 K it was established that the lithium distribution is largely temperature independent indicating that the time and spatially averaged structural picture represents static disorder of lithium ions, rather than a dynamic motion of cations inside the octahedron.

The presence of shared polyhedral faces between the oxide tetrahedra and octahedra necessarily leads to the presence of short $Li \cdots Li$ distances in the structure. It was initially assumed that by locally ordering the lithium cations these short inter-lithium separations could be minimised and so the repulsion between cations would be avoided. This would lead to local regions of the material which contained lithium exclusively in the oxide octahedra and other regions in which only the tetrahedra are occupied and so suggested a cluster model for the lithium distribution reminiscent of the complex cation arrangements observed in LISICON.[67] However, subsequent studies of compounds containing a higher concentration of lithium cations have identified the presence of a distance of only 2.4 Å between two positions which have a majority occupation.[175] The high occupancy of these sites means that for a majority of lithium cations in the phase $Li_{6.6}Ba_{1.6}La_{1.4}Ta_2O_{12}$ there is a neighbouring lithium cation in the adjacent polyhedron at a distance of 2.4 Å as shown in Figure 3.26.

The garnet structure has shown itself to be remarkably flexible in accommodating not only a compositionally diverse range of elements and oxidation states, but also the ability to incorporate considerable

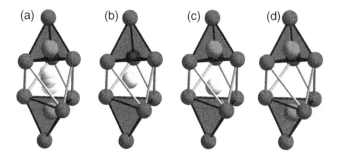

Figure 3.26 The lithium positions observed crystallographically in the lithium-stuffed garnet $Li_{6.6}Ba_{1.6}La_{1.4}Ta_2O_{12}$. Li^+ is found in the centre of two tetrahedra, shown in grey, that are linked *via* sharing the opposing faces of an octahedron. The octahedron contains lithium on three positions; close to the centre and displaced away from one shared face and towards the other showing the short separation between lithium cations in adjacent octahedral and tetrahedral sites. For this composition the majority of both the octahedra and the tetrahedra are occupied indicating that simultaneous occupation of the face-sharing octahedra and tetrahedra must exist locally.[175] Chemically reasonable distributions of Li^+ are shown with (b) the octahedron occupied with Li^+ at the central position, (c) the octahedron occupied with Li^+ displaced away from the occupied tetrahedron and towards a vacant tetrahedron and (d) both tetrahedra occupied and the octahedron vacant. These arrangements avoid extremely short $Li \cdots Li$ distances, but the $Li \cdots Li$ distance in (c) is only 2.4 Å and must exist locally for most of the lithium in the material

excess lithium beyond that anticipated for a conventional garnet stoichiometry. By doping with alkaline earth cations in place of the lanthanum cation the lithium content can be readily increased, as in the series $Li_{5+x}Ba_xLa_{3-x}Ta_2O_{12}$,[171, 175, 178, 179] but more startling still is the replacement of Ta^{5+} with Zr^{4+} to yield a composition, $Li_7La_3Zr_2O_{12}$, that contains a remarkable 133% more lithium than could be accommodated in the tetrahedra of a conventional garnet structure.[180] This compound displays a room temperature total conductivity, of 3×10^{-4} S cm^{-1}, that is higher than any other crystalline lithium ion conductor and Zr^{4+} provides excellent stability under reducing conditions.

This wide-range of cation content in the garnet structure may seem surprising, especially so when it is noted that the vast majority of garnets crystallise with the same space group, $Ia\bar{3}d$. Refinements of the lithium garnet structures against neutron diffraction data have shown that this high symmetry is retained for the lithium-rich compositions discussed above. The cubic structure can be considered to be composed of a continuous, three-dimensional body centred lattice of LnO_8 units linked by shared edges. However, this substructure does not impose strict demands on the size of the unit cell and so the lattice can demonstrate considerable expansion without disrupting the connectivity of the LnO_8 sublattice or introducing distortions that violate the space group symmetry. Examination of a series of compounds that follow the conventional garnet stoichiometry, $Li_3Ln_3Te_2O_{12}$, shows that although the LnO_8 units show a reduction in volume expected from the lanthanide contraction, the tetrahedral interstices that accommodate lithium remain more regular in size due to alterations in the structure that are compatible with the $Ia\bar{3}d$ space group symmetry.[181] Structural adjustments of this kind are responsible for a large variation in the density of different garnet phases; $Si_3Co_3Al_2O_{12}$[166] and $Li_5La_3Ta_2O_{12}$[173] both crystallise in the space group $Ia\bar{3}d$ but consideration of the ionic radii[182] shows a considerable variation in the packing efficiency; the unit cells of these compounds are 71% and 58% filled respectively. This surprising observation indicates that the lithium-stuffed garnet has a larger amount of free volume than found in a conventional garnet.

Variable temperature neutron diffraction experiments have shown that the $[La_3Ta_2O_{12}]^{5-}$ sublattice is fully ordered and appears typical for a robust, dense, framework material, thus indicating that a hopping model for lithium ion mobility is operative in these compounds, rather than a more cooperative model, as observed in some systems such as Li_2SO_4. However, the presence of a complex distribution of lithium cations in these fast ion conducting phases has made it difficult to identify which lithium sites contribute to the lithium conduction pathway. A study of compounds with the conventional garnet stoichiometry, such as $Li_3Nd_3Te_2O_{12}$ found that in these compounds the lithium filled the tetrahedral interstices and there was no lithium in the oxide octahedra,[181] despite the coordination environment appearing as suitable for lithium as in the lithium-rich phases where this site is occupied.[173]

In $Li_3Nd_3Te_2O_{12}$ the lithium on the filled tetrahedral site showed minimal mobility, with a conductivity several orders of magnitude less than $Li_5La_3Ta_2O_{12}$ and an activation energy *ca* 1.3 eV. Whilst this suggests that the tetrahedrally coordinated lithium plays no role the fast lithium conducting phases, it should be remembered that in $Li_5La_3Ta_2O_{12}$ the lithium sites contain *ca* 20% vacancies and so the behaviour of the tetrahedrally coordinated lithium in $Li_3Nd_3Te_2O_{12}$ and $Li_5La_3Ta_2O_{12}$ may be dramatically different. A ^7Li NMR study of $Li_5La_3Nb_2O_{12}$ observed two distinct lithium signals.[183] However, the intensity of these resonances showed considerable variation and did not correlate with the ratio of lithium on the octahedral and tetrahedral sites derived consistently from a number of neutron diffraction studies on a range of compounds.[173–175, 184] Moreover, the signals showed a large variation with temperature that does not reflect the minimal temperature dependence of the lithium distribution observed crystallographically.[173, 177]

A study of the solid solution that bridges the gap between the insulating $Li_3Nd_3Te_2O_{12}$ and the fast ion conducing phase $Li_5La_3Ta_2O_{12}$ has identified the pathway for lithium conductivity.[185] On introducing as little as 2% lithium into the oxide octahedra, the conductivity increases by two orders of magnitude compared with $Li_3Nd_3Te_2O_{12}$ as shown in Figure 3.27. This is accompanied by a step change in the activation energy from *ca* 1.3 eV to a value, *ca* 0.6 eV, that is similar to that reported in the fast ion conducting phases. NMR shows that there is no exchange between the lithium cations occupying the oxide octahedra and tetrahedra and, as the tetrahedral lithium has been shown to be relatively immobile in the $Li_3Nd_3Te_2O_{12}$ end member of the series, it necessarily follows that the lithium conductivity in these compounds occurs by cation migration exclusively through the octahedral interstices. These are linked by edge sharing to produce the three-dimensional conduction pathway shown in Figure 3.28.

The identification of the pathway for lithium migration through the garnet structure provides the necessary information for optimising the conductivity of these phases. However, a number of additional features have been observed in some garnets that are incompletely understood and

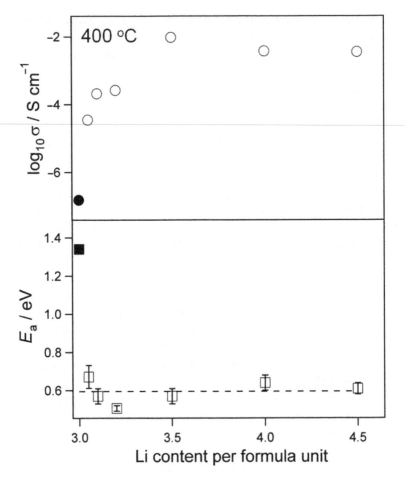

Figure 3.27 The conductivity at 400 °C and activation energy as a function of lithium content in the series $Li_{3+x}Nd_3Te_{2-x}Sb_xO_{12}$.[185] The solid symbols indicate data from the composition $Li_3Nd_3Te_2O_{12}$ that contains only tetrahedrally coordinated Li^+ cations. The dashed line indicates the average value for the activation energy for all compositions that contain $> 3\ Li^+$ per formula unit

may have a considerable impact on the performance of these materials. Whilst it has been demonstrated that the Ta^{5+}-based materials show excellent thermal stability and resistance to reduction it is noteworthy that the lattice parameters reported for the initial phase, $Li_5La_3Ta_2O_{12}$, show a significant variation. In some cases the samples have not been subjected to structural refinement and so it is possible that this arises from lithium nonstoichiometry or some other unsuspected departure from the published composition. But a high resolution neutron diffraction study of some Sb^{5+}-based analogues shows these samples contained a mixture of

a) b)

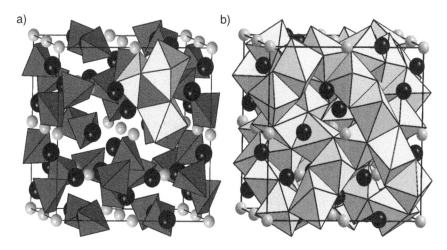

Figure 3.28 The Li^+ cations found in the oxide tetrahedra (a) that are occupied in the conventional garnet structure are immobile. The pathway for lithium conductivity in lithium garnets uses the three-dimensional network of edge-sharing octahedra shown in (b)[185]

garnet phases with significantly different lattice parameters but with no detectable difference in the composition or lithium distribution between the two phases.[186] It should be noted that this does not appear to be an intrinsic feature of all garnets and indeed the initial structural determination of $Li_5La_3Sb_2O_{12}$ clearly identified a single phase of this composition.[184]

In addition to showing this unusual structural effect, the compounds $Li_5Ln_3Sb_2O_{12}$ ($Ln=$ La, Pr, Nd, Sm, Eu) also showed a temperature dependence in the activation energy with a gradual change in gradient in the Arrhenius plot between 200 °C and 300 °C.[187, 188] It has been shown that at least in part this nonlinearity is due to experimental artifacts,[187] but it is possible that the structure of the material is varying as a function of temperature. A structural analysis of $Li_6La_2SrNb_2O_{12}$ shows the same complex distribution of lithium across octahedra and tetrahedra but identified a small reduction in the occupancy of the tetrahedra, from 0.59 at room temperature to 0.55 at 350 °C.[177] Whilst this change is at the detection limits of the structural refinement, conductivity data show a clear change in the performance of this material depending on the sample history. If the material is quenched from high temperature then the low temperature conductivity is increased by a factor of five compared with a sample that has been slowly cooled to room temperature as shown in Figure 3.29.

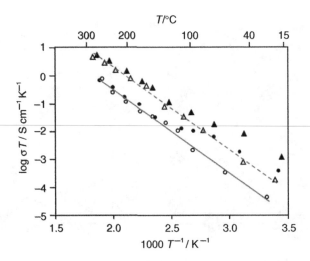

Figure 3.29 The bulk conductivity of $Li_6ALa_2Nb_2O_{12}$ (A = Ca, Sr) shows a significant history dependence below *ca* 120 °C. Triangles and circles represent data collected from A = Ca and Sr respectively. Filled and empty circles indicate data collected from samples that were quenched or slow cooled from 700 °C[177]

It is unclear what the origin of this effect is. Clearly it is highly unusual to kinetically trap a compound in a more mobile state as the mobility should allow it to relax to the thermodynamically stable configuration. It may be that this observation of history dependence is related to micro-structural effects in the sample and that these can trap the lithium cations in a more mobile configuration. NMR studies of two samples of $Li_5La_3Nb_2O_{12}$ have observed substantial differences depending on the final annealing temperature but it is unclear whether these samples were quenched or slow cooled and, in any event, the temperatures employed of 850 °C or 900 °C are in a regime considerably higher than the temperatures that lead to recovery in the $Li_6La_2SrNb_2O_{12}$ system.

Thus it can be concluded that in this ongoing area of research there remain some substantive aspects of fast ion conductivity that are incompletely understood. There is general agreement that the garnet structure can realise total lithium ion conductivity that rivals that of the fastest crystalline lithium ion conducting phases due to small inter-grain resistance. Diffraction studies show a consistent complex distribution of lithium cations across both oxide tetrahedra and distorted octahedra and it is the latter, connected through a three-dimensional, edge-sharing network, that provide the route for fast ion conductivity. Whilst these points are reasonably well established it is unclear what is the origin of the history dependence of the conductivity in some of these samples. It is possible that this phenomenon is widespread in the garnets but many of the reported syntheses do not detail how the sample is cooled from the final synthesis temperature and so make it difficult to reliably compare conflicting results. It is also striking that the conductivity does not vary dramatically in response to quite substantial changes in lithium content and the number of vacant sites. It may be that in these lithium-rich materials the high concentration of mobile cations leads to a degree of local ordering and percolation that is perturbed by the thermal history of the material.

REFERENCES

[1] J.M. Garcia-Ruiz, R. Villasuso, C. Ayora, A. Canals and F. Otalora, *Geology*, 35, 327 (2007).

[2] J.-M. Tarascon and M. Armand, *Nature*, 414, 359 (2001).

[3] K. Xu, *Chem. Rev.*, 104, 4303 (2004).

[4] E.J. Plichta and W.K. Behl, *J. Power Sources*, 88, 192 (2000).

[5] K. Xu, S. Zhang, T.R. Jow, W. Xu, C.A. Angell, *Electrochem. Solid-State Lett.*, 5, A26 (2002).

[6] C.E. Webster, R.S. Drago and M.C. Zerner, *J. Phys. Chem. B*, 103, 1242 (1999).

[7] A. Dyer, *An Introduction to Zeolite Molecular Sieves*, John Wiley & Sons, New York, 1988.

[8] A.R. West, in *Solid State Electrochemistry*, P.G. Bruce (Ed.), Cambridge University Press, Cambridge, 1997, p. 7.

[9] J.B. Goodenough, in *Solid State Electrochemistry*, P.G. Bruce (Ed.), Cambridge University Press, Cambridge, 1997, p. 43.

[10] A.K. Cheetham and P. Day, (Eds), *Solid State Chemistry: Techniques*, Oxford University Press, Oxford, 1990.

[11] G.C. Farrington and J.L. Briant, *Science*, 204, 1371 (1979).

[12] W.L. Roth and G.C. Farrington, *Science*, 196, 1332 (1977).

[13] J. Tegenfeldt, M. Underwood and G.C. Farrington, *Solid State Ionics*, 18/19, 668 (1986).

[14] G.A. Rankin and H. E. Merwin, *J. Am. Chem. Soc.*, 38, 568 (1916).

[15] C.W. Stillwell, *J. Phys. Chem.*, 30, 1441 (1926).

[16] C.A. Beevers and M.A.S. Ross, *Z. Kristallogr.*, 97, 59 (1937).

[17] Y.-F.Y. Yao and J.T. Kummer, *J. Inorg. Nucl. Chem.*, 29, 2453 (1967).

[18] J.D. Jorgensen, F.J. Rotella and W.L. Roth, *Solid State Ionics*, 5, 143 (1981).

[19] J.L. Briant and G.C. Farrington, *J. Electrochem. Soc.*, 128, 1830 (1981).

[20] M.S. Wittingham and R.A. Huggins, *J. Chem. Phys.*, 54, 414 (1971).

[21] R.H. Radzilowski and J.T. Kummer, *J. Electrochem. Soc.*, 118, 714 (1971).

[22] K. Edstrom, T. Gustafsson, J.O. Thomas and G.C. Farrington, *Acta Crystallogr.*, *Sect. B*, **53**, 631 (1997).

[23] B.C. Tofield and G.C. Farrington, *Nature*, **278**, 438 (1979).

[24] A. Marcelli, A. Mottana and G. Cibin, *J. Appl. Crystallogr.*, **33**, 234 (2000).

[25] M.D. Ingram, *J. Am. Ceram. Soc.*, **63**, 248 (1980).

[26] R.H. Radzilowski, Y.F. Yao and J.T. Kummer, *J. Appl. Phys.*, **40**, 4716 (1969).

[27] J.L. Briant and G.C. Farrington, *J. Solid State Chem.*, **33**, 385 (1980).

[28] J.E. Garbarczk, W. Jakubowski and M. Wasiucionek, *Solid State Ionics*, **18/19**, 653 (1986).

[29] J.G. Albright, *Z. Kristallogr.*, **84**, 150 (1932).

[30] N.W. Alcock, D.A. Evans and H.D.B. Jenkins, *Acta Crystallogr.*, *Sect. B*, **29**, 360 (1973).

[31] L. Nilsson, J.O. Thomas and B.C. Tofield, *J. Phys. C: Solid State Phys.*, **13**, 6441 (1980).

[32] R.Kaber, L.Nilsson, N.H. Andersen, A. Lundén and J.O. Thomas, *J. Phys.: Condens. Matter*, **4**, 1925 (1992).

[33] A. Lundén, *Solid State Commun.*, **65**, 10 (1988).

[34] R.W. Impey, M.L. Klein and I.R. McDonald, *J. Chem. Phys.*, **82**, 4690 (1985).

[35] E.A. Secco, *Solid State Ionics*, **60**, 233 (1993).

[36] E.A. Secco, *Solid State Commun.*, **66**, 921 (1988).

[37] M. Jansen, *Angew. Chem. Int. Ed.*, **30**, 1547 (1991).

[38] D. Wilmer, H. Feldmann, J. Combet and R.E. Lechner, in *5th International Conference on Quasi-Elastic Neutron Scattering (QENS 2000)*, Edinburgh, Scotland, 2000, p. 99.

[39] M. Ferrario, M.L. Klein and I.R. McDonald, *Mol. Phys.*, **86**, 923 (1995).

[40] M. Dissanayake, M.A. Careem, P. Bandaranayake and C.N. Wijayasekera, *Solid State Ionics*, **48**, 277 (1991).

[41] D.C. Parfitt, D.A. Keen, S. Hull, W.A. Crichton, M. Mezouar, M. Wilson and P.A. Madden, *Phys. Rev. B*, **72**, 054121 (2005).

[42] L. Nilsson, N.H. Andersen and A. Lundén, *Solid State Ionics*, **34**, 111, (1989).

[43] R. Tarneberg and A. Lundén, *Solid State Ionics*, **90**, 209 (1996).

[44] B. Graneli, P. Fisher, J. Roos, D. Brinkmann and A.W. Hewat, *Physica B*, **180**, 612 (1992).

[45] J. Mata, X. Solans, M.T. Calvet, J. Molera and M. Font-Bardia, *J. Phys.: Condens. Matter*, **14**, 5211 (2002).

[46] B.E. Mellander, B. Granéli and J. Roos, *Solid State Ionics*, **40/41**, 162 (1990).

[47] L. Karlsson and R.L. McGreevy, *Solid State Ionics*, **76**, 301 (1995).

[48] N. Bagdassarov, H.C. Freiheit and A. Putnis, *Solid State Ionics*, **143**, 285 (2001).

[49] M. Zhang, A. Putnis and E.K.H. Salje, *Solid State Ionics*, **177**, 37 (2006).

[50] H.A. Øye, *Acta Chem. Scand.*, **18**, 361 (1964).

[51] L. Nilsson, N.H. Andersen and J.K.Kjems, *Solid State Ionics*, **6**, 209 (1982).

[52] A. Kvist, *Z. Naturforsch.*, *A: Phys. Sci.*, **22**, 208 (1967).

[53] H. Voellenkle and A. Wittmann, *Z. Kristallogr.*, **128**, 66 (1969).

[54] H. Voellenkle, A. Wittmann and H. Nowotny, *Monatsh. Chem.*, **99**, 1360 (1968).

[55] W.H. Baur and T. Ohta, *J. Solid State Chem.*, **44**, 50 (1982).

[56] B.H.W.S. deJong, D. Ellerbroek and A.L. Spek, *Acta Crystallogr.*, *Sect. B*, **50**, 511 (1994).

[57] I.M. Hodge, M.D. Ingram and A.R. West, *J. Am. Ceram. Soc.*, **59**, 360 (1976).

[58] A. Khorassani and A.R. West, *Solid State Ionics*, **7**, 1 (1982).

[59] L. Chen, L. Wang, G. Che, G. Wang and Z. Li, *Solid State Ionics*, **9/10**, 149 (1983).

[60] A. Khorassani and A.R. West, *J. Solid State Chem.*, **53**, 369 (1984).

[61] A. Elfakir, G. Wallez, M. Quarton and J. Pannetier, *Phase Transitions*, **45**, 281 (1993).

[62] A.R. West and F.P. Glasser, *J. Solid State Chem.*, **4**, 20 (1972).

[63] I. Abrahams, P.G. Bruce, W.I.F. David and A.R. West, *J. Solid State Chem.*, **110**, 243 (1994).

[64] J. Zemann, *Acta Crystallogr.*, **13**, 863 (1960).

[65] J. Kuwano and A.R. West, *Mater. Res. Bull.*, **15**, 1661 (1980).

[66] H.Y.P. Hong, *Mater. Res. Bull.*, **13**, 117 (1978).

[67] I. Abrahams, P.G. Bruce, W.I.F. David and A.R. West, *Acta Crystallogr., Sect. B*, **45**, 457 (1989).

[68] L. Sebastian, R.S. Jayashree and J. Gopalakrishnan, *J. Mater. Chem.*, **13**, 1400 (2003).

[69] M. Bose, A. Basu and D. Torgenson, *Solid State Ionics*, **18-9**, 539 (1986).

[70] P.G. Bruce and A.R. West, *J. Solid State Chem.*, **44**, 354 (1982).

[71] Y. Saito, T. Asai, K. Ado, H. Kageyama and O. Nakamura, in *7th International Conference on Solid State Ionics*, Hakone, Japan, 1989, p. 34.

[72] H.H. Sumathipala, M. Dissanayake and A.R. West, in *10th International Conference on Solid State Ionics (SSI-10)*, Elsevier Science BV, Singapore, Singapore, 1995, p. 719.

[73] J.B. Chavarria, P. Quintana and A. Huanosta, *Solid State Ionics*, **83**, 245 (1996).

[74] K. Jackowska and A.R. West, *J. Mater. Sci.*, **18**, 2380 (1983).

[75] C. Masquelier, M. Tabuchi, T. Takeuchi, W. Soizumi, H. Kageyama and O. Nakamura, in *20th Commemorative Symposium on Solid State Ionics in Japan*, Tokyo, Japan, 1994, p. 98.

[76] A. Robertson and A.R. West, *Solid State Ionics*, **58**, 351 (1992).

[77] B.J. Neudecker and W. Weppner, *J. Electrochem. Soc.*, **143**, 2198 (1996).

[78] A.K. Ivanov-Shits and S.E. Sigaryov, *Solid State Ionics*, **27**, 89 (1988).

[79] F. Stewner and R. Hoppe, *Acta Crystallogr., Sect. B*, **27**, 616 (1971).

[80] F. Stewner and R. Hoppe, *Z. Anorg. Allg. Chem.*, **380**, 241 (1971).

[81] F. Stewner and R. Hoppe, *Z. Anorg. Allg. Chem.*, **381**, 140 (1971).

[82] R.M. Biefeld and R.T. Johnson Jr, *J. Electrochem. Soc.*, **126**, 1 (1979).

[83] T. Esaka and M. Greenblatt, *Solid State Ionics*, **21**, 255 (1986).

[84] T. Esaka and M. Greenblatt, *J. Solid State Chem.*, **71**, 164 (1987).

[85] L.O. Hagman and P. Kierkegaard, *Acta Chem. Scand.*, **22**, 1822 (1968).

[86] F. Sudreau, D. Petit and J.P. Boilot, *J. Solid State Chem.*, **83**, 78 (1989).

[87] G.J. Long, G. Longwith, P.D. Battle, A.K. Cheetham, R.V. Thundathil and D. Beveridge, *Inorg. Chem.*, **18**, 624 (1979).

[88] M. Casciola, U. Costantino, L. Merlini, I.G.K. Andersen and E.K. Andersen, *Solid State Ionics*, **26**, 229 (1988).

[89] D. Petit, P. Colomban, G. Collin and J. P. Coilot, *Mater. Res. Bull.*, **21**, 365 (1986).

[90] H.Y.P. Hong, *Mater. Res. Bull.*, **11**, 173 (1976).

[91] M. Catti, A. Comotti and S. DiBlas, *Chem. Mater.*, **15**, 1628 (2003).

[92] J.E. Iglesias and C. Pecharroman, *Solid State Ionics*, **112**, 309 (1998).

[93] M. Catti, S. Stramare and R. Ibberson, *Solid State Ionics*, **123**, 173 (1999).

[94] M. Catti, N. Morgante and R.M. Ibberson, *J. Solid State Chem.*, **152**, 340 (2000).

[95] F. Brunet, N. Bagdassarov and R. Miletich, *Solid State Ionics*, **159**, 35 (2003).

[96] M.A. Subramanian, R. Subramanian and A. Clearfield, *Solid State Ionics*, **18/19**, 562 (1986).

[97] J. Kuwano, N. Sato, M. Kato and K. Takano, in *9th International Conference on Solid State Ionics*, Elsevier Science BV, The Hague, Netherlands, 1993, p. 332.

[98] K. Nomura, S. Ikeda, K. Ito and H. Einaga, *Solid State Ionics*, **61**, 293 (1993).

[99] A. Aatiq, M. Menetrier, L. Croguennec, E. Suard and C. Delmas, *J. Mater. Chem.*, **12**, 2971 (2002).

[100] E.R. Losilla, M.A.G. Aranda, M. MartinezLara and S. Bruque, *Chem. Mater.*, **9**, 1678 (1997).

[101] E. Morin, J. Angenault, J.C. Couturier, M. Quarton, H. He and J. Klinowski, *Eur. J. Solid State Inorg. Chem.*, **34**, 947 (1997).

[102] M. Alami, R. Brochu, J.L. Soubeyroux, P. Gravereau, G. Leflem and P. Hagenmuller, *J. Solid State Chem.*, **90**, 185 (1991).

[103] M.A. Parìs, A. Martìnez-Juárez, J.E. Iglesias, J.M. Rojo and J. Sanz, *Chem. Mater.*, **9**, 1430 (1997).

[104] M.A. Parìs and J. Sanz, *Phys. Rev. B*, **55**, 14270 (1997).

[105] A. Martinez-Juarez, C. Pecharroman, J.E. Iglesias and J.M. Rojo, *J. Phys. Chem. B*, **102**, 372 (1998).

[106] H. Aono, E. Sugimoto, Y. Sadaoka, N. Imanaka and G. Adachi, *J. Electrochem. Soc.*, **140**, 1827 (1993).

[107] D.T. Qui, S. Hamdoune, J.L. Soubeyroux and E. Prince, *J. Solid State Chem.*, **72**, 309 (1988).

[108] K. Arbi, M. Tabellout, M.G. Lazarraga, J.M. Rojo and J. Sanz, *Phys. Rev. B*, **72**, 8 (2005).

[109] K. Arbi, J. M. Rojo and J. Sanz, *J. Eur. Ceram. Soc.*, **27**, 4215 (2007).

[110] F.E. Mouahid, M. Bettach, M. Zahir, P. Maldonado-Manso, S. Bruque, E.R. Losilla and M.A.G. Aranda, *J. Mater. Chem.*, **10**, 2748 (2000).

[111] J. Fu, *Solid State Ionics*, **96**, 195 (1997).

[112] H. Aono, E. Sugimoto, Y. Sadaoka, N. Imanaka and G. Adachi, *J. Electrochem. Soc.*, **137**, 1023 (1990).

[113] S. Wong, P.J. Newman, A.S. Best, K.M. Nairn, D.R. MacFarlane and M. Forsyth, *J. Mater. Chem.*, **8**, 2199 (1998).

[114] M. Cretin and P. Fabry, *J. Eur. Ceram. Soc.*, **19**, 2931 (1999).

[115] K. Ado, Y. Saito, T. Asai, H. Kageyama and O. Nakamura, *Solid State Ionics*, **53–6**, 723 (1992).

[116] P. Maldonado-Manso, E.R. Losilla, M. Martinez-Lara, M.A.G. Aranda, S. Bruque, F.E. Mouahid and M. Zahir, *Chem. Mater.*, **15**, 1879 (2003).

[117] K. Ado, Y. Saito, T. Asai, H. Kageyama and O. Nakamura, in *8th International Conference on Solid State Ionics (SSI-8)*, Elsevier Science BV, Lake Louise, Canada, 1991, p. 723.

[118] Y. Saito, K. Ado, T. Asai, H. Kageyama and O. Nakamura, *J. Mater. Sci. Lett.*, **11**, 888 (1992).

[119] H. Aono, E. Sugimoto, Y. Sadaoka, N. Imanaka and G.Y. Adachi, *J. Electrochem. Soc.*, **136**, 590 (1989).

[120] B.V.R. Chowdari, K. Radhakrishnan, K.A. Thomas and G.V.S. Rao, *Mater. Res. Bull.*, **24**, 221 (1989).

[121] I.A. Stenina, I.Y. Pinus, A.I. Rebrov and A.B. Yaroslavtsev, in *14th International Conference on Solid State Ionics*, Monterey, CA, 2003, p. 445.

[122] J.P. Boilot, G. Collin and P. Colomban, *Mater. Res. Bull.*, **22**, 669 (1987).

[123] P.J. Squattrito, P.R. Rudolf, P.G. Hinson, A. Clearfield, K. Volin and J.D. Jorgensen, *Solid State Ionics*, **31**, 31 (1988).

[124] A. Clearfield, *Eur. J. Solid State Inorg. Chem.*, **28**, 37 (1991).

[125] F. d'Yvoire, M. Pintard-Scrépel, E. Bretey and M. de la Rochère, *Solid State Ionics*, **9/10**, 851 (1983).

[126] A.B. Bykov, A.P. Chirkin, L.N. Demyanets, S.N. Doronin, E.A. Genkina, A.K. Ivanov-shits, I.P. Kondratyuk, B.A. Maksimov, O.K. Melnikov, L.N. Muradyan, V.I. Simonov and V.A. Timofeeva, *Solid State Ionics*, **38**, 31 (1990).

[127] T. Suzuki, K. Yoshida, K. Uematsu, T. Kodama, K. Toda, Z.G. Ye, M. Ohashi and M. Sato, in *11th International Conference on Solid State Ionics (SSI-11)*, Elsevier Science Bv, Honolulu, Hi, 1997, p. 89.

[128] J. Gaubicher, C. Wurm, G. Goward, C. Masquelier and L. Nazar, *Chem. Mater.*, **12**, 3240 (2000).

[129] L.S. Cahill, R.P. Chapman, J.F. Britten and G.R. Goward, *J. Phys. Chem. B*, **110**, 7171 (2006).

[130] R.H. Mitchell, *Perovskites: Modern and Ancient*, Almaz Press, Thunder Bay, Ontario, 2002.

[131] A.M. Glazer, *Acta Crystallogr., Sect. B*, **28**, 3384 (1972).

[132] P.M. Woodward, *Acta Crystallogr., Sect. B*, **53**, 32 (1997).

[133] P.M. Woodward, *Acta Crystallogr., Sect. B*, **53**, 44 (1997).

[134] C.J. Howard, B.J. Kennedy and P.M. Woodward, *Acta Cryst., Sect. B*, **59**, 463 (2003).

[135] P. Woodward, R.-D. Hoffman and A.W. Sleight, *J. Mater. Res.*, **9**, 2118 (1994).

[136] A.G. Belous, G.N. Novitskaya, S.V. Polyanetskaya and Y.I. Gomikov, *Izv. Akad. Nauk SSSR, Neorg. Mater.*, **23**, 470 (1987).

[137] Y. Inaguma, C. Liquan, M. Itoh, T. Nakamura, T.U.H. Ikuta and M. Wakihara, *Solid State Commun.*, **86**, 689 (1993).

[138] P.D. Battle, C.P. Grey, M. Hervieu, C. Martin, C.A. Moore and Y. Paik, *J. Solid State Chem.*, **175**, 20 (2003).

[139] W.R. Gemmill, M.D. Smith and H.C. zur Loye, *J. Solid State Chem.*, **179**, 1750 (2006).

[140] O. Bohnke, C. Bohnke and J.L. Fourquet, *Solid State Ionics*, **91**, 21 (1996).

[141] C. León, J. Santamarìa, M.A. Parìs, J. Sanz, J. Ibarra and L.M. Torres, *Phys. Rev. B: Condens. Matter*, **56**, 5302 (1997).

[142] Y. Inaguma, L. Chen, M. Itoh and T. Nakamura, *Solid State Ionics*, **70/71**, 196 (1994).

[143] S. Stramare, V. Thangadurai and W. Weppner, *Chem. Mater.*, **15**, 3974 (2003).

[144] M. Itoh, Y. Inaguma, W.-H. Jung, L. Chen and T. Nakamura, *Solid State Ionics*, **70/71**, 203 (1994).

[145] J.M.S. Skakle, G.C. Mather, M. Morales, R.I. Smith and A.R. West, *J. Mater. Chem.*, **5**, 1807 (1995).

[146] A. Varez, M.T. Fernández-Dìaz, J.A. Alonso and J. Sanz, *Chem. Mater.*, **17**, 2404 (2005).

[147] A.D. Robertson, S.G. Martin, A. Coats and A.R. West, *J. Mater. Chem.*, **5**, 1405 (1995).

[148] J.A. Alonso, J. Sanz, J. Santamaria, C. Leon, A. Varez and M.T. Fernandez-Diaz, *Angew. Chem. Int. Ed.*, **39**, 619 (2000).

[149] M. Sommariva and M. Catti, *Chem. Mater.*, **18**, 2411 (2006).

[150] S. Garcìa-Martìn, M. Á. Alario-Franco, H. Ehrenberg, J. Rodriguez-Carvajal and U. Amador, *J. Am. Chem. Soc.*, **126**, 3587 (2004).

[151] M. Catti, *Chem. Mater.*, **19**, 3963 (2007).

[152] M. Yashima, M. Itoh, Y. Inaguma and Y. Morii, *J. Am. Chem. Soc.*, **127**, 3491 (2005).

[153] Y. Inaguma, Y. Matsui, J. Yu, Y.-J. Shan, T. Nakamura and M. Itoh, *J. Phys. Chem. Solids*, 58, 843 (1997).
[154] M. Catti, *J. Phys. Chem. C*, 112, 11068 (2008).
[155] M.L. Sanjuán, M.A. Laguna, A.G. Belous and O. I. V'Yunov, *Chem. Mater.*, 17, 5862 (2005).
[156] B.S. Guiton, H. Wu and P.K. Davies, *Chem. Mater.*, 20, 2860 (2008).
[157] O. Bohnke, C. Bohnke, J. O. Sid'Ahmed, M.P. Crosnier-Lopez, H. Duroy, F. Le Berre and J.L. Fourquet, *Chem. Mater.*, 13, 1593 (2001).
[158] A.G. Belous, *Ionics*, 4, 360 (1998).
[159] A. Morata-Orrantia, S. Garcìa-Martìn and M.Á. Alario-Franco, *Chem. Mater.*, 15, 3991 (2003).
[160] S. Garcìa-Martìn, A. Morata-Orrantia, M.Á. Alario-Franco, J. Rodriguez-Carvajal and U. Amador, *Chem. Eur. J.*, 13, 5607 (2007).
[161] V. Thangadurai, A. K. Shukla and J. Gopalakrishnan, *Chem. Mater.*, 11, 835 (1999).
[162] F. LeBerre, M.-P. Crosnier-Lopez, Y. Laligant, E. Suard, O. Bohnke, J. Emery and J.-L. Fourquet, *J. Mater. Chem.*, 14, 3558 (2004).
[163] M. Vijayakumar, Y. Inaguma, W. Mashiko, M.-P. Crosnier-Lopez and C. Bohnke, *Chem. Mater.*, 16, 2719 (2004).
[164] C. Bohnke, H. Duroy and J.-L. Fourquet, *Sens. Actuators, B*, 89, 240 (2003).
[165] V. Thangadurai, H. Kaack and W.J.F. Weppner, *J. Am. Ceram. Soc.*, 86, 437 (2003).
[166] S.C. Abrahams and S. Geller, *Acta Crystallogr.*, 11, 437 (1958).
[167] H.M. Kasper, *Inorg. Chem.*, 8, 1000 (1968).
[168] D. Mazza, *Mater. Lett.*, 7, 205 (1988).
[169] H. Hyooma and K. Hayashi, *Mater. Res. Bull.*, 23, 1399 (1988).
[170] J. Isasi, M.L. Veiga, R. Saez-Puche, A. Jerez and C. Pico, *J. Alloys Compd.*, 177, 251 (1991).
[171] V. Thangadurai and W. Weppner, *J. Power Sources*, 142, 339 (2005).
[172] V. Thangadurai, S. Adams and W. Weppner, *Chem. Mater.*, 16, 2998 (2004).
[173] E.J. Cussen, *Chem. Commun.*, 412 (2006).
[174] J. Percival, E. Kendrick and P.R. Slater, *Mater. Res. Bull.*, 43, 765 (2008).
[175] M.P. O'Callaghan and E.J. Cussen, *Chem. Commun.*, 2048 (2007).
[176] M.P. O'Callaghan and E.J. Cussen, *Solid State Sci.*, 10, 390 (2008).
[177] J. Percival, D. Apperley and P.R. Slater, *Solid State Ionics*, 179, 1693 (2008).
[178] V. Thangadurai and W. Weppner, *Adv. Funct. Mater.*, 15, 107 (2005).
[179] R. Murugan, V. Thangadurai and W. Weppner, *Ionics*, 13, 195 (2007).
[180] R. Murugan, V. Thangadurai and W. Weppner, *Angew. Chem. Int. Ed.*, 46, 7778 (2007).
[181] M.P. O'Callaghan, D.R. Lynham, G.Z. Chen and E.J. Cussen, *Chem. Mater.*, 18, 4681 (2006).
[182] R.D. Shannon, *Acta Crystallogr., Sect. A*, 32, 751 (1976).
[183] L. vanWüllen, T. Echelmeyer, H.-W. Meyer and D. Wilmer, *Phys. Chem. Chem. Phys.*, 9, 3298 (2007).
[184] E.J. Cussen and T.W.S. Yip, *J. Solid State Chem.*, 180, 1832 (2007).
[185] M.P. O'Callaghan, J.T. Titman, G.Z. Chen and E.J. Cussen, *Chem. Mater.*, 20, 2360 (2008).
[186] J. Percival and P.R. Slater, *Solid State Commun.*, 142, 355 (2007).
[187] R. Murugan, W. Weppner, P. Schmid-Beurmann and V. Thangadurai, *Mater. Res. Bull.*, 43, 2579 (2008).
[188] J. Percival, E. Kendrick and P.R. Slater, *Solid State Ionics*, 179, 1666 (2008).
[189] P.G. Bruce and A.R. West, *J. Electrochem. Soc.*, 130, 662 (1983).

4

Thermoelectric Oxides

Sylvie Hébert and Antoine Maignan
Laboratoire CRISMAT, UMR6508 CNRS et ENSICAEN, Caen Cedex, France

4.1 INTRODUCTION

The rich properties of oxides have been objects of desire for chemists and physicists. In 1986, the discovery of high T_c superconductors created a rush to the layered cuprates for which the low dimensionality of the structure together with the copper mixed-valency were found to play a crucial role. From the fundamental physics point of view, this has been also the beginning of a new era devoted to the strongly correlated systems. As compared with conventional degenerate semiconductors, in these oxides the role of electronic correlations has been shown to play a crucial role. After the hour of glory of cuprates, the possibility to use magnetic oxides as storage media by using the concept of spintronics has attracted a lot of researchers, especially after the discovery of the so-called magneto-resistance in perovskite manganites. More recently, much effort has been focused on electronic materials (transistors, dielectrics, superconductors...) but also on catalysts, batteries, fuel cells, luminescent materials, photovoltaics and thermoelectrics. For societal impact reasons, the environmental risk of new materials is again pushing forward the oxide-based materials. This is especially timely if one considers the importance of transparent conducting oxides in photovoltaic cells, solid oxide fuel cells or lithium batteries based on layer cobaltites. In that respect,

Functional Oxides   Edited by Duncan W. Bruce, Dermot O'Hare and Richard I. Walton
© 2010 John Wiley & Sons, Ltd

thermoelectricity is one field where oxides are also worth studying. There exists a large amount of waste-heat on our planet: a majority of the consumed fossil energies is discharged in the environment as waste-heat. For instance in a car, it is about 70% of the consumed energy which is released as waste-heat. In such individual vehicles, the waste-heat recovery by using classical systems – as vapour turbines – is not possible and thus, the partial conversion of waste-heat into electricity by using thermoelectric generators (TEG) has attracted much attention, even if their efficiency is only of 5–10% (percentage of heat crossing the TEG that is transformed into electricity). One explanation for this poor efficiency comes from the poor performance of the thermoelectric materials, which are qualified by the figure of merit Z (this will be detailed later on). The most commonly used thermoelectrics are Bi_2Te_3 and $PbTe$. Although these materials exhibit high enough Z values, their use for waste-heat recovery is limited by their chemical stability at high temperatures as they start to decompose at low T ($T > 200$ °C) precluding the use of conventional thermoelectrics at higher temperatures. A lot of interesting materials have been identified: clathrates, half-Heusler, Zintl phases, skutterudites, Re-Te *etc.*[1] Nonetheless, it appears that most of them are not molten at 1000 °C, but their surface tends to oxidise in ambient air applications with sublimation of chemical elements, as Sb in the partially filled skutterudites. The toxicity of elements such as Te or Sb is also a severe constraint for application. In that context, oxides which are prepared in air or oxidising conditions at high temperatures are very attractive candidates (754000 occurrences on Google for thermoelectric oxides, for example in May 2009).

In this chapter, we report on advances made in the last ten years in the field of thermoelectric oxides.

4.2 HOW TO OPTIMISE THERMOELECTRIC GENERATORS (TEG)

4.2.1 Principle of a TEG

Figure 4.1 presents a schematic description of a TEG. n- and p-type thermoelectric (TE) materials are electrically connected in series, and thermally connected in parallel between two plates submitted to a temperature gradient $\Delta T = T_h - T_c$.

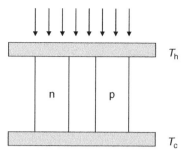

Figure 4.1 Schematic description of a thermoelectric generator with n- and p-type materials connected, electrically in series and thermally in parallel, between the hot source (T_h) and the heat sink (T_c)

To determine the efficiency of such a device, let us first consider a simpler model, with only one leg (see Figure 4.2), of electrical resistivity ρ, thermal conductivity, κ and Seebeck coefficient, S.

The efficiency of such a device is defined by the ratio between the output power generated by the material (P) with the thermal flux through the bar (Q).

$$\eta = P/Q \tag{4.1}$$

When considering the energy balance of the system, one has:

$$\eta = \frac{R_{\text{load}}I^2}{\kappa\Delta T - ST_cI - \frac{1}{2}RI^2} \tag{4.2}$$

with R_{load} the resistance of the load and R the resistance of the bar.

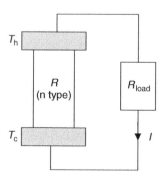

Figure 4.2 A one-leg device, with internal material resistance R, adapted on a load resistance R_{load}

By optimising η with respect to R_{load}, the maximum efficiency is obtained:[1]

$$\eta_{max} = \frac{T_h - T_c}{T_h} \frac{\sqrt{1 + ZT_m} - 1}{\sqrt{1 + ZT_m} + \dfrac{T_h}{T_c}} = \eta_{Carnot} \frac{\sqrt{1 + ZT_m} - 1}{\sqrt{1 + ZT_m} + \dfrac{T_h}{T_c}} \quad (4.3)$$

where the dimensionless coefficient ZT has been introduced:

$$ZT_m = \frac{S^2}{\rho\kappa} T_m \quad (4.4)$$

and

$$T_m = \frac{T_c + T_h}{2} \quad (4.5)$$

ZT is called the figure of merit of the thermoelectric material and should be as large as possible to optimise η.

In the case of a thermoegenerator as presented in Figure 4.1, with both n- and p-type materials, it can be shown that the efficiency of the TEG defined by the ratio:

$$\eta = P/Q \quad (4.6)$$

with P the power from the TEG and Q the thermal flux through the TEG, is maximum for:

$$\eta_{max} = \frac{T_h - T_c}{T_h} \frac{\sqrt{1 + Z_{np}T_m} - 1}{\sqrt{1 + Z_{np}T_m} + \dfrac{T_h}{T_c}} = \eta_{Carnot} \frac{\sqrt{1 + Z_{np}T_m} - 1}{\sqrt{1 + Z_{np}T_m} + \dfrac{T_h}{T_c}} \quad (4.7)$$

where

$$T_m = \frac{T_c + T_h}{2} \quad (4.8)$$

and

$$Z_{np} = \frac{(S_p - S_n)^2}{[(\rho_p\kappa_p)^{1/2} + (\rho_n\kappa_n)^{1/2}]} \approx \frac{Z_n + Z_p}{2} \quad (4.9)$$

if the properties of n- and p-type materials are similar.

The efficiency of the thermogenerator will therefore strongly depend on three parameters of the materials: the electrical resistivity, ρ, the thermal conductivity, κ and the Seebeck coefficient, S. These three parameters which are strongly interlinked through the carrier density should be optimised for the two materials, and this is the major challenge for material scientists.

Apart from these fundamental restrictions on the choice of materials, the major technological limiting factor for the efficiency of thermogenerators is the quality of electrical and thermal contacts. In the formulae given above, all the contacts are supposed to be ideal, but all the equations should be rewritten taking into account thermal and electrical resistances. When performing such a treatment, the manufacturing factor MF which reflects the quality of the contacts, can be introduced.[2] The output power is defined by:

$$P_{\text{out}} = MF \times P \qquad (4.10)$$

with

$$MF = R_{\text{ideal}}/R_{\text{int}} <1 \qquad (4.11)$$

R_{ideal} being linked to the n- and p-type materials resistances and R_{int} including the contact resistances.

4.2.2   The Figure of Merit

The figure of merit is given by

$$ZT = \frac{S^2}{\rho\kappa}T = \frac{S^2}{\rho(\kappa_e + \kappa_l)}T \qquad (4.12)$$

with κ_e and κ_l the electronic and lattice part of the thermal conductivity respectively. To maximise ZT, a material with a large Seebeck coefficient, small thermal conductivity and small electrical resistivity is therefore required. Concerning the electronic part, S, ρ and κ_e are strongly interdependent. On the other hand, the lattice part κ_l can be in principle minimised independently of the other parameters.

Figure 4.3 presents the evolution of S, ρ and κ as a function of the carrier density.

S and ρ are small in metals, leading to small power factors. On the other hand, S is large for semiconductors or insulators but ρ is too large, so that S^2/ρ is again too small. As shown in Figure 4.3, the maximal value for Z can be obtained for intermediate values of the carrier density, close to 10^{18}–10^{19} cm^{-3}, typical of semiconductors. Three different families have been traditionally used: bismuth-based alloys (BiSb, Bi_2Te_3...) are used for T ranging from $100\,K$ to $450\,K$, lead telluride materials are

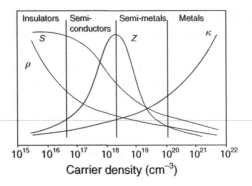

Figure 4.3 Evolution of S, ρ, κ and the figure of merit Z as a function of the carrier density[1]

used for intermediate T up to $850\,K$, and for higher T, silicium germanium can be used up to $1300\,K$ (Figure 4.4). The ZT of all these materials reach maximum values of 1–1.2.

In the presence of an electric field, E and thermal gradient, ∇T, the general transport equations can be written in the following manner:[3]

$$J = L_{11}E + L_{12}\nabla T$$
$$U = L_{21}E + L_{22}\nabla T \tag{4.13}$$

with J the electrical current and U the thermal current. It can be shown that the three important parameters for thermoelectricity are:

$$\sigma = L_{11} \tag{4.14}$$

$$S = \frac{-L_{12}}{L_{11}} \tag{4.15}$$

and

$$\kappa_e = -\left(L_{22} - \frac{L_{12}L_{21}}{L_{11}}\right) \tag{4.16}$$

κ_e represents the electronic part of the total thermal conductivity.

In the case of metals, from these equations, the thermopower can be written as:

$$S = \frac{\pi^2 k_B^2}{3e}T\left[\frac{\partial \ln\sigma(E)}{\partial E}\right]_{E\,=\,E_F} \tag{4.17}$$

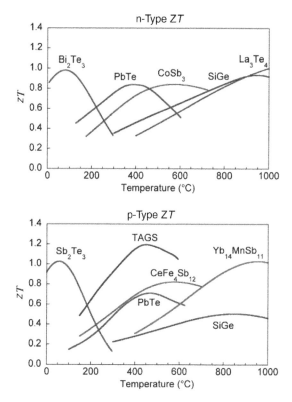

Figure 4.4 *ZT* as a function of *T* of the classical thermoelectric materials, after Snyder and Toberer, *Nat. Mater.*, 7, 105 (2008). Copyright (2008) Nature Publishing Group.

with $\sigma(E) = en(E)\mu(E)$ is the electronic conductivity, $n(E)$ the carrier density and $\mu(E)$ the carrier mobility. This formula, often referred to as the Mott formula, shows that S varies linearly with T, and that the measure of S is a good probe of the density of states at the Fermi level. The Wiedemann law can also be derived from these equations, with:

$$\frac{\kappa_e}{\sigma T} = \frac{\pi^2 k_B^2}{3e^2} = L_0 \tag{4.18}$$

Taking into account the Wiedemann–Franz law, the following formula is obtained:

$$ZT = \frac{S^2}{L_0} \left(\frac{1}{1 + \dfrac{\kappa_l}{\kappa_e}} \right) \tag{4.19}$$

Considering only the electronic part of the thermal conductivity, the Seebeck coefficient is described by Equation 4.20:

$$S = 156\sqrt{ZT} \quad \mu V/K \qquad (4.20)$$

For $ZT > 1$, S should be larger than 156 $\mu V/K$. As the lattice thermal conductivity is never zero, this condition is even more drastic.

In the case of semiconducting materials, the thermopower can be written as:

$$S = \frac{k_B}{e} \frac{E_g}{k_B T} \qquad (4.21)$$

with E_g the band gap.

4.2.3 Beyond the Classical Approach

4.2.3.1 Nanostructuration

The thermopower strongly depends on the shape of the density of states (DOS) as shown by the Mott formula:

$$S = \frac{\pi^2 k_B^2}{3e} T \left[\frac{\partial \ln \sigma(E)}{\partial E} \right]_{E = E_F} \qquad (4.22)$$

To obtain large values of S, one possibility is therefore to modify the shape of the DOS at the Fermi level, to induce large derivatives of $\sigma(E)$. It has been proposed[4] that by reducing the dimensionality of the materials, as the DOS can be sharpened, large S can be obtained. A smooth $\sigma(E)$ in 3D could be in principle transformed into spikes for 0D materials, leading to theoretically infinite Seebeck coefficients. This approach has been tested by designing superlattices based in most cases on classical thermoelectrical materials, such as Bi_2Te_3 or PbTe. Enhancement of ZT has already been reported,[5, 6] but the effect is not due to thermoelectric enhancement but rather to a strong reduction of the thermal conductivity. In these superlattices, the electrical properties are almost not changed, but the thermal conductivity can be drastically reduced due to diffusion of the phonons at the interfaces.

Another way to enhance ZT is by decoupling the electronic and thermal properties to design 'PGEC', i.e. a 'Phonon Glass and Electrical

Crystal' as proposed by Slack.[7] The material should have lattice thermal conductivities close to those of amorphous materials, but electronic properties associated with crystalline materials. As the characteristic length scales for phonon and electron diffusion can be different, it could in principle be possible to independently optimise these two properties. Some clathrates indeed show a glass-like thermal conductivity.[8] Also, in the case of skutterudites, it has been proposed that rattling by introducing one heavy atom in vacant sites could reduce the thermal conductivity. Thermal conductivity is indeed reduced, even if the rattling mechanism is not at the origin of this phenomenon as shown recently by inelastic neutron diffraction.[9]

4.2.3.2   Strong Correlations

As shown in Figure 4.4, ZT is maximal for a carrier density close to $\sim 10^{18} - 10^{19} \, \text{cm}^{-3}$. However, it has been shown by Terasaki et al.[10] that a power factor as large as that of Bi_2Te_3 at 300 K can be obtained in the metallic cobalt oxide Na_xCoO_2, for a larger carrier density close to $\sim 10^{22} \, \text{cm}^{-3}$. This family of oxides is part of the so-called 'strongly correlated materials' in which strong correlation effects exist between the charge carriers, which can drastically modify the physics of the carriers (mobility, shape of the density of states...). In these strongly correlated systems, calculations of the Seebeck coefficients have been derived from the Hubbard model, taking into account these narrow bands. Using the Kubo formalism, it can be shown that S is composed of two terms:[11]

$$S = \frac{-S^{(2)}/S^{(1)} + \mu/e}{T} \quad (4.23)$$

where $S^{(1)}$ and $S^{(2)}$ are integrals depending on velocity and energy flux operators and μ is the chemical potential. When $T \to \infty$, Equation 4.23 tends to

$$S \to \frac{\mu}{eT} = \frac{-1}{|e|} \left(\frac{\partial \sigma}{\partial N} \right) = \frac{-k_B}{e} \frac{\partial \ln g}{\partial N} \quad (4.24)$$

where σ is the system entropy, N the number of particles and g the degeneracy of the system ($\sigma = k_B \ln g$). The Seebeck coefficient is in this case a direct measurement of the entropy per carrier of the system. For these strongly correlated systems, maximising the entropy will induce large Seebeck coefficients at high T. In the case of spinless fermions, the Heikes formula is obtained:

$$S = \frac{-k_B}{|e|} \ln \left(\frac{x}{1-x} \right) \tag{4.25}$$

with x the carrier concentration.

Extensions of the Heikes formula have been proposed taking into account the other possible origins of entropy which are linked to the spins and orbitals. First, by considering a system containing mixed-valent cations $M^{(n)+}/M^{(n+1)+}$ with spin values S_n and S_{n+1}, an extra entropy term coming from spins can be added[12] and:

$$S = \frac{-k_B}{|e|} \ln \left(\beta \frac{x}{1-x} \right) \text{ with } \beta = \frac{2S_n + 1}{2S_{n+1} + 1} \tag{4.26}$$

More recently, such a calculation has been performed for the Na_xCoO_2 case, taking into account the orbital and spin degeneracy term.[13] In these layered cobaltites, Co^{3+} and Co^{4+} are supposed to be in low spin states, *i.e.* only with t_{2g} orbitals. S would then be equal to

$$S = \frac{-k_B}{|e|} \ln \left(\frac{g_3}{g_4} \frac{x}{1-x} \right) \tag{4.27}$$

with g_3 and g_4 the spin and orbital degeneracies associated with Co^{3+} and Co^{4+}, respectively, and x the Co^{4+} concentration. In the low spin states, this g_3/g_4 term is very small, equal to 1/6, which leads to huge values of S. For a given x, an extra term of 154 μV/K is added (Figure 4.5).

The Heikes formula is valid only at high T, and other calculations including the transport term have been performed at low T (for example, for Na_xCoO_2[14]) which show the dominant part of the Heikes formula on the measured thermopower.

In the last 15 years, the Dynamical Mean Field Theory (DMFT) has been developed to investigate the properties for strongly correlated materials.[15] Using this technique, the thermopower has been calculated, in particular in the case of a triangular lattice.[16] Starting from the Hubbard model, the thermopower can be written at low T, as:

$$S = \frac{-k_B}{|e|} \left[\frac{\partial D_0(E)/\partial E}{D_0(E)} \right]_{E=E_F} \frac{I_{21}}{I_{01}} \frac{T}{Z} \tag{4.28}$$

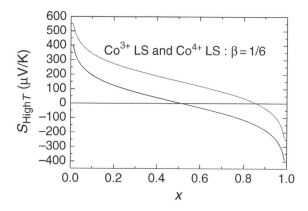

Figure 4.5 S as a function of the carrier concentration x for the different models in the case of Co^{3+}/Co^{4+} with the Heikes formula (in black), and extra contribution due to spin degeneracy and spin and orbital degeneracy for low spin states (LS) (in grey)

with $I_{21}/I_{01} = 2.65$, Z is the inverse of the effective mass and D_0 is the density of states with no correlation.

The ratio between S and γ, the electronic part of specific heat, can be written as:

$$\frac{S}{\gamma T} = \frac{-1}{|e|} \frac{3}{2\pi^2} \left[\frac{\partial D_0(E)/\partial E}{D_0(E)^2} \right]_{E=E_F} \frac{I_{21}}{I_{01}} \qquad (4.29)$$

For a triangular network, the value of $\left[\frac{\partial D_0(E)/\partial E}{D_0(E)^2}\right]_{E=E_F}$ is large, and as γ is also large,[17] this could explain why large thermopower can be found, with a large slope at low T, in these strongly correlated oxides.

The link between specific heat and S has been experimentally investigated in different systems (heavy fermions, oxides . . .).[18] When $T \to 0$, the quantity $q = \frac{S}{T} \frac{N_{Av}e}{\gamma}$ tends to a constant value with $0.5 < |q| < 2$. For free electrons, $q = -1$. This means that this ratio is not strongly modified in the case of strong correlations. Moreover, the thermopower at low T will be larger when γ is large, *i.e.* in the case of strong correlations.

4.3 THERMOELECTRIC OXIDES

Until 1997, a few oxides were investigated both for n- and p-type materials. The best ZT values reported were 0.14 for p-type $LaCrO_3$ at $1600\,K$[19]

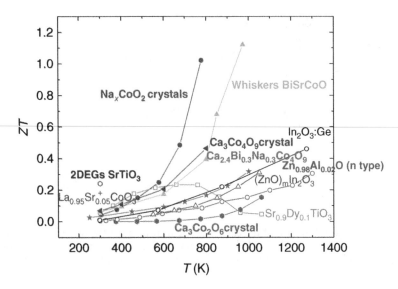

Figure 4.6 The best ZT for oxides. Data taken from Fujita *et al.*, 2001 [21]; Muta *et al.*, 2003 [22]; Xu *et al.*, 2002 [23]; Funahashi and Shikano, 2002 [24]; Mikami *et al.*, 2003 [25]; Ohta *et al.*, 2007 [26]; Shikano and Funahashi, 2003 [27]; Androulakis *et al.*, 2004 [28]; Ohtaki *et al.*, 1996 [20]; Guilmeau *et al.*, 2008 [100]

and 0.3 for n-type, Al-doped ZnO at 1272 K.[20] The discovery of a large thermopower in the metallic Na_xCoO_2[10] has induced an upsurge of research on thermoelectric oxides. Figure 4.6 summarises the best ZT values, as a function of T, which have been obtained, both for p- and n-type oxides.

In the last few years, ZT has been continuously improved, but a striking difference is observed between n- and p-type, with a ZT systematically smaller in the case of n-type oxides. The origin of this asymmetry will be discussed in the following sections. First, the results obtained on semiconducting oxides will be presented and compared with the Heikes formula, to evaluate the relevancy of this formula. The second part will focus on the misfit oxides which are metallic oxides with CdI_2 type layers similar to Na_xCoO_2. In this family, the best p-type oxides can be found, and the origin of these properties is fascinating from a fundamental point of view, as one has to take into account both the triangular nature of the lattice and the strong correlations between carriers. However, only p-type oxides have been found so far in this family, and the search for new n-type materials, with enhanced ZT, has to be pursued. The results obtained on degenerate semiconductors, among which the best n-type oxides can be found, will thus be detailed, emphasising the recent improvements obtained. The possible benefits of nanostructuration in these materials will be presented. To conclude this part, results obtained with all oxides modules will be briefly described.

4.3.1 Semiconducting Oxides and the Heikes Formula

For localised carriers, the Heikes formula states that:

$$S = \frac{-k_B}{|e|} \ln \left(\frac{x}{1-x} \right) \tag{4.30}$$

with x the carrier concentration. As shown previously, large enhancement of S can be achieved by introducing spin and orbital degeneracies. Also, a change of the sign of S can be obtained by doping with electrons or holes, Figure 4.7.

These two properties have been investigated in the case of chromium and cobalt oxides. For the first family, the aim was to investigate the possible spin and orbital degeneracy term and its impact on the Seebeck coefficient, while for cobalt oxides, the purpose of the investigation was to obtain large Seebeck coefficient, both positive and negative.

4.3.1.1 $Pr_{1-x}Ca_xCrO_3$

Orthochromites $Pr_{1-x}Ca_xCrO_3$ with $0 \leq x \leq 0.5$ crystallise in the orthorhombic perovskite $GdFeO_3$ type structure, Figure 4.8.[30] Their properties are shown in the Figure 4.9.

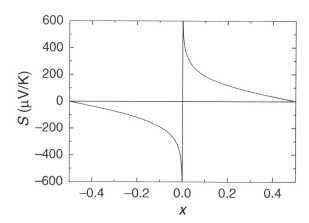

Figure 4.7 The Heikes formula for holes and electrons

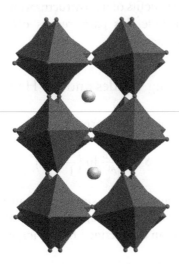

Figure 4.8 The perovskite structure (in the case of $Pr_{1-x}Ca_xCrO_3$, Pr^{3+} and Ca^{2+}, are the large grey spheres and Cr^{3+}/Cr^{4+} are at the centre of the octahedra that share corners through oxygen atoms, shown by the small spheres)

They all present semiconducting behaviour with large values of ρ for the undoped compound. When introducing Ca^{2+}, a mixed valency Cr^{3+}/Cr^{4+} is introduced, leading to a decrease in the resistivity, even if the materials remain semiconductors. Due to the large resistivity values, S can be measured only down to 100 K. S is positive, large for the undoped compound ($+1000$ $\mu V/K$), and decreases as x increases. For $T > 150 - 200$ K, a plateau in S is observed, and high T measurements for $x = 0.05$ and $x = 0.3$ show that this plateau extends up to 700 K. The

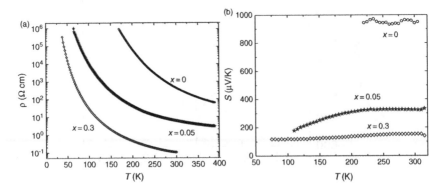

Figure 4.9 $\rho(T)$ (a) and $S(T)$ (b) of $Pr_{1-x}Ca_xCrO_3$

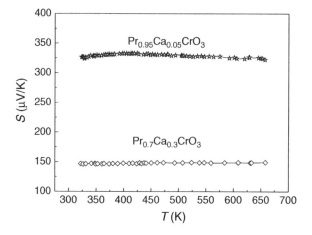

Figure 4.10 High T measurements of S

values measured for $T > 150\,\mathrm{K}$ therefore correspond to the high T limit of S, Figure 4.10.

Theoretical calculations have been performed for these t_{2g} systems,[31] using the Hubbard model. In this paper, the whole Seebeck coefficient, and not only the high temperature limit, has been calculated. It was shown that, taking into account the spin and orbital terms, the Heikes formula has to be modified by adding an extra contribution:

$$\Delta S_d = \frac{k_B}{|e|}\ln\left[\Gamma_{orb}\Gamma_{spin}\right] \tag{4.31}$$

with $\Gamma_{orb} = 3$ for Cr^{4+} $(3d^2)$ and $\Gamma_{orb} = 1$ for Cr^{3+} $(3d^3)$, and in the case of weak magnetic coupling, $\Gamma_{spin} = 2\sigma+1$ (with σ is the spin), or, in the case of strong magnetic coupling, $\Gamma_{spin} = 1$. For weak magnetic coupling,

$$\Delta S_d = \frac{k_B}{|e|}\ln\left(\left[\Gamma_{orb}\Gamma_{spin}\right]^{Cr^{4+}} / \left[\Gamma_{orb}\Gamma_{spin}\right]^{Cr^{3+}}\right)$$
$$= \frac{k_B}{|e|}\ln\left(\frac{3\times 3}{1\times 4}\right) = 69.9\ \mu V/K. \tag{4.32}$$

In Figure 4.11, the values of S at $300\,\mathrm{K}$ are reported as a function of x. Also shown is the evolution of $S(x)$ calculated with the Heikes formula, and with the spin and orbital degeneracy calculated by Marsh et al.[31]

The data are perfectly fitted by the Marsh and Parris formula,[31] showing that an enhancement of thermopower can be obtained by optimising the spin and orbital degeneracies. Combining the resistivity and Seebeck

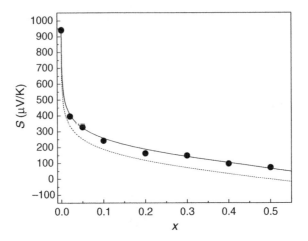

Figure 4.11 S as a function of carrier doping in $Pr_{1-x}Ca_xCrO_3$, calculated from the Heikes formula (dashed line) or taking into account the spin and orbital degeneracy (solid line)

coefficients, a power factor 2×10^{-4} W/m/K^2 is obtained at 300 K, close to the values obtained in misfit cobaltate polycrystals.

4.3.1.2   LaCoO$_3$

One of the major tasks for the development of thermoelectric materials is to find and optimise the thermoelectric properties of both n- and p-type legs. As explained in the Introduction, in the case of oxides, the ZT of n-type materials are always smaller, and it is therefore important to find new ways to generate n-type materials. From the simple Heikes formula, with only the x dependence, and no spin and orbital degeneracy term, it can be shown that n- and p-type materials should be symmetrically obtained by doping an insulator with a small amount of carriers or holes (Figure 4.12).

This strategy has been used in the case of LaCoO$_3$ perovskites.[32] For this undoped compound, S should be very large ($x \sim 0$ in Equation 4.29). As shown in Figure 4.13, S is equal to +600 μV/K at 300 K, with S decreasing as a function of T as expected for semiconductors. Due to the different oxidation states of Co, it is possible to introduce different mixed valency, Co^{2+}/Co^{3+} (electron doping) with Ti^{4+} or Ce^{4+} such as in La$^{3+}_{1-x}$Ce$^{4+}_x$CoO$_3$, or Co^{3+}/Co^{4+} in La$^{3+}_{1-x}$Sr$^{2+}_x$CoO$_3$ (hole doping). As shown in Figure 4.13, with hole doping, the Seebeck coefficient of the naturally hole-doped LaCoO$_3$ is decreased to +300 μV/K at 300 K for

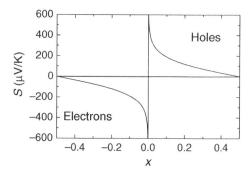

Figure 4.12 S calculated from the Heikes formula for electron or hole doping

$La_{0.98}Sr_{0.02}CoO_3$. On the other hand, in $La_{0.99}Ce_{0.01}CoO_3$, S is negative, with $S \sim -300$ μV/K at 300 K. By using these values, and considering the simplest Heikes formula (equation 4.25), a good agreement is obtained for the p-type between the nominal Sr^{2+} content (2×10^{-2}) and the calculated one (2.2×10^{-2}). For n-type, the agreement is not so good (3.6×10^{-2} from the Heikes formula to be compared with the nominal 10^{-2}).

Nevertheless, as demonstrated in $Pr_{1-x}Ca_xCrO_3$, the spin and orbital degeneracy terms should be added to the simple Heikes formula, and this excess of Seebeck coefficient will shift the x value. A detailed analysis of the spin and orbital degeneracies in the case of Co^{2+}/Co^{3+} or Co^{3+}/Co^{4+} has been performed by Taskin et al.[33] Due to the larger degeneracy of the high-spin Co^{2+} compared with the low-spin Co^{4+}, larger Seebeck coefficients can be obtained for a given x in the case of electron doped cobalt oxides.

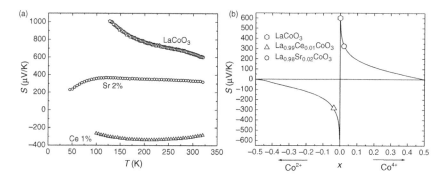

Figure 4.13 $S(T)$ for $LaCoO_3$ and related compounds (a) and the theoretical $S(x)$ with the experimental points (b)

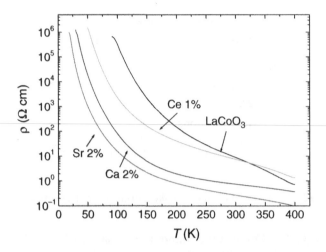

Figure 4.14 $\rho(T)$ for $LaCoO_3$ and related p- (Ca and Sr doping) and n-type (Ce doping) compounds

Large Seebeck coefficients can therefore be obtained in semiconducting oxides, both for p- and n-types, by maximising the spin and orbital degeneracies. However, the power factor of these materials remains small due to the large resistivity. In Figure 4.14, a strong asymmetry is observed between p- and n-type doped $LaCoO_3$, the value of ρ being much larger for n-type $La_{0.99}Ce_{0.01}CoO_3$.

It was proposed that this difference of resistivity is due to the difference in the filling of the t_{2g}/e_g orbitals, depending on the spin states of Co.[34] The orbital filling for Co^{2+}, Co^{3+} and Co^{4+} is shown in Figure 4.15. In the case of Co^{3+}/Co^{4+}, a hole can hop in the t_{2g} orbitals. On the other hand, the hopping of an electron through the e_g level is forbidden due to an impossible orbital filling.

Even if the Seebeck coefficient can be larger for n-type materials, the power factor remains much smaller due to the spin blockade effect.

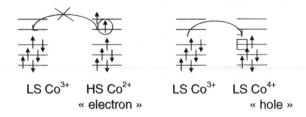

Figure 4.15 The spin blockade phenomenon

Also, a transition to a metallic state associated with a strong decrease of S is observed in $LaCoO_3$ at $\sim 600\,K$,[35] whose origin is still a matter of debate. The $LaCoO_3$ compound will therefore not be used for thermoelectric applications even if good thermoelectric properties are reported at room temperature.[28] The results presented here are nevertheless very interesting as they show that by a simple doping, n- and p-type thermoelements can be obtained. Similar results have been obtained in the 1D compound $Ca_3Co_2O_6$ where a mixed-valence Co^{2+}/Co^{3+} or Co^{3+}/Co^{4+} can be obtained, with, respectively, negative and positive Seebeck coefficients.[36] The ZT value depends on the square of the Seebeck coefficient, and due to the large resistivity in these compounds, it is now crucial, as shown already in $Pr_{1-x}Ca_xCrO_3$, to maximise the spin and orbital degeneracy term in order to enhance S to obtain potential materials for application.

Finally, it should also be kept in mind that the use of the Heikes formula is theoretically valid only at high T, in the limit of 'infinite' temperature, $i.e.$ when all the relevant energies become negligible with respect to kT. The calculation has been performed at any temperature range only for the case of orthochromites[31] and in manganites.[37] This distinction is important as shown for example, in $La_{1-x}Ca_xMnO_3$,[37] where S has been calculated and measured in different temperature ranges. The calculation of S has to be adjusted depending on the temperature range, the larger values for small doping being obtained at low T ($T < 400\,K$). High temperature measurements of these semiconducting oxides are therefore required to compare the S value with the theoretical value calculated from the Heikes formula.

4.3.2 Na_xCoO_2 and the Misfit Cobaltate Family

4.3.2.1 Na_xCoO_2

Na_xCoO_2 and related compounds such as Li_xCoO_2 have been investigated in detail in the last 30 years due to their potential use in batteries.[38] Na_xCoO_2 is a layered oxide, with CoO_2 layers of CdI_2 type, made of edge-shared CoO_6 octahedra, and separated by a layer filled with Na^+ (Figure 4.16). Compared with the perovskites previously described, the geometry is completely different as, for example, the angle for $O - Co - O$ is no longer 180° but 90°. The Co ions are now located on a triangular lattice.

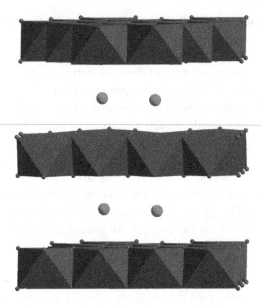

Figure 4.16 Structure of Na_xCoO_2 (here the case of $x = 0.5$, with Na^+ as grey spheres, and the CoO_2 layers shown as polyhedra)

In 1997, Terasaki et al.[10] showed that for $x = 0.7$, a metallic behaviour coexists with a large thermopower at 300 K, making this material interesting for thermoelectric applications. A few years after this report, the discovery of superconductivity below 30 K in the hydrated phase $Na_{0.3}CoO_2 \cdot yH_2O$[39] stimulated a lot of research from a broader solid state community.

For single crystals, the combination at 300 K of $\rho \sim 0.2$ mΩ cm together with $S \sim 80\,\mu$V/K leads to a power factor of 50×10^{-4} W/m/K^2, slightly larger than the one of the classical thermoelectric materials Bi_2Te_3. Owing to the lamellar structure of Na_xCoO_2, the resistivity strongly depends on the microstructure and a good texturation is required to obtain small resistivity. For example, Fujita et al. have shown that ρ increases from 0.29 mΩ cm for single crystals to 2 mΩ cm for polycrystals.[21] The thermal conductivity of this material is small, and Na_xCoO_2 belongs to the 'PGEC' family, i.e. 'Phonon Glass and Electrical Crystal' as defined by Slack.[7] For single crystals, $\kappa \sim 20$ W/m/K at 300 K, decreases to 5 W/m/K at 800 K, and is reduced to 2 W/m/K for polycrystals between 300 K and 800 K.[40,21] The Seebeck coefficient increases as T increases. Combining these different values, $ZT \sim 1$ is obtained in single crystals at 800 K.[21]

Na_xCoO_2 could also be used for thermoelectric applications at low T. For $x > 0.70$, Z presents a peak at $T \sim 100\,K$, due to a peak in the Seebeck coefficient.[41] The values of Z obtained are equal to $1.8 \times 10^{-3}\,K^{-1}$ at $100\,K$, one of the largest Z among thermoelectric materials.

The Na_xCoO_2 family has been investigated in detail both theoretically and experimentally due to the large thermopower for $x \sim 0.7$ and superconductivity for $x \sim 0.3$ in the hydrated phase. Also, charge ordering is observed for $x = 0.5$[42] suggesting a strong influence of the Na layer on the transport properties. One of the most surprising result is that the correlations are maximum for $x \sim 0.7$, $i.e.$ for a rather small concentration of carriers,[42,43] with a Curie-Weiss susceptibility, turning to Pauli susceptibility for $x \sim 0.3$. Considering only the thermoelectric properties, the fascinating coexistence of metallicity and large thermopower for $x \sim 0.7$, and this large ZT, has motivated a lot of theoretical investigations. Two different approaches have first been proposed. As previously explained in Section 4.2.3.2, the Heikes formula has been extended for the Co^{3+} $(3d^6)$/ Co^{4+} $(3d^5)$ low-spin states $(t_{2g}^6$ and $t_{2g}^5)$.[13] This formula which takes into account the spin and orbital degeneracies leads to large values of the Seebeck coefficient, close to $100\,\mu V/K$ if $x \sim 0.6$. The correct value for the Seebeck coefficient is thus obtained, but this model should be applied only in the case of localised carriers (narrow band systems in the Hubbard model), and at high T. The metallicity can thus not be explained in such a framework.

Simultaneously, the band structure of $NaCo_2O_4$ has been calculated by Singh[44] taking into account the rhombohedral symmetry of the CdI_2 layers. Due to this symmetry, the t_{2g} orbitals are split in two, the a_{1g} narrow band, with localised carriers close to the Fermi level responsible for large Seebeck coefficient, and a broad band of light carriers, e'_g, responsible for metallicity.[44] This band structure has then been experimentally investigated by different techniques. The search for the e'_g pockets is a crucial point, which could play a role also for superconductivity.[45] Since this calculation by Singh, there is a debate on the existence of these pockets, as the ARPES experiments and Shubnikov de Haas oscillations have found no evidence so far for them.[46–48] The difference between the calculations from Singh and the ARPES results could come from strong correlation effects,[49] not taken into account in the DOS determination, or from Na disorder.[50] Beyond this debate, it appears that most of the properties and calculations for this family of oxides have to be discussed by considering the coexistence of localised moments and itinerant carriers.[48]

To account for the magnetic properties of these materials, the model of 'spin polarons' has been proposed.[51,52] In this model, due to the peculiar geometry of the CdI_2 like layers, the Co^{3+}/Co^{4+} are not only in the t_{2g} orbitals, but excitations from t_{2g} to the e_g levels induce the existence of 'spin polarons', with $S = 1$, instead of $S = 0$. As shown later, these possible spin polarons could explain also the ARPES results in these materials.

Finally, in this framework of strongly correlated material, the transport properties of Na_xCoO_2 have been calculated using the Kubo formalism. The complexity of the Kubo formalism can be avoided by using certain limits of this formalism and several quantities such as the Hall coefficient can be calculated.[53] The Seebeck coefficient, Lorenz number and figure of merit have been calculated considering this t–J model on a 2D triangular lattice.[14] The values of S for a given T have been compared with the values extracted from the Heikes formula, and it was shown that the high T limit where the two curves merge is obtained for $T \sim 5 - 6t$, with t the hopping parameter. In Na_xCoO_2, t, calculated from the Hall constant was found to be rather small, close to $25\,K$,[54] which shows that the thermopower could already be in the infinite T limit at $300\,K$. In these conditions, the use of the Heikes formula would thus be justified, even for these metallic oxides.

One of the major issues in these materials is also the role played by the separating layer of randomly filled Na^+. The origin of the magnetic correlations and the paramagnetic susceptibility have been ascribed to the electrostatic potential associated with the Na layer.[43] Also, the Na layer, when disordered, could be efficient to reduce the thermal conductivity. It is therefore very important to be able to investigate other families of layered cobaltates, with the same CdI_2 layer, but different block layer. This can be done by investigating the misfit oxide family.

4.3.2.2 The Misfit Cobalt Oxides

Na_xCoO_2 has been intensively investigated since 1997 due to its thermoelectric and superconductivity properties. All the properties of this material come from the presence of the CoO_2 layer of the CdI_2 type. Other systems possess the same kind of CdI_2 type layers, as for example the misfit cobalt oxides, the difference with Na_xCoO_2 being the separating layer, which consists of a rock-salt like layer instead of the randomly filled Na^+ layer.[55] Furthermore, the two different monoclinic sublattices have common a, c and β parameters, but two different b parameters with an incommensurate ratio b_1/b_2.

Figure 4.17 The misfit structure of 'Ca$_3$Co$_4$O$_9$'. The plane of edge-shared octahedral corresponds to a CdI$_2$-type CoO$_2$ layer. The second sublattice (in the centre) is made of three layers, 2 CaO layers surrounding a central layer made of corner-shared CoO$_6$ octahedral

The misfits have been investigated in the sulfur family,[56] and have first been found in oxides in the CRISMAT laboratory in 1996.[55] The NaCl-like block can be made of $n = 2$,[57] 3,[55] or 4[58] separating layers.

The misfit formula of Ca$_3$Co$_4$O$_9$ (Figure 4.17) can be written as [Ca$_2$CoO$_3$][CoO$_2$]$_{1.62}$, with the first bracket corresponding to the NaCl-like block layer, the second bracket the CdI$_2$-type layer and the b_1/b_2 parameter linking the two sublattices. As for Na$_x$CoO$_2$, the two sublattices are electrostatically bound, with a positive charge in the NaCl-like layers which act as charge reservoir block, and the CoO$_2$ layer with a Co^{3+}/Co^{4+} mixed valency. Different substitutions can be made mainly in the NaCl-like layers, and also in the CdI$_2$-like layer as shown in [Bi$_{1.95}$Ba$_{1.95}$Rh$_{0.1}$O$_4$][RhO$_2$]$_{1.8}$.[59,60] The oxygen stoichiometry can also be tuned.[61] The result of all these substitutions is to modify the b_1/b_2 ratio, which is found to lie in the range 1.62–2, with 1.62 in [Ca$_2$CoO$_3$][CoO$_2$]$_{1.62}$[62] and 2 in [Bi$_2$Ba$_{1.8}$Co$_{0.2}$O$_4$][CoO$_2$]$_2$.[63]

Transport Properties The properties of misfit cobaltites are very close to those of Na$_x$CoO$_2$, with a large, positive thermopower together with small, metallic-like resistivity at 300 K. Nevertheless, there exist differences, especially at low T. More precisely, two different behaviours are observed at $T < 100$ K. In most of the materials, as in [Ca$_2$CoO$_3$][CoO$_2$]$_{1.62}$, below $T \sim 100$ K, a large increase in the resistivity is observed at low T, together with large negative magnetoresistance.[62] Strong magnetothermopower, as in Na$_x$CoO$_2$,[64] is also reported for example in $[Bi_{1.7}Co_{0.3}Ca_2O_4][CoO_2]_{1.67}$.[65] In only two misfits, another behaviour has been reported so far: down to 2 K, the resistivity is metallic, following a T^2 behaviour characteristic of a Fermi liquid.[66,67] The negative magnetoresistance is replaced by a small and positive magnetoresistance.[66] For these two metallic misfits, the thermopower is slightly smaller with $S \sim 70$–80 $\mu V/K$ at 300 K. Figure 4.18 presents an example of this difference with the comparison of S and ρ between $[Ca_2CoO_3][CoO_2]_{1.62}$ and $[Tl_{0.81}Co_{0.2}Sr_{1.99}O_3][CoO_2]_{1.79}$, two misfits with three separating block layers.

One of the most important questions to optimise the figure of merit of these materials is to understand the evolution of S vs T, correlate the value of S at 300 K with doping, and compare this value with the Heikes formula. To investigate the doping effect, several misfit compounds have been synthesised and characterised.

Optimisation of the Seebeck Coefficient by Doping   Following the Heikes formula or the generalised Heikes formula, S should increase when the Co^{4+} concentration decreases. Following the electroneutrality equation between the two different sublattices, the Co valency in the CdI_2 plane can be written as

$$v_{Co} = 4 - \alpha/(b_1/b_2) \qquad (4.33)$$

with α the positive charge from the block layer, and b_1/b_2 the misfit ratio. To decrease the Co valency, it is therefore necessary to increase α and/or decrease b_1/b_2.

In the three separating layer $[Ca_2CoO_3][CoO_2]_{1.62}$, the substitution by Ti^{4+} has been attempted. The results are shown in Figure 4.19.

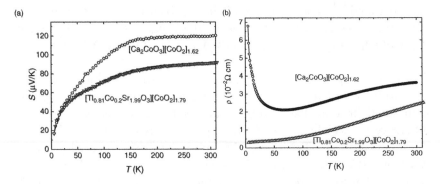

Figure 4.18 $S(T)$ (a) and $\rho(T)$ (b) of $[Ca_2CoO_3][CoO_2]_{1.62}$ and $[Tl_{0.81}Co_{0.2}Sr_{1.99}O_3]$ $[CoO_2]_{1.79}$

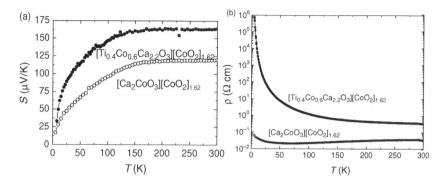

Figure 4.19 $S(T)$ (a) and $\rho(T)$ (b) of $[Ca_2CoO_3][CoO_2]_{1.62}$ undoped and substituted with Ti^{4+}

The introduction of Ti^{4+} leads to a semiconducting-like behaviour for the substituted compound. More interestingly, in the Ti^{4+} substituted sample, the Seebeck coefficient has been increased from 120 μV/K to 165 μV/K. It should be noted that even if the resistivities are strongly modified by the substitution, the Seebeck coefficients present the same T dependence, only shifted in magnitude. This enhancement of S at 300 K shows that the increase in α by the introduction of Ti^{4+} in the rock-salt layer has a positive impact as expected from the Heikes formula.

A second proof of this doping effect is by tuning b_1/b_2. With this aim, the $n = 4$ Bi-based family is the most interesting family to investigate, as it spans a large range of b_1/b_2 from 1.67 for 'BiCaCoO' to the commensurate value of 2 for 'BiBaCoO'. The resistivity and Seebeck coefficients of these different compounds, for which the major difference is the b_1/b_2 parameter considering a fixed oxygen stoichiometry, are presented in Figure 4.20. Note that the resistivities have been measured on single

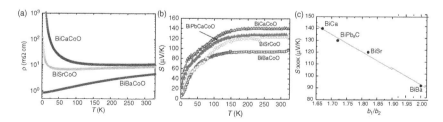

Figure 4.20 The transport properties in the $n = 4$ Bi-based family, with the $\rho(T)$ measured on single crystals (a), $S(T)$ measured on polycrystals (b) and S as a function of the misfit ratio b_1/b_2 (c)

crystals, while the thermopower has been measured on polycrystals (the thermopower measurements performed in the case of BiCaCoO, BiSrCoO and BiBaCoO on single crystals confirm that the thermopower is the same as the one of polycrystals).

The modification of b_1/b_2 has a drastic impact on the transport properties. The resistivity changes from a metallic behaviour in almost the whole T range for 'BiBaCoO', to a more resistive behaviour for 'BiCaCoO', strongly localised at low T. Together with this metal to insulator transition, the thermopower increases in the whole temperature range when b_1/b_2 decreases from 2 ('BiBaCoO') to 1.67 ('BiCaCoO'). The same phenomenon is observed in the $[Pb_{0.7}Sr_{2-x}Ca_xCo_{0.3}O_3][CoO_2]b_1/b_2$ compounds (with $b_1/b_2 = 1.62$ for $x = 2$, and 1.79 for $x = 0$). The Seebeck coefficient is strongly enhanced from 120 μV/K for $x = 0$ to 165 μV/K for $x = 2$, i.e. when b_1/b_2 decreases.[68]

The oxygen content in $[Ca_2CoO_3][CoO_2]_{1.62}$ can be modified by different annealing. Using iodometric titration, Karppinen et al. have shown that oxygen vacancies can be introduced in this compound, and when using the $Ca_3Co_{3.95}O_{9+\delta}$ formula, δ can change from 0.07 to 0.29. As δ decreases, the mean Co valency decreases and the associated Seebeck coefficient is increased.[61]

Two major conclusions can be derived from all these experiments. First the Seebeck coefficient temperature dependence is rather insensitive to the valency of cobalt and to the resistivity curve ($d\rho/dT > 0$ or $d\rho/dT < 0$). For all the compounds described here, all the $S(T)$ curves present the same shape, with a plateau from $\sim$100 K to 300 K. Second, qualitatively, all these results show that the evolution of the Seebeck coefficient at 300 K fits with the Heikes formula. As the Co valency decreases through the b_1/b_2 parameter, the oxygen content or the cationic substitutions, the Seebeck coefficient increases. A more quantitative analysis of the Heikes formula cannot be seriously performed considering the number of unknown parameters, as for example the oxygen content or the valency of Co in the RS layer. More experiments are needed to directly correlate the value of S with the doping parameter 'x' of the Heikes formula, and investigate in particular the importance of the spin and orbital degeneracies term. Recently, such an analysis was performed in the case of 'BiCaCoO' by combining thermopower, magnetic measurements and chemical analysis,[69] and it was concluded that, in the case of misfits, the Heikes formula, should be used with the degeneracy term of ½:

$$S = \frac{-k_B}{|e|} \ln \left(\frac{1}{2} \frac{x}{1-x} \right) \qquad (4.34)$$

This factor of two (instead of $6^{[13]}$) is due to the special filling of the t_{2g} orbitals, with the e'_g level below the a_{1g} due to electronic correlations.[70]

Before concluding, it should be noted that the Heikes formula should be used also only if a plateau is reached for the $S(T)$ curve. This is not always the case as an increase in S can be observed for some misfits[68] or in Na_xCoO_2.[21] This increase has not been investigated in detail so far, and it would be very important to understand if this is really an intrinsic increase or if it is related to some oxygen reduction as T increases, depending on the measuring atmosphere.

In the following, we will show that beyond the Heikes formula, other contributions to thermopower can be added, giving more information on the nature of the carriers in these materials and the origin of correlations.

Magnetothermopower and Magnetoresistance At high T ($T > 100\,K$), the optimisation of the Seebeck coefficient can be achieved by reducing the number of carriers. At low T, another source of enhancement of the Seebeck coefficient has been observed in the case of 'BiCaCoO'. Below $100\,K$, S strongly depends on the magnetic field, and the application of a magnetic field of 9 T leads to a decrease of the Seebeck coefficient by a factor of 2 (Figure 4.21).[65]

The same effect has been observed in Na_xCoO_2, with $x \sim 0.7$.[64] This is not a universal effect in the case of misfits, as, for example, a flat dependence of $S(H)$ is observed in the case of 'BiBaCoO'. In the case of 'BiCaCoO', the strong magnetothermopower correlates with the presence of a very large negative magnetoresistance (Figure 4.22).

The magnetic properties of these misfits are very difficult to measure macroscopically as the signals related to Co^{3+}/Co^{4+} in low-spin states are

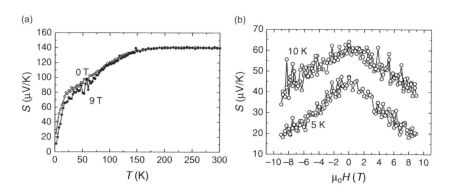

Figure 4.21 $S(T)$ under 0 T and 9 T and $S(H)$ at low T in 'BiCaCoO'

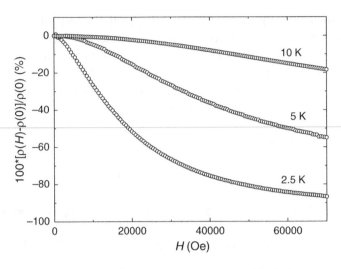

Figure 4.22   The magnetoresistance in 'BiCaCoO'

always very small, and can rapidly be polluted by the signal of magnetic impurities. NMR has been used as a powerful tool to investigate the microscopic environment of Co, and measure the spin susceptibility.[71] A clear correlation has been established between Na_xCoO_2 and the misfit family, using the thermopower as scaling parameter. Within the Bi-based misfit family, by going from 'BiBaCoO' to 'BiCaCoO', it is possible to span the equivalent $x \sim 0.6$–0.9 region of Na_xCoO_2. The NMR measurements show that the susceptibility is maximum, and Curie–Weiss like, for the 'BiCaCoO' compound with a Seebeck coefficient of $\sim 140\ \mu V/K$ when $x \sim 0.75$. This maximum in the susceptibility corresponds to the material for which the magnetothermopower and negative magnetoresistance are huge.

The analysis of magnetothermopower has been performed by scaling the $S(H)$ curves on a single curve, which can be fitted by a Brillouin function.[72] This was also observed in Na_xCoO_2[64] for $x \sim 0.7$, the x value in the range of maximal spin susceptibility detected by NMR. This scaling was attributed to a spin entropy contribution to the thermopower. A detailed investigation of magnetoresistance has been performed and used to analyse the magnetothermopower in more detail.[73] A variable range hopping (VRH) mechanism has been proposed to be at the origin of the low T upturn and strong increase of resistivity. The VRH mechanism would be due to the diffusion of carriers on paramagnetic spins $S = 1/2$. Due to strong Hund's coupling, the resistivity is minimum when the spin of the carriers and of the localised spins are parallel, this mechanism being at the origin of the strong negative magnetoresistance.

The difficulty in this model is to be able to define two different types of carriers, the mobile ones and the localised ones. In a purely ionic model, with Co^{3+} and Co^{4+} both in low spin states (t_{2g}^6 and t_{2g}^5), only localised carriers are considered, with carriers in the d_{xy}, d_{xz} and d_{yz} orbitals. Assuming some structural distorsion, the d_{z^2} level could be lower than the d_{xy} level, therefore enabling the e_g orbitals to be filled. In this case, as schematically depicted in Figure 4.23, a 'virtual' spin state '$S = 1$' could be realised, and the Hund's coupling would then be established, justifying the VRH mechanism at the origin of magnetothermopower and magnetoresistance.

Following this model, the thermopower has then been analysed considering two terms. First, a spin entropy term associated with the paramagnetic $S = 1/2$ localised spins is included, at the origin of the magnetothermopower. The second term comes from the mobile carriers, within the e_g band, which can be treated as renormalised quasiparticles. At low T ($T < T_F$, T_F being the Fermi energy, here $T_F \sim 260\,K$), S is proportional to T while for $T > T_F$, S saturates and recovers its purely entropic form, following the Heikes formula.[73]

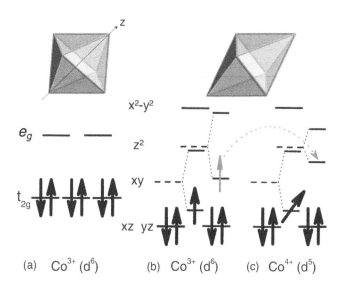

Figure 4.23 Possible filling of the t_{2g} and e_g levels in the CoO_2 layers. Reprinted with permission from Limelette *et al.*, 2008 [73]. Copyright (2008) American Physical Society

The origin of the '$S = 1$' state is not clear at present. This possible '$S = 1$' virtual spin state has also been recently proposed in the model of 'spin polaron'.[51] A possible transfer from the t_{2g} to e_g orbitals can happen, at the origin of the spin polaron formation. This transfer is due to the fact that the introduction of Co^{4+} reduces the cubic symmetry of the 6 Co^{3+} centres surrounding it. These spin polarons can explain the unexpectedly large values of susceptibility obtained in Na_xCoO_2 for $x \sim 0.7$ when there are only a few diluted Co^{4+},[51] and can also explain other experimental results such as the shape of the ARPES spectra.[74,75]

Na_xCoO_2 *and the Misfits: Concluding Remarks*   The discovery of a large thermopower in Na_xCoO_2 is at the origin of an intense experimental and theoretical investigation of these systems. Twelve years after the paper by Terasaki *et al.*[10] the phase diagrams of these materials have turned out to be very rich, with superconductivity for the hydrated phase with $x \sim 0.3$, and strong correlations between Co^{4+} for $x \sim 0.7$, *i.e.* surprisingly for a small concentration of diluted spins. Also, as for Na_xCoO_2, the electronic structure of misfit cobaltates determined by ARPES experiments seems different from that calculated in Na_xCoO_2, with no pockets observed.[74,75] The physical interpretation of these properties thus requires new approaches, taking into account the peculiarities of these materials: a layered structure of CdI_2 type with a triangular symmetry for the Co lattice, a separating block layer acting as charge reservoir and as a possible source of disorder, and the Co^{3+}/Co^{4+} in low spin states (t_{2g}^5 and t_{2g}^6), with possible excitations to the e_g states. For thermoelectric properties, these unique materials combine a small, metallic resistivity, together with a thermopower more characteristic of localised systems. The use of the Heikes formula can be justified due to the fact that the characteristic energies are small, so that the high-temperature limit is reached at $T \sim 250$ K. The thermopower at high T can thus be adjusted by doping and optimising the carrier concentration. Moreover, the localised carriers can induce an extra spin entropy term, which increases the thermopower at 300 K. Following these conclusions, and the unique properties coming from these CdI_2-type layers, the physical properties of related compounds have been investigated.

4.3.2.3   Other Materials with the CdI_2 Type Layer

Rh Misfits   Owing to its isoelectronic configuration (t_{2g}^5/t_{2g}^6 for Rh^{3+}/Rh^{4+}) with cobalt cations, Rh is a good candidate for substitution in the misfits. $4d$ orbitals have to be considered for Rh, instead of $3d$ and this could have some influence on the transport properties (the larger extension of $4d$ orbitals could favour a better overlap, and thus reinforce metallicity). The CoO_2 layers in cobalt misfits can be partially[76] or completely substituted by Rh as shown in the Bi-based family.[60] The transport properties of these rhodates are shown in Figure 4.24.

Due to the larger ionic size of rhodium compared with cobalt, the misfit ratio is reduced when comparing 'BiBaRhO' to 'BiBaCoO'. Otherwise, the characteristic results obtained for cobalt misfits are found in the case of rhodium misfits, with a metallic behaviour ($20\,\text{m}\Omega$ cm in polycrystals at $300\,\text{K}$) together with a large thermopower of $95\,\mu\text{V/K}$ at $300\,\text{K}$. Also, magnetoresistance is reported in these materials below $20\,\text{K}$, positive for large magnetic fields and negative at low fields,[59] together with a small negative magnetothermopower. Due to the larger crystal field for $4d$ than $3d$ orbitals, Rh is always in low-spin states, and the similarity of results between the Co and Rh misfits was used as a first clue of the importance of t_{2g} orbitals to understand the physics of these materials. Moreover, there exists to our knowledge no report on magnetoresistance and magnetothermopower in the case of rhodium oxides, and the CdI_2 nature of the RhO_2 layers is most probably again responsible for this, due to a possible splitting of the carriers into mobile carriers and localised spins. These results should now be reconsidered and analysed following the spin polaron model, and related models, which have been developed since then.

BiSrRhO: A New Layered Structure with a Different Separating Block Layer Among the different models which have been developed to understand the transport and magnetic properties of Na_xCoO_2, one of

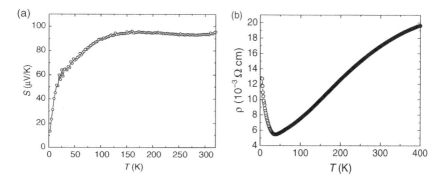

Figure 4.24 $S(T)$ (a) and $\rho(T)$ (b) of Rh misfits

the relevant parameters which has been analysed is the impact of the Na^+ layer and its disorder. The potential created by the Na^+ layer could for example be at the origin of the magnetic correlations.[43] Also, comparisons have been performed on the thermal conductivity of these materials, and a possible reduction in the thermal conductivity induced by the misfit between the CoO_2 and NaCl-like layers has been claimed.[77] The influence of the block layer is important both from fundamental and experimental points of view. The discovery of new family of compounds is therefore very stimulating.

By attempting to grow single crystals of BiSrRhO misfits, a new family of layered rhodates has been discovered. It is composed of the same RhO_2 layers of CdI_2 types, but separated this time by wurzite type layers.[78,79]

Figure 4.25 presents the transport properties of these crystals. A metallic behaviour is observed from 2 K to 300 K, with $\rho \sim 1.5$ mΩ cm at 300 K, and $\rho \sim T^2$ characteristic of a Fermi liquid. A very small positive magnetoresistance is observed (+3% at 2.5 and 7 T). As for misfits and Na_xCoO_2, the thermopower presents the same T dependence, with a rather large value of 63 μV/K at 300 K, increasing to 110 μV/K at 800 K.[80] The properties of this material can be compared with those of 'TlSrCoO', with one of the smaller TEP values reported so far in misfits.

The p-type Delafossite Thermoelectric Oxides The triangular network of cobalt cations in the CoO_2 layers for Na_xCoO_2 or 'misfit' structures has been the object of intense research in thermoelectricity. Interestingly, there exists another structure, called delafossite with the AMO_2 formula, which also contains MO_2 layers similar to the CoO_2 one. The delafossite

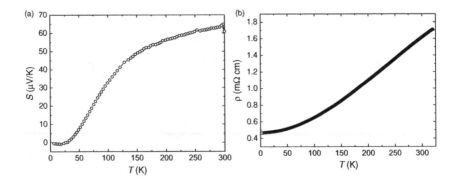

Figure 4.25 $S(T)$ (a) and $\rho(T)$ (b) of 'BiSrRhO'

mineral corresponds to $CuFeO_2$. Its rhombohedral structure (space group $R\bar{3}m$) corresponds to the stacking of CdI_2-type FeO_2 layer separated by monovalent copper in dumbbell coordination (Figure 4.26).[82] In fact, there exist numerous compounds of AMO_2 formula crystallising in that layer structure (2H hexagonal delafossites, such as crednerite $CuMnO_2$, also exist), with monovalent A^+ cation and trivalent M^{3+} cations.

Furthermore, depending on the ionic radius of the M cations, the delafossite structure may incorporate extra oxygen atoms in the A plane leading to oxygen overstoichiometry $AMO_{2+\delta}$. For example, this situation is encountered in the $CuYO_{2+\delta}$ case (see Singh[82] and references therein), for which the electrical conduction results in the CuO_δ plane and not in the YO_2 plane. This is a very different situation

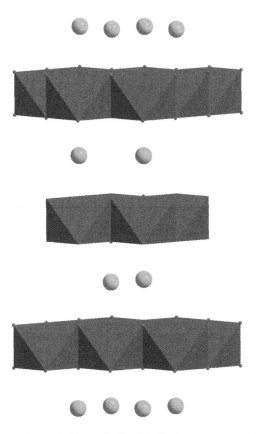

Figure 4.26 Illustration of the $CuCrO_2$ delafossite structure. The CrO_6 layers (shown as polyhedra) are linked through O–Cu–O dumbbells

as compared with the Na_xCoO_2 phases. Among the oxygen stoichiometric delafossites, those containing cobalt, $PdCoO_2$ and $PtCoO_2$, are known for their lower electrical resistivities than those of the lamellar cobaltites (Na_xCoO_2 or misfit types). From the chemical aspect, the Co^{3+} oxidation state of these delafossites is not favourable to charge delocalisation. However, at room temperature, their electrical resistivity is only of few $\mu\Omega$ cm and is associated with low S values (~ 1 $\mu V/$ K) (see Eyert et al.[83] and references therein). In order to understand the very metallic behaviour of these cobalt-based delafossites, electronic structure calculations are useful. They show that for both $PdCoO_2$ and $PtCoO_2$, the in-plane $4d$ or $5d$ orbitals of Pd^+ or Pt^+, respectively, are contributing at the Fermi level (E_F). In contrast, the CoO_2 layers are insulating acting as charge reservoirs for the Pd or Pt metal-like layers.

Considering that result, the low electrical resistivity ($1-10$ mΩ cm at 300 K) in bulk samples of Mg-doped $CuCrO_2$ and $CuRhO_2$ is intriguing especially if one considers also the insulating behaviour of the undoped compound $CuCrO_2$ ($\sim$ kΩ cm at room temperature).[81] In fact, two interpretations were proposed: as Mg^{2+} is substituted for Cr^{3+}, it induces either Cu^{2+} or Cr^{4+}, the transport being thus made either in the Cu or CrO_2 planes.[83,84] Both situations supposed a p-type character consistent with the positive values of the Seebeck coefficient. But electronic structure calculations show that the most important contribution at the Fermi level comes from the d-orbitals of the high-spin Cr^{3+} ($S = 3/2$) near E_F with a polarisation of the spins. Such a prediction has been validated by measurements versus applied magnetic field showing the existence of a negative magnetoresistance and magnetothermopower (Figure 4.27). Thus, this situation contrasts with that found in $CuYO_{2+\delta}$ where transport is made in the CuO_δ planes.

As shown by the $S(T)$ curves of $CuCr_{1-x}Mg_xO_2$,[85] for $x \geq 0.02$, the S values are almost unchanged with x, the Mg^{2+} doping effect increasing with x for $x \leq 0.02$ (Figure 4.28).

This is explained by inspecting the X-ray diffraction patterns. They reveal the presence of $MgCr_2O_4$ spinel impurities as soon as $x \sim 0.01$, indicating the very limited Mg solubility. The too large cation size mismatch between Cr^{3+} ($r = 0.0615$ nm) and Mg^{2+} ($r = 0.0720$ nm) could explain this limited range of solubility. Best power factors, $PF \sim 2 \times 10^{-4}$ W/m/K^2, are found for $x \sim 0.02$ in $CuCr_{1-x}Mg_xO_2$.[83,84] Such low doping levels contrast with the high cobalt oxidation state reachable in the Na_xCoO_2 compounds by varying x. However, considering

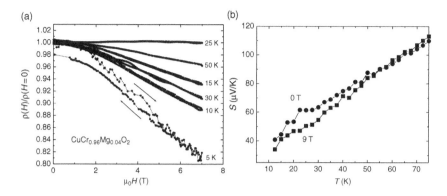

Figure 4.27 (a) Isothermal magnetic field dependent resistivity of CuCr$_{0.96}$Mg$_{0.04}$O$_2$ and (b) enlargement of the $S(T)$ curves for the same compound collected upon cooling from 315 K in 0 T and then in 9 T

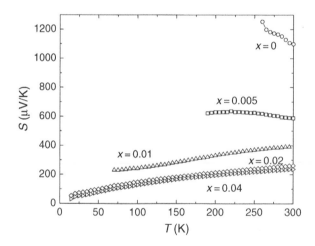

Figure 4.28 T dependence of the Seebeck coefficient, S, for the CuCr$_{1-x}$Mg$_x$O$_2$ series

the physical similarity of cobaltates and rhodates with the misfit structures, the CuRh$_{1-x}$Mg$_x$O$_2$ delafossite series is worth studying. Furthermore, as the Rh^{3+} ionic size, $r = 0.0665$ nm, is closer to that of Mg^{2+}, up to 12% Mg^{2+} can be substituted for Cr^{3+} in CuRhO$_2$.[86] Also, the starting material, CuRhO$_2$, is already much more conducting than CuCrO$_2$ (at room temperature, $\rho_{\text{CuRhO}_2} \sim 0.5$ Ω cm and $\rho_{\text{CuCrO}_2} \sim 1$ kΩ cm)

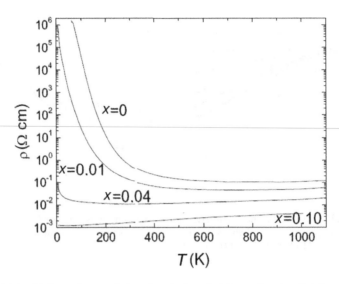

Figure 4.29 T dependence of the electrical resistivity ρ for the $CuRh_{1-x}Mg_xO_2$ series

(Figure 4.29). This could be attributed to the more covalent character of Rh-O bonds resulting from the more extended d-orbitals.

The magnetism of $CuRhO_2$ and $CuCrO_2$ is also very different (Figure 4.30), the former exhibiting a Pauli-like paramagnetism contrasting with the antiferromagnetic ordering ($T_N = 25$ K) of the latter.

Electronic structure calculations support the low-spin state of rhodium in $CuRhO_2$. The t_{2g} $4d$ orbitals of Rh lie near E_F, the e_g orbital being separated by an optical band gap of 0.75 eV. A low-spin state of rhodium was also observed in the misfit rhodates.[86]

The large Mg^{2+} solubility and the lack of strong Rh^{3+} paramagnetism allow a metal-like behaviour to be induced for all T in the 40–1000 K range as shown in Figure 4.29 for $CuRh_{0.9}Mg_{0.1}O_2$. At 100 K, the electrical resistivity drops by seven orders of magnitude as one goes from the $CuRhO_2$ curve to the one of $CuRh_{0.9}Mg_{0.1}O_2$. The $\rho(T)$ curves exhibit a metal-to-insulator transition as T decreases from $\sim$1000 K. The characteristic temperature of the ρ minimum decreases from 800 K down to 38 K for $CuRhO_2$ and for $CuRh_{0.9}Mg_{0.1}O_2$, respectively. These T dependences differ from the $d\rho/dT < 0$ coefficient found in $CuCr_{1-x}Mg_xO_2$. In these delafossites chromites, the $\rho(T)$ curves are fitted with a polaronic model. The latter is not working in the case of $CuRh_{1-x}Mg_xO_2$. Instead, Fermi-liquid behaviours, $\rho \propto T^2$, are experimentally observed up to unusually large T, 900 K in $CuRh_{0.96}Mg_{0.04}O_2$. As expected from the

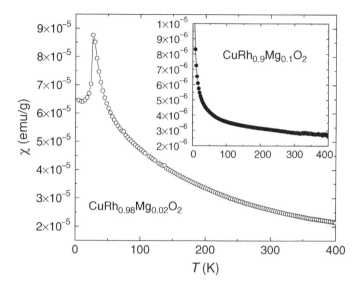

Figure 4.30 T-dependent magnetic susceptibility of $CuCr_{0.98}Mg_{0.02}O_2$ and $CuRh_{0.9}Mg_{0.1}O_2$ (under 0.3 T after zero field cooling)

$CuRh^{3+}_{1-2x}Rh^{4+}_xMg_xO_2$ formula, showing the creation of 'Rh^{4+}' holes with the Mg^{2+} substitutions for Rh^{3+}, the Seebeck coefficient values, measured between 300 K and ~ 1000 K, are positive with a magnitude that decreases as x increases (Figure 4.31). The T dependence shows an almost $S \propto T$ linear behaviour for all compositions such as $x > 0.01$. At 1000 K, S

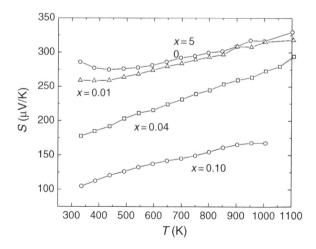

Figure 4.31 T dependence of the Seebeck coefficient, S, for the $CuRh_{1-x}Mg_xO_2$ samples

decreases from ~ 320 μV/K to ~ 170 μV/K as the Mg^{2+} content increases from $x = 0.00$ to $x = 0.10$, respectively.

By combining S and ρ, which depend on T and T^2, respectively, one obtains a S^2/ρ power factor which is rather T-independent. The highest value is obtained for $CuRh_{0.9}Mg_{0.1}O_2$, with $PF \sim 7 \times 10^{-4}$ W/m/K^2 for all T between 400 K and 1000 K. Such a PF value is much higher than the highest value of the $CuCr_{1-x}Mg_xO_2$ series at 800 K, reaching only $PF = 1.4 \times 10^{-4}$ W/m/K^2. Among the delafossite oxides, this value is also greater than the value reached in $CuFe_{1-x}Ni_xO_2$ series exhibiting a maximum of 5.1×10^{-4} W/m/K^2 at 1100 K.[87]

These delafossites, especially those with $M = Cr$, Fe, Rh in $CuMO_2$, exhibit quite large PF values. But their thermal conductivity values are large, with κ values in the range 6 to 10 W/m/K at 300 K.[84] Additionally, chromium and rhodium are not good candidates for TE applications, due to their toxicity and high cost, respectively. Of course, delafossites based on $CuFeO_2$ could be attractive candidates, especially if one takes into account the promising value $ZT = 0.14$ reported for $CuFe_{0.99}Ni_{0.01}O_2$.[87] However, a severe reduction of the κ values has to be made. In fact, the excellent fitting between the A and MO_2 layers in these materials of AMO_2 formula must favour the phonon propagation. One possible way to reduce κ could be to create some disorder on these sublattices. With the existence of several kinds of AMO_2 delafossites with the same M trivalent cations but different A monovalent cations (for instance, fixing M to Co, the following delafossites cobaltites exist: $PtCoO_2$, $PdCoO_2$, $CuCoO_2$ and $AgCoO_2$), it could be interesting to prepare AMO_2 delafossites containing a mixture of A cations to try to reduce κ.

4.3.3 Degenerate Semiconductors

From the inspection of the $ZT(T)$ graphs showing the data reported for p- and n-type oxides (Figure 4.6), it is obvious that the p-type layer cobaltites as the Na_xCoO_2 or misfist cobaltites exhibit larger ZT values than the n-type such as the perovskite titanates or the Al-doped ZnO oxide. Such a difference is linked to the electron-hole asymmetry generally found in oxides. As mentioned in the case of the layer p-type cobaltites, electron correlations create a different physics as compared with the classical models of degenerate semiconductors. This does not hold in the case of the n-type oxides mentioned in Figure 4.6.

4.3.3.1 Perovskite Magnetic Oxides

At first, such an asymmetry can be first explained by the different electronic configurations of the cations corresponding to electrons and holes in magnetic oxides. As previously explained in Section 4.3.1, this can be illustrated by the $LaCoO_3$ or $Ca_3Co_2O_6$ cobaltites. For both, the cobalt cations are trivalent ($3d^6$), and at room temperature or below, the electronic configuration is low spin, with $t_{2g}^6 e_g^0$ orbital filling. On the n-type side, as high-spin state Co^{2+} cations are created in the Co^{3+} matrix, the corresponding $t_{2g}^5 e_g^2$ configuration is not favourable to delocalise an electron from the e_g^2 orbital to an e_g^0 orbital of a Co^{3+} neighbour as this would create two species – a low-spin $t_{2g}^6 e_g^1$ Co^{2+} and an intermediate spin state $t_{2g}^5 e_g^1$ Co^{3+} – different from the starting species. In contrast, the process of moving a hole, from low-spin state Co^{4+} ($t_{2g}^5 e_g^0$) to Co^{3+} ($t_{2g}^6 e_g^0$) can be thermally activated in the t_{2g} orbitals (Figure 4.15). This explains the lower electrical resistivity values measured in hole-doped cobaltites as compared with their n-doped counterparts. Such an effect, called 'spin blockade', has been also used to explain the change of electrical resistivity at the spin-state transitions of ordered perovskite cobaltites. However, the direct consequence of it for the search for TE candidates is that it is hardly conceivable to use the same starting compound to form the n- and p-type pairs of TE legs in all oxide TEG. This would have been of great technological interest as in that case the respective dilatation coefficients of n- and p-legs would have been similar, avoiding the possible crack formation due to thermomechanical strains.

Keeping in mind the problem linked to the cobalt oxidation states on each side of the trivalent cobalt, it is rather interesting to explore the oxides in which the d^n magnetic cations allow an electron conduction at the level of the e_g orbitals. Such an idea is inspired by the n-type metallicity of tetravalent perovskite cobaltite such as $SrCoO_{3-\delta}$.[88] In the latter, the oxygen nonstoichiometry, $\delta \sim 0.05$, is responsible for the electron creation ('Co^{3+}') in the Co^{4+} matrix. The much lower resistivity, $\rho^{300\,K}$ < 1 mΩ cm, is explained by the higher cobalt spin states than those in $LaCoO_3$ or $Ca_3Co_2O_6$ cobaltites containing trivalent cobalt cations. In fact, the higher cobalt spin states in $SrCoO_{3-\delta}$ are responsible for the charge delocalisation in the e_g orbitals rather than in the t_{2g} ones. In that respect it was tempting to study the isostructural $Ca_{1-x}Ln_xMnO_3$ perovskite manganites. Starting from $CaMnO_3$, the motion of an electron in this matrix of Mn^{4+} consists in moving an e_g^1 electron in the empty e_g orbitals of the $t_{2g}^3 e_g^0$ Mn^{4+} neighbours.

Although such a doping is very efficient for instance in the $Ca_{1-x}Sm_xMnO_3$ series to decrease the electrical resistivity,[89] we have to also consider the Jahn-Teller effect of the trivalent manganese cations in the $Mn^{3+}O_6$ octahedral coordination, which renders the electronic transport polaronic. Nevertheless, at low doping, when the polaron concentration is low enough, the temperature dependence of the electric resistivity is metal like and also that of the Seebeck coefficient showing an $S \propto T$ behaviour (Figure 4.32).

Accordingly, their physics can be explained by the model used for classical TE materials, using the Mott formula (Equation 4.17). Such T dependences contrast with the rather T-independent behaviour observed for the hole-doped cobaltites, rhodates and ruthenates. Obviously, for the TE applications, the metal-like behaviour of the electron-doped manganites is interesting. Measurements up to high temperature ($>1000\,K$) reveal that the linear regimes are sustained, yielding an increasing PF with T. The highest ZT values for such manganites reach $ZT = 0.3$ at $1000\,K$ for $Ca_{0.97}La_{0.03}MnO_3$. This explains why dense ceramics of these compositions have been used as n-legs of TEG[90] as discussed in the last part of this chapter.

One of the limiting factors for the use of these electron-doped manganites is their too high thermal conductivity values, which is near $\kappa = 10$ W/m/K at 300 K for the $x = 0.05$ best composition in the $Ca_{1-x}Sm_xMnO_3$ series.[91] In order to reduce κ, several strategies could be used to limit the phonon propagation, such as creating the electron doping by substitutions at the manganese site by using M cations with oxidation state greater than four. Examples are given by the two series $CaMn_{1-x}Mo_xO_3$ or $CaMn_{1-x}Nb_xO_3$ in which molybdenum and niobium are hexavalent and pentavalent, respectively.[92] Along such

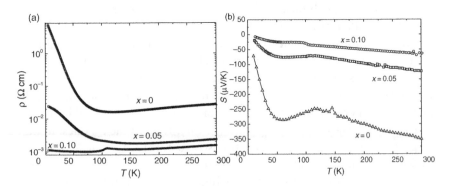

Figure 4.32 $Ca_{1-x}Sm_xMnO_3$ perovskite manganites: $\rho(T)$ (a) and $S(T)$ (b) curves

substitutions, the oxygen content is kept unchanged so that Mn^{3+} species are created. The presence of scattering centres at the metal site is believed to reduce the phonon propagation. As shown for the Nb-doped series, this effect can be concomitant to that induced by grain size reduction.[93] Both factors have been invoked to explain their low thermal conductivity $\kappa \sim 1$ W/m/K at 800 K, leading to ZT values reaching 0.3 at 800 K.

From that study of electron-doped perovskite manganites, we learn that oxides with empty electronic bands which can be partially filled by doping are in principle the best candidates as n-type legs. In the following, several examples among the best n-type oxides have been chosen to illustrate that point.

4.3.3.2 From Transparent Conducting Oxides to Thermoelectric Oxides

Doped In_2O_3 For the development of new display screens or photo-voltaic cells, the search for conducting and transparent oxides is an important challenge. Among these materials, indium tin oxide $In_{2-x}Sn_xO_3$, crystallising in the bixbyite structure, is one of the most famous.[94] The key physical property of this class of materials is the existence of a large bandgap necessary for the optical properties, between the $5s^0$ empty conduction band from In^{3+} and the $2p^6$ filled valence band formed by the oxygen anions O^{2-}.[95] Doping In^{3+} by Sn^{4+} (the so-called ITO) creates donors located in an impurity band below the $5s^0$ conduction band. This leads to a degenerate semiconductor. Accordingly, such materials can also be viewed as classical TE materials. A similar doping effect can be achieved by creating oxygen defects following the formula In_2O_{3-2x}.[94] This is known to induce, at room temperature, an electrical resistivity reduction by three orders of magnitude from ~ 1 Ω cm to ~ 1 mΩ cm. Nevertheless, one major chemical problem with ITO lies in the preparation requiring reducing conditions such as treatments in H_2 flow. As reduction to the metal state is difficult to avoid by such reducing conditions, the synthesis reproducibility of these ceramics is made diffi-cult. This difficulty can be avoided by using solid-state reaction in air for co-substituted indium oxides, such as $In_{2-x}M_{x/2}Sn_{x/2}O_3$ where the divalent M cations ($M = Zn$, Cu, Ni, Mg, Ca) compensate the tetravalent state of tin.[96] Although such substitution is supposed to preserve the electroneutrality, the synthesis in air of such ceramics has been shown to be an efficient way to dope In_2O_3. At room temperature, an electrical resistivity as low as 0.4 mΩ cm has been measured for $In_{1.9}Cu_{0.05}Sn_{0.05}O_3$. Such an effect can be explained only by nonstoichiometry effects: the divalent and tetravalent cations having different solubility, the larger amount of Sn^{4+} compared with Cu^{2+} is supposed to create the electron doping.

In the following, we will see that the existence of the solubility limit for the doping cations at the indium site of In_2O_3 is advantageous to enhance the thermoelectric properties in air-prepared ceramics. This is particularly true for $In_{2-x}Ge_xO_3$, in which tetravalent germanium is substituted for trivalent indium.[97] In contrast to the large Sn solubility at the In site of ITO, the solubility of Ge is limited to $x = 0.01$. Nevertheless, a significant level of doping can be achieved. By increasing x from $x = 0.000$ to $x = 0.002$, the electrical resistivity at 300 K is decreased from 15 mΩ cm to 2 mΩ cm. A lowest value, $\rho \sim 0.7$ mΩ cm, is even reached for $x = 0.015$ (Figure 4.33). However, beyond this x value, as the solubility limit is overshot, ρ increases with x. This means that for such Ge nominal contents, the samples are no longer single phase. Thus, they can be described as composite materials with a secondary phase, $In_2Ge_2O_7$.

Correspondingly, as the charge-carrier concentration (n) does not change for $x \geq 0.015$, the Seebeck coefficient, which is proportional to $n^{-2/3}$, tends to become x independent. As the T dependence of ρ is linear, and the $|S|$ magnitude increases with T, the combination of S^2 and ρ, leads to power factor values which are increasing linearly with T. The best composition corresponds to $x = 0.01$, with $PF \sim 8 \times 10^{-4}$ W/m/K at 1000 K against $PF \sim 2.2 \times 10^{-4}$ W/m/K for In_2O_3 prepared in the same conditions.

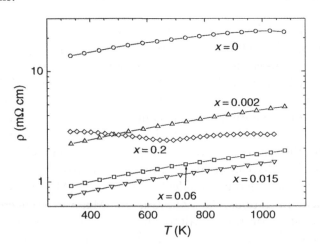

Figure 4.33 T-dependent electrical resistivity of the compounds with nominal composition $In_{2-x}Ge_xO_3$

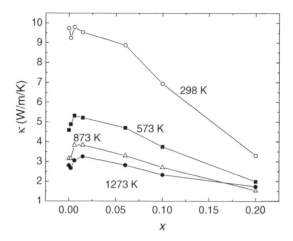

Figure 4.34 x dependence of the thermal conductivity for the $In_{2-x}Ge_xO_3$ composite compounds

One other important consequence of the composite nature of $In_{2-x}Ge_xO_3$ for $x > 0.01$ is the thermal conductivity reduction accompanying the x increase (Figure 4.34).

This has to be compared with the microstructure studied by scanning electron microscopy. It reveals the existence of microdefects, voids and $In_2Ge_2O_7$ inclusions whose amount increases with x, dispersed in the Ge-doped In_2O_3 matrix. Such defects limit the phonon propagation and are responsible for the thermal conductivity decrease at room temperature from $\lambda \sim 10$ W/m/K for $x \leq 0.01$ to $\lambda \sim 3.5$ W/m/K for the composite of nominal composition $In_{1.8}Ge_{0.2}O_3$. Thus, the highest ZT value reaching $ZT = 0.46$ at 1273 K is obtained for the composite $In_{1.8}Ge_{0.2}O_3$. As shown in the $ZT(T)$ graphs compiling the data of oxides (Figure 4.6), it is the highest value reached for a n-type oxide. When compared with the TE materials of higher ZT, such as skutterudites or 'LAST' alloys, although the ZT values of the indium doped oxides are smaller, their chemical stability in air at high T (up to 1200 K) makes them attractive.

Doped ZnO   ZnO is a simple oxide crystallising in the hexagonal wurtzite structure. This oxide shows a broad spectrum of properties and has been the subject of intensive research for applications as a photocalyst, a magnetic semiconductor and also very recently as a piezoelectric.[98] For electronic properties, the doped ZnO oxides exhibit a n-type character. It is also a TCO with a bandgap of ~ 3.5 eV which is used in applications such as photovoltaic cells or flat displays. Basically, its physics can be compared with that of In_2O_3: Zn^{2+} has $3d^{10}$ filled orbitals and the pseudogap is in between the O^{2-} $2p$ orbitals (valence band) and the empty $4s^0$ orbitals (conduction band) as compared with $5s^0$ for In^{3+}.

The most famous study of doped ZnO for TE purpose was reported by Tsubota et al.,[99] who showed that in the Al^{3+}-doped ZnO series, a $ZT \sim$ 0.3 at 1273 K is reached for air-prepared ceramics $Zn_{0.98}Al_{0.02}O$. As shown in Figure 4.35, the Al^{3+} doping, expected to create donor levels (electrons) to the conduction band, strongly reduces the electrical resistivity (ρ) as compared with ZnO. At RT, ρ decreases by more than 3 orders of magnitude, the semiconducting behaviour of ZnO being replaced by almost T independent ρ values.

Accordingly, the absolute value of $|S|$ decreases from $\sim 345\,\mu V/K$ to $\sim 100\,\mu V/K$ for ZnO and $Zn_{0.98}Al_{0.02}O$, respectively.[100] As in the case of ITO, the $|S|$ values increase with T leading to high power factor values, PF reaching 15×10^{-4} W/m/K for $Zn_{0.98}Al_{0.02}O$ at 1273 K.[99] Focusing on the small x values in $Zn_{1-x}Al_xO$, it can be observed that the $x = 0.01$ composition exhibits minimum values of ρ (Figure 4.35) and S (not shown). This nonmonotonous variation of the properties with x is related to the Al^{3+} solubility limit. Indeed, the X-ray diffraction patterns show the presence of a $ZnAl_2O_4$ spinels impurity as soon as $x = 0.01$ (Figure 4.36).

This very reduced Al^{3+} solubility limit can be compared with the Ge^{4+} doping in In_2O_3.[97] Hall coefficient measurements support this limit of solubility as the maximum electron concentration, $7.7 \times 10^{25}\,m^{-3}$, is

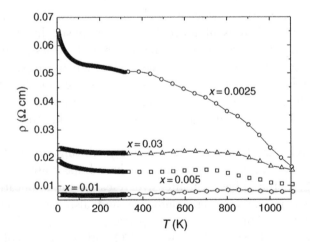

Figure 4.35 $\rho(T)$ curves of the Al^{3+} doped ZnO (nominal composition $Zn_{1-x}Al_xO$)

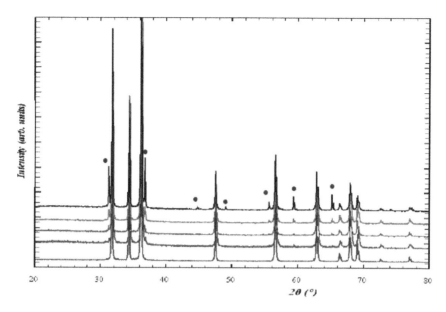

Figure 4.36 Powder X-ray diffraction of samples with $Zn_{1-x}Al_xO$ nominal compositions. The circular symbol corresponds to $ZnAl_2O_4$ impurity, the latter is detected even for $x = 0.01$. From top to bottom, the spectra correspond to $x = 0.00$, 0.01, 0.02, 0.03 and 0.05

reached for $Zn_{0.99}Al_{0.01}O$.[99] The limiting factor for having a high ZT is coming from the too high thermal conductivity of these oxides, as high as $\kappa \sim 40$ W/m/K at room temperature for $Zn_{0.98}Al_{0.02}O$. Nonetheless, these values decrease as T increases, reaching $\kappa \sim 5$ W/m/K at 1273 K. Combining the power factor and κ values of $Zn_{0.98}Al_{0.02}O$, one obtains $ZT = 0.3$ at that temperature. It must be emphasised that the Al^{3+} efficiency to create electrons in ZnO is not unique. A study of doped thin films showed that in the same group of the periodic table, Ga^{3+} and In^{3+} are also interesting dopants.[101] The comparison of their $\rho(T)$ and $S(T)$ curves measured for ceramic samples prepared in air confirms this similarity. As in the In_2O_3 case, the comparable electronic configurations of Ga^{3+} and In^{3+} with Zn^{2+}, with empty or filled d-orbitals, but with s or p empty orbitals, appear to be an important factor. Furthermore, as these cations can also adopt a tetrahedral coordination as Zn^{2+} in the wurtzite structure, their substitution for zinc is, in principle, possible. This doping effect is much less pronounced for magnetic transition metal elements such as $M = Cr$, Fe, Co $(3d)$ or Rh $(4d)$, these compounds for $Zn_{0.99}M_{0.01}O$ nominal compositions exhibiting much higher ρ values than those induced by $M = Al$, Ga and In.

In the future, a special care would have to be taken in order to tune more finely the doping level in these ZnO oxides. And, as the solubility limit of these doping elements appears to be very limited, the possibility to reduce the thermal conductivity by using the composite nature for doping contents beyond the solubility limit would deserve a special attention in order to try to enhance the ZT values of doped ZnO.

4.3.3.3 Back to Perovskite $3d$ Oxides: Doped SrTiO$_3$

As concluded in the section devoted to perovskite magnetic oxides, the use of empty t_{2g} or e_g orbitals to reach both high S value and low electrical resistivity is an interesting route. This is illustrated by the doped SrTiO$_3$ perovskite titanates.[102] SrTiO$_3$ crystallises in a cubic structure and its insulating state, characterised by a large dielectric constant, is used in high-voltage capacitors. This oxide is also extensively used as substrate for epitaxial growth of superconducting cuprate thin films, as the unit-cell parameters of the latter fit well with those of the former. It must also be pointed out that, for charge injection purposes, conductive substrates obtained by doping SrTiO$_3$ can also be used. For instance, the Nb^{5+} ($4d^0$) cation can be substituted for Ti^{4+} ($3d^0$) to create 'Ti^{3+}' electrons ($3d^1$), i.e. to inject electrons in the empty t_{2g} orbitals.[103] Moving to bulk perovskite titanates, in the series Sr$_{1-x}$La$_x$TiO$_3$, the trivalent lanthanum substitution for divalent strontium has been also shown to be an efficient way to create 'Ti^{3+}' electrons in the threefold degenerated t_{2g} orbitals.[102] Measurements made on crystals showed that PF values up to 36×10^{-4} W/m/K^2 could be reached at room temperature for compositions corresponding to $x \sim 0.08$ in Sr$_{1-x}$La$_x$TiO$_3$. At such 'electron' carrier density ($\sim 10^{21}$ cm^{-3}), the resistivity is metallic with $\rho^{RT} \sim 1$ mΩ cm and the thermopower is $S^{300\,K} \sim -150$ μV/K with absolute values that increase with T. Although the PF values at room temperature of these perovskite titanates are similar to the classical (non-oxide) TE materials such as Bi$_2$Te$_3$, their performances are limited by their too high thermal conductivity values ($\kappa \sim 12$ W/m/K for Sr$_{0.95}$La$_{0.05}$TiO$_3$, $ZT = 0.09$ and 300 K). As the weak point of these materials is their too high κ values, different strategies have been developed to increase their Seebeck coefficient values.

Following the ideas developed by Dresselhaus and others, which are based on the fact that the dimension reduction of a TE material induces an increase of the slope of the density of states at the conduction band edge responsible for a S increase, nanosize intergrowths of conducting layers of doped SrTiO$_3$ with insulating layers have been studied. Such structures are natural in the case of the Ruddlesden–Popper series $Sr_{n+1}Ti_nO_{3n+1}$ or can be artificially created by layer-by-layer thin film deposition techniques. Superlattices have been realised by growing intergrowths of insulating $SrTiO_3$ and conducting $Sr_0Ti_{0.8}Nb_{0.2}O_3$ perovskite layers.[103] When the thickness of the conducting layers was reduced to the nanoscale ($\sim$1.56 nm corresponding to four unit cells thickness of doped $SrTiO_3$), the absolute value of |S| was enhanced to reach up to 1300 μV/K for a charge carrier concentration of 10^{21} cm^{-3} against 200 μV/K in single crystals of $SrTiO_3$ with the same level of doping.

Although very promising, such epitaxial heterostructures cannot be used in TEG under very large T gradients ($\Delta T >> 100$ °C) as the T gradient cannot be set for such thin films. At present, no high ZT values have been reached in bulk 2D natural structures as in the Ruddlesden-Popper titanate phases, the latter having $ZT \sim 0.15$ at 1000 K as for $Sr_{2.85}La_{0.05}Ti_2O_7$ (see Ohta et al.[103] and references therein).

However, there remain numerous oxide candidates to be studied with low-dimensional structures where the electron doping can be tuned via chemical substitutions, as in vanadates, molybdates etc., by adding an electron in empty d-orbitals of the metal.

4.3.4 All-Oxide Modules

The main goal of the thermoelectric oxides research is their use as pairs of n- and p-type legs in TEG to convert a part of the waste-heat going through the module into electricity. As given in Section 4.2.1, the efficiency of a TEG is the ratio of the produced electrical power to the heat flux entering at the hot side and passing through the TEG. It depends not only on the figure of merit of the materials, but also on the quality of the electrical and thermal contacts, the shape of the legs, the properties of the frames used at the hot and cold side, etc.[104] Knowing the physical parameters of the legs soldering and frames, one can calculate a theoretical efficiency which can be compared with the measured one. Their ratio allows definition of the manufacturing factor (MF). Thus, MF is the important quantity qualifying a TEG assembly. Due to the problems linked to the electrical contacts to be made in between the legs and to the presence of thermomechemical strains, the manufacturing factors of home-made TEG are usually not very high ($< 60\%$) and also not very reproducible from module to module, with MF values varying from $\sim$10% to 60%. For instance, a module of 18 pairs with a misfit cobaltite

Figure 4.37 Picture of an all-oxide module with misfit cobaltite and perovskite manganite as p- and n-legs. A one euro coin is also shown for sake of size comparison

$Ca_{2.75}Gd_{0.25}Co_4O_3$ (as p-type) and a perovskite manganite $Ca_{0.92}La_{0.08}MnO_3$ (as n-type), using Pt paste and wire as electrical soldering and connections (Figure 4.37), led to a manufacturing factor of only 53% due to a too high internal resistance coming from the bad quality of the electrical contacts. Such a module under a gradient of 390 K allowed the production of 39 mW.

Other pairs of cobaltite and perovskite nickelate as n-type have been assembled in modules where the electrical contacts were made by using silver paste charged with a powder of these oxides to improve the electrical contacts. With one pair of these materials, a power of 80 mW was delivered for a 500 K gradient. Larger output power values, reaching 340 mW have been also obtained by using more pairs.

At present, the low output power values obtained by using all-oxide TE modules come first from the too low ZT values of the oxide legs. For instance ZT values near ~ 1 can be reached in the case of layered cobaltites but such materials need a texturation process to align the crystallographic planes in order to benefit from the superior in-plane properties. Using such anisotropic materials without grain alignment yield $ZT \sim 0.2$ values near 1000 K which strongly reduce the possibilities of electricity generation. The other important limitation lies in the resistance contacts at the oxide–metal interfaces which are in the range of several $m\Omega\,cm^2$ for the specific resistance of these contacts. Such values are two orders of magnitude higher than in classical, commercial-based TEG.

Although these points create severe limitations for short-term applications, the test of all-oxide modules has proved that large ΔT gradients up to $\sim 600\,K$ can be applied on such modules, and, interestingly, that the performances of these TEG are reproducible after several cycles. Many research groups are putting efforts in this field.

4.4 CONCLUSION

In this chapter, we have tried to give an overview on the emerging field of thermoelectric (TE) oxides. Dealing with this class of materials, the first important conclusion lies in the physics difference between the p- and n-type TE oxides. As a large part of that research has been devoted to the p-type Na_xCoO_2 layer cobaltates and derived phases, one major output lies in the important role played by the electronic correlations. Usually the physics of conventional TEs is based on models of degenerate semiconductors and thus strongly correlated materials have opened new perspectives for the search of TE materials.

The cobaltates behave as mixed-valent heavy fermions, with two bands, broad and narrow, responsible for the metallicity and large Seebeck coefficients, respectively. The origin of the coexistence of these localised spins and itinerant carriers is still under investigation, but this possible decoupling is very promising to optimise the TE properties.

This situation contrasts with the results obtained for the n-type TE oxides: their physics is very similar to that of degenerate semiconductors. Beating the current ZT values of ≈ 0.3 for this class of oxides necessitates developing new strategies. As for the composites $In_{2-x}Ge_xO_3$, a decrease in the thermal conductivity can be induced by the presence of nano- or micro-inclusions that limit the phonon propagation. A second route consists in using the spin/orbital degeneracy term, working in the p-type oxides, to add an extra contribution to S. This could be achieved by mixing the d^0 or d^{10} cations of the n-type with magnetic cations.

Finally, the possibility to create artificial superlattices to control the thickness of the conducting layer in order to take advantage of the S increase with the structure dimensionality decrease is also a promising route to enhance the figure of merit of oxides. This should help the solid state scientist in designing bulk TE oxides to use in TEG of the future.

ACKNOWLEDGEMENTS

The authors would like to acknowledge their coworkers in CRISMAT laboratory who have strongly collaborated with this work. Collaborations with Jiri Hejtmanek (Czech Academy of Science, Prague), Julien Bobroff and Véronique Brouet (LPS, Orsay), Patrice Limelette (LEMA, Tours), Volker Eyert (Augsburg University), Bogdan Dabrowski (DeKalb, Illinois) and Wataru Kobayashi (Waseda University, Tokyo) are acknowledged. Financial support has been given through ANR OCTE.

REFERENCES

[1] D.M. Rowe, in *Thermoelectrics Handbook: Macro to Nano*, D.M. Rowe (Ed.), CRC Press, Boca Raton, 2005.

[2] D.M. Rowe, *J. Power Sources*, **73**, 193 (1998).

[3] J.M. Ziman, *Principles of the Theory of Solids*, 2nd Edn, Cambridge University Press, Cambridge, 1972.

[4] L.D. Hicks and M.S. Dresselhaus, *Phys. Rev. B*, **47**, 12727 (1993); *Phys. Rev. B*, **47**, 16631 (1993).

[5] R. Venkatasubramanian, E. Siivola, T. Colpitts and B. O'Quinn, *Nature*, **413**, 597 (2001).

[6] L.D. Hicks, T.C. Harman, X. Sun and M.S. Dresselhaus, *Phys. Rev. B*, **53**, R10493 (1996).

[7] G.A. Slack, in *CRC Handbook of Thermoelectrics*, (Ed.) D.M. Rowe, CRC Press, Boca Raton, 1995, p. 407.

[8] J.L. Cohn, G.S. Nolas, V. Fessatidis, T.H. Metcalf and G.A. Slack, *Phys. Rev. Lett.*, **82**, 779 (1999).

[9] M.M. Koza, M.R. Johnson, R. Viennois, H. Mukta, L. Girard and D. Ravot, *Nat. Mater.*, **7**, 805 (2008).

[10] I. Terasaki, Y. Sasago and K. Uchinokura, *Phys. Rev. B*, **56**, R12685 (1997).

[11] P.M. Chaikin and G. Beni, *Phys. Rev. B*, **13**, 647 (1976).

[12] J.P. Doumerc, *J. Solid State Chem.*, **110**, 419 (1994).

[13] W. Koshibae, K. Tsutsui and S. Maekawa, *Phys. Rev. B*, **62**, 6869 (2000).

[14] M.R. Peterson, B.S. Shastry and J.O. Haerter, *Phys. Rev. B*, **76**, 165118 (2007).

[15] A. Georges, G. Kotliar, W. Krauth and M. J. Rozenberg, *Rev. Mod. Phys.*, **68**, 13 (1996).

[16] J. Merino and R.H. McKenzie, *Phys. Rev. B*, **61**, 7996 (2000).

[17] Y. Ando, N. Miyamoto, K. Segawa, T. Kawata and I. Terasaki, *Phys. Rev. B*, **60**, 10580 (1999).

[18] K. Behnia, D. Jaccard and J. Flouquet, *J. Phys.: Condens. Matter*, **16**, 5187 (2004).

[19] W.J. Weber, C.W. Griffin and J. L. Bates, *J. Am. Ceram. Soc.*, **70**, 265 (1987).

[20] M. Ohtaki, T. Tsubota, K. Egushi and H. Arai, *J. Appl. Phys.*, **79**, 1816 (1996).

[21] K. Fujita, T. Mochida and K. Nakamura, *Jpn. J. Appl. Phys.*, **40**, 4644 (2001).

[22] H. Muta, K. Kurosaki and S. Yamanaka, *J. Alloys Comp.*, **350**, 292 (2003).

[23] G. Xu, R. Funahashi, M. Shikano, I. Matsubara and Y. Zhou, *Appl. Phys. Lett.*, **80**, 3760 (2002).

[24] R. Funahashi and M. Shikano, *Appl. Phys. Lett.*, **81**, 1459 (2002).

[25] M. Mikami, R. Funahashi, M. Yoshimura, Y. Mori and T. Sasaki, *J. Appl. Phys.*, **94**, 6579 (2003).

[26] H. Ohta, S. Kim, Y. Mume, T. Mizogachi, K. Nomura, S. Ohta, T. Nomura, Y. Nakamishi, Y. Ikuhara, M. Hirano, H. Hosono and K. Koumoto, *Nat. Mater.*, **6**, 129 (2007).

[27] M. Shikano and R. Funahashi, *Appl. Phys. Lett.*, **82**, 1851 (2003).

[28] J. Androulakis, P. Migiakis and J. Giapintzakis, *Appl. Phys. Lett.*, **84**, 1099 (2004).

[29] D. Bérardan, E. Guilmeau, A. Maignan and B. Raveau, *Solid State Commun.*, **146**, 97 (2008).

[30] S. Pal, S. Hébert, C. Yaicle, C. Martin and A. Maignan, *Eur. Phys. J. B*, **53**, 5 (2006).

[31] D.B. Marsh and P.E. Parris, *Phys. Rev. B*, **54**, 7720 (1996).

[32] A. Maignan, D. Flahaut and S. Hébert, *Eur. Phys. J. B*, **39**, 145 (2004).

[33] A.A. Taskin, A.N. Lavrov and Y. Ando, *Phys. Rev. B*, **73**, 121120 (2006).

[34] A. Maignan, V. Caignaert, B. Raveau, D. Khomskii and G. Sawatzky, *Phys. Rev. Lett.*, **93**, 026401 (2004).

[35] J.-Q. Yan, J.-S. Zhou and J.B. Goodenough, *Phys. Rev. B*, **69**, 134409 (2004).

[36] S. Hébert, D. Flahaut, C. Martin, S. Lemonnier, J. Noudem, C. Goupil, A. Maignan and J. Hejtmanek, *Progr. Solid State Chem.*, **35**, 457 (2007).

[37] D.B. Marsh and P.E. Parris, *Phys. Rev. B*, **54**, 16602 (1996).

[38] C. Fouassier, G. Matejka, J.M. Reau and P. Hagenmuller, *J. Solid State Chem.*, **6**, 532 (1973).

[39] K. Takada, H. Sakurai, E. Takayama-Muromachi, F. Izumi, R.A. Dilanian and T. Sasaki, *Nature*, **422**, 53 (2003).

[40] K. Takahata, Y. Iguchi, D. Tanaka, T. Itoh and I. Terasaki, *Phys. Rev. B*, **61**, 12551 (2000).

[41] M. Lee, L. Viciu, L. Li, Y. Wang, M.L. Foo, S. Watauchi, R.A. Pascal Jr, R.J. Cava and N.P. Ong, *Nat. Mater.*, **5**, 537 (2006).

[42] M.L. Foo, Y. Wang, S. Watauchi, H.W. Zandbergen, T. He, R.J. Cava and N.P. Ong, *Phys. Rev. Lett.*, **92**, 247001 (2004).

[43] C.A. Marianetti and G. Kotliar, *Phys. Rev. Lett.*, **98**, 176405 (2007).

[44] D.J. Singh, *Phys. Rev. B*, **61**, 133397 (2000).

[45] M.D. Johannes, I.I. Mazin, D.J. Singh and D.A. Papacostantopoulos, *Phys. Rev. Lett.*, **93**, 097005 (2004).

[46] L. Balicas, M. Abdel-Jawad, N.E. Hussey, F.C. Chou and P.A. Lee, *Phys. Rev. Lett.*, **94**, 236402 (2005).

[47] L. Balicas, J.G. Analytis, Y.J. Jo, K. Storr, H. Zandbergen, Y. Xin, N.E. Hussey, F.C. Chou and P.A. Lee, *Phys. Rev. Lett.*, **97**, 126401 (2006).

[48] L. Balicas, Y.J. Jo, G.J. Shu, F.C. Chou and P.A. Lee, *Phys. Rev. Lett.*, **100**, 126405 (2008).

[49] K.W. Lee, J. Kunes and W.E. Pickett, *Phys. Rev. B*, **70**, 045104 (2004).

[50] D.J. Singh and D. Kasinathan, *Phys. Rev. Lett.*, **97**, 016404 (2006).

[51] M. Daghofer, P. Horsch and G. Khaliullin, *Phys. Rev. Lett.*, **96**, 216404 (2006).

[52] J. Chaloupka and G. Khaliullin, *Phys. Rev. Lett.*, **99**, 256406 (2007).

[53] B. Sriram Shastry, B.I. Shraiman and R.R.P. Singh, *Phys. Rev. Lett.*, **70**, 2004 (1993).

[54] Y. Wang, N.S. Rogado, R.J. Cava and N.P. Ong, arXiv: cond-mat/0305455v1.
[55] P. Boullay, B. Domengès, M. Hervieu, D. Groult and B. Raveau, *Chem. Mater.*, **8**, 1482 (1996).
[56] J. Rouxel, A. Meerschaut and G.A. Wiegers, *J. Alloys Compd.*, **229**, 144 (1995).
[57] M. Shizuya, M. Isobe, Y. Baba, T. Nagai, M. Osada, K. Kosuda, S. Takenouchi, Y. Matsui and E. Takayama-Muromachi, *J. Solid State Chem.*, **180**, 249 (2006).
[58] H. Leligny, D. Grebille, O. Pérez, A.C. Masset, M. Hervieu and B. Raveau, *Acta Crystallogr. B*, **56**, 173 (2000).
[59] Y. Klein, S. Hébert, D. Pelloquin, V. Hardy and A. Maignan, *Phys. Rev. B*, **73**, 165121 (2006).
[60] S. Okada and I. Terasaki, *Jpn. J. Appl. Phys.*, **44**, 1834 (2005).
[61] M. Karppinen, H. Fjellvag, T. Konno, Y. Morita, T. Motohashi and H. Yamauchi, *Chem. Mater.*, **16**, 2790 (2004).
[62] A.C. Masset, C. Michel, A. Maignan, M. Hervieu, O. Toulemonde, F. Studer, B. Raveau and J. Hejtmanek, *Phys. Rev. B*, **62**, 166 (2000).
[63] M. Hervieu, A. Maignan, C. Michel, V. Hardy, N. Créon and B. Raveau, *Phys. Rev. B*, **67**, 045112 (2003).
[64] Y. Wang, N.S. Rogado, R.J. Cava and N.P. Ong, *Nature*, **423**, 425 (2003).
[65] A. Maignan, S. Hébert, M. Hervieu, C. Michel, D. Pelloquin and D. Khomskii, *J. Phys.: Condens. Matter.*, **15**, 2711 (2003).
[66] S. Hébert, S. Lambert, D. Pelloquin and A. Maignan, *Phys. Rev. B*, **64**, 172101 (2001).
[67] S. Ishiwata, I. Terasaki, Y. Kusano and M. Takano, *J. Phys. Soc. Jpn.*, **75**, 104716 (2006).
[68] A. Maignan, S. Hébert, D. Pelloquin, C. Michel and J. Hejtmanek, *J. Appl. Phys.*, **92**, 1964 (2002).
[69] M. Pollet, J.P. Doumerc, E. Guilmeau, D. Grebille, J.F. Fagnard and R. Cloots, *J. Appl. Phys.*, **101**, 083708 (2007).
[70] S. Landron and M.-B. Lepetit, *Phys. Rev. B*, **74**, 184507 (2006).
[71] J. Bobroff, S. Hébert, G. Lang, P. Mendels, D. Pelloquin and A. Maignan, *Phys. Rev. B*, **76**, 100407 (2007).
[72] P. Limelette, S. Hébert, V. Hardy, R. Frésard, Ch. Simon and A. Maignan, *Phys. Rev. Lett.*, **97**, 046601 (2006).
[73] P. Limelette, S. Hébert, H. Muguerra, R. Frésard and Ch. Simon, *Phys. Rev. B*, **77**, 235118 (2007).
[74] J. Chaloupka and G. Khaliulinn, *Phys. Rev. Lett.*, **99**, 256406 (2007).
[75] V. Brouet, A. Nicolaou, M. Zacchigna, A. Tejeda, L. Patthey, S. Hébert, W. Kobayashi, H. Muguerra and D. Grebille, *Phys. Rev. B*, **76**, 100403 (2008).
[76] D. Pelloquin, S. Hébert, A. Maignan and B. Raveau, *J. Solid Stat. Chem.*, **178**, 769 (2005).
[77] A. Satake, H. Tanaka, T. Ohkawa, T. Fujii and I. Terasaki, *J. Appl. Phys.*, **96**, 931 (2004).
[78] W. Kobayashi, S. Hébert, D. Pelloquin, O. Pérez and A. Maignan, *Phys. Rev. B*, **76**, 245102 (2007).
[79] D. Pelloquin, O. Pérez, W. Kobayashi, S. Hébert and A. Maignan, unpublished results.
[80] W. Kobayashi, S. Hébert, D. Pelloquin, O. Pérez, A. Maignan and I. Terasaki, unpublished results.
[81] J.P. Doumerc, A. Wichainchai, A. Ammar, M. Pouchard and P. Hagenmuller, *Mater. Res. Bull.*, **21**, 745 (1986).

[82] D.J. Singh, *Phys. Rev. B*, **77**, 205126 (2008).

[83] V. Eyert, R. Frésard and A. Maignan, *Chem. Mater.*, **20**, 2370 (2008); T. Okuda, N. Jufuku, S. Hikada and N. Terada, *Phys. Rev. B*, **72**, 144403 (2008).

[84] Y. Ono, T. Satoh, T. Nozaji and T. Kajitari, *Jpn. J. Appl. Phys.*, **46**, 1071 (2007).

[85] A. Maignan, C. Martin, R. Frésard, V. Eyert, E. Guilmeau, S. Hébert, M. Poienar and D. Pelloquin, *Solid State Commun.*, **149**, 962 (2009).

[86] A. Maignan, V. Eyert, C. Martin, S. Kremer, R. Frésard and D. Pelloquin, *Phys. Rev. B*, **80**, 115103 (2009).

[87] T. Nozaki, K. Hayashi and T. Kajitani, *J. Chem. Eng. Jpn.*, **40**, 1205 (2007).

[88] P. Bedzdicka, A. Wattiaux, J.C. Grenier, M. Pouchard and P. Hagenmuller, *Z. Anorg. Allg. Chem.*, **619**, 7 (1993).

[89] A. Maignan, C. Martin, F. Damay, B. Raveau and J. Hejtmanek, *Phys. Rev. B*, **58**, 2758 (1998).

[90] I. Matsubara, R. Funahashi, T. Takeuchi, S. Sodeoka, T. Shimizu and K. Ueno, *Appl. Phys. Lett.*, **78**, 3627 (2001).

[91] J. Hejtmanek, Z. Jirak, M. Marysko, C. Martin, A. Maignan, M. Hervieu and B. Raveau, *Phys. Rev. B*, **60**, 14057 (1999).

[92] B. Raveau, Y.M. Zhao, C. Martin, M. Hervieu and A. Maignan, *J. Solid State Chem.*, **149**, 203 (2000).

[93] L. Bocher, M.H. Aguirre, D. Logvinovich, A. Shkabko, R. Robert, M. Truttman and A. Weidenkaff, *Inorg. Chem.*, **47**, 8077 (2008).

[94] P.P. Edwards, A. Porch, M.O. Jones, D.V. Morgan and R.M. Perks, *Dalton Trans.*, 2995 (2004).

[95] J.C.C. Fan and J.B. Goodenough, *J. Appl. Phys.*, **48**, 3524 (1977).

[96] L. Bizo, J. Choisnet, R. Retoux and B. Raveau, *Solid State Commun.*, **136**, 163 (2008).

[97] D. Bérardan, E. Guilmeau, A. Maignan and B. Raveau, *Solid State Commun.*, **146**, 97 (2008).

[98] S. Singh, P. Thiyagerajan, K. Mohan Kant, D. Anita, S. Thirupathiah, N. Rama, B. Tiwari, M. Kottaisamty and M.S. Ramachandra Rao, *J. Phys, D: Appl. Phys.*, **40**, 6312 (2007).

[99] T. Tsubota, M. Ohtaki, K. Eguchi and H. Arai, *J. Mater. Chem.*, **7**, 85 (1997).

[100] E. Guilmeau, A. Maignan and C. Martin, *J. Electron. Mater.*, **38**, 1104 (2009).

[101] V. Bhosle, A. Tiwari and J. Narayan, *Appl. Phys. Lett.*, **88**, 032106 (2006).

[102] T. Okuda, K Nakanishi, S. Miyasaka and Y. Tokura, *Phys. Rev. B*, **63**, 113104 (2001).

[103] S. Ohta, T. Nomura, H. Ohta, M. Hirano, H. Hosono and K. Koumoto, *Appl. Phys. Lett.*, **87**, 092108 (2005).

[104] S. Lemonnier, *PhD Thesis*, Caen University, France (2008).

5

Transition Metal Oxides: Magnetoresistance and Half-Metallicity

Tapas Kumar Mandal and Martha Greenblatt
Department of Chemistry and Chemical Biology, Rutgers University, New Jersey, USA

5.1 INTRODUCTION

Transition metal oxides constitute an interesting and widely investigated class of compounds in solid state and materials chemistry.[1, 2] They form the basis of solid state chemistry for the study of structure, composition and various physical and chemical properties[2] that are not only of fundamental importance, but have technological applications. The important properties of these materials include superconductivity, ferromagnetism, magnetoresistance, half-metallicity, catalysis, nonlinear optical property, ferroelectricity and multiferroic property; all are of current immense interest.

This chapter will provide the readers a brief overview of the phenomena of magnetoresistance (MR) and half-metallicity (HM). First, we will introduce the concepts and development of the phenomenon of MR. In describing MR more specific terminologies like the giant magnetoresistance, colossal magnetoresistance and tunnelling magnetoresistance effects will be introduced. Particular mention will be made of the class of materials

Functional Oxides   Edited by Duncan W. Bruce, Dermot O'Hare and Richard I. Walton
© 2010 John Wiley & Sons, Ltd.

where the discovery or the prediction of the phenomenon was made for the first time. The second part of the chapter will outline the first theoretical prediction of HM in Heusler alloys and later its occurrence in various transition metal oxides including the double perovskites. In the third part of the chapter, the double perovskite family will be emphasised in the light of their half-metallic and magnetoresistive properties. The crystal and electronic structure of double perovskites, their half-metallic property along with ferro/ferrimagnetism (with characteristic Curie temperature, T_C) and the relationship of high transition temperature (T_C) with the MR will be discussed. In this section, we also make an attempt to draw interrelationships between high T_C, HM and room temperature MR with a futuristic view of the field towards potential materials discovery. The chapter concludes with a new emerging field of electronics, namely, 'spintronics'[3] where the half-metallic and magnetoresistive properties are capitalised for the new generation of electronic devices and technological applications.

5.2   MAGNETORESISTANCE: CONCEPTS AND DEVELOPMENT

5.2.1   Phenomenon of Magnetoresistance: Metallic Multilayers and Anisotropic Magnetoresistance (AMR)

The MR can simply be defined as change in electrical resistance of a material on application of an external magnetic field. Thus, MR can broadly be defined as the relative change in resistance of a material on application of magnetic field according to Equation 5.1:

$$MR = [\rho(H) - \rho(0)]/\rho(0) \qquad (5.1)$$

where $\rho(H)$ and $\rho(0)$ are the resistivities in the presence of a magnetic field (H) and at zero magnetic field, respectively. It is usually represented as a percentage ratio. Therefore, the MR in Equation 5.1 is also presented as $MR = 100 \times [\rho(H) - \rho(0)] / \rho(0)$. According to this definition the MR can have a maximum value of 100%.

The MR effect was discovered by Lord Kelvin in 1857,[4] first in iron and then in nickel. The change in resistance was initially very low, but later he showed that the change in resistance can be higher when the applied magnetic field direction is switched between parallel and perpendicular to the direction of electrical current. The respective resistivities were denoted as $\rho_{\parallel}$ and $\rho_{\perp}$. This effect was referred to as anisotropic magnetoresistance (AMR), while the previously observed small magnetoresistance was called an ordinary magnetoresistance effect, which is exhibited by many metals. The name 'anisotropic' signifies the directional differences in resistivities ($\rho_{\parallel}$ vs $\rho_{\perp}$) for currents flowing parallel and perpendicular to the direction of magnetisation. The read head sensors based on the AMR effect were first introduced in 1992 and until 1997[5] it was the most widely used technology in the fabrication of magnetic read heads of hard disk drives and magnetic sensors.

5.2.2   Giant Magnetoresistance (GMR) Effect

The discovery of the phenomenon of giant magnetoresistance (GMR) was made independently by the French scientist Albert Fert and the German scientist Peter Grünberg at the same time in 1988.[6, 7] In their experiments with metallic multilayers (Fe/Cr/Fe) they both demonstrated a huge change in resistance while aligning the magnetisation of adjacent Fe layers by application of magnetic fields and simultaneously measuring the transmission of conduction electrons through the thin Cr-layers. Since the change in resistance was huge in this case, *i.e.* much greater than what was previously observed in the AMR effect, it was therefore named the GMR effect. For this achievement Fert and Grünberg received the 2007 Nobel Prize in Physics.

In their work, Baibich *et al.*[6] considered spin-dependent transmission of conduction electrons through the thin Cr-layers between the ferromagnetically coupled Fe layers as the origin of the GMR effect in the superlattice structures (Figure 5.1). The MR as observed in Fe/Cr multilayers is shown in Figure 5.2. The figure significantly shows the decrease in resistance as the magnetic field strength is increased and it becomes almost constant when the field reaches the respective saturation magnetisation value. At the same time, Binasch *et al.*[7] investigated the resistivity behaviour of multilayers made of the same metal combinations as used by Baibich *et al.* but with antiferromagnetically coupled Fe-double layers separated by Cr spacer layer. Figure 5.3 depicts the relative change of resistivity when the field is scanned throughout the

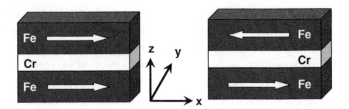

Figure 5.1 A schematic diagram of ferromagnetic (parallel) and antiferromagnetic (antiparallel) exchange coupled (Fe/Cr/Fe) multilayer structure. For the application of field and measurement of current, the x- and y-directions are considered as in-plane (IP) and z-direction as perpendicular to the plane of the multilayers

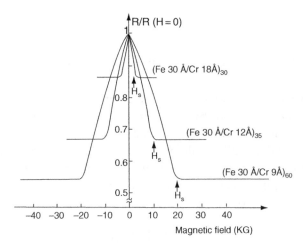

Figure 5.2 Magnetoresistance of Fe/Cr(001) multilayers at 4.2 K. The direction of current and magnetic field both are along (110) axis in the plane of the layers. Reprinted with permission from Baibich *et al.*, 1988 [6]. Copyright (1988) American Physical Society

hysteresis loop. The resistivity difference, $\Delta\rho$, is the $(\rho - \rho_{\parallel})$, where $\rho_{\parallel}$ is the resistivity for saturation magnetisation parallel to the (100) direction (which was the long axis of the sample/easy axis of magnetisation for their film). The figure also shows the AMR effect for a 25 nm thick single Fe film. While the single Fe film showed only the AMR effect (negative MR values), the Fe-Cr-Fe film showed MR both due to the AMR (negative) and antiparallel alignment (positive) of Fe layers. The zero in the plot indicates the MR at $\rho = \rho_{\parallel}$.

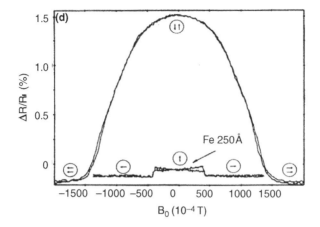

Figure 5.3   Magnetoresistance of Fe double layers with antiferromagnetic coupling compared with the anisotropic MR of a 250 Å thick Fe film. Reprinted with permission from Binasch *et al.*, 1989 [7]. Copyright (1989) American Physical Society

Soon after their discovery there was phenomenal development in the field of magneto-electronics, which revolutionised the field of electronics especially in the miniaturisation of hard disk drive read heads based on magnetoresistive effect. In the technological front, the AMR sensor based read heads of hard disk drives (HDDs) were replaced by GMR sensors in 1997 and paved the way for a huge increase in the areal recording density by more than two orders of magnitude (from ≈ 1 to ≈ 600 Gbit/in^2 in 2007).[3] The discovery and development of the GMR effect introduced the emerging concept of 'spintronics'; where the electron spins of the spintronic material are manipulated for application in new quantum computers.[8]

5.2.3   Colossal Magnetoresistance (CMR) in Perovskite Oxomanganates

After the discovery of GMR in metallic multilayers in the late 1980s, MR effects even larger than GMR were reported in 1993 by von Helmolt *et al.*[9] and others[10] and a little later (in 1994) by Jin *et al.*[11] in a class of compounds known as the perovskite oxomanganates with the general formula $Ln_{1-x}A_xMnO_3$ (where $Ln =$ lanthanide and $A =$ alkaline earth metal). In addition to thin films, these effects are also observed in poly-crystalline[12] and single crystal[13] materials. This phenomenal increase of MR in oxomanganates was called colossal magnetoresistance (CMR) to distinguish it from GMR in granular and metallic multilayered materials. In oxomanganates, this dramatic decrease of resistance when exposed to magnetic fields gives a negative MR value according to Equation 5.1. It should be noted that the colossal values of MR reported in the literature with the MR defined as $[\rho(H)-\rho(0)]/\rho(H)$, or $100 \times \rho(0)/\rho(H)$ instead of the equation, $MR = 100 \times [\rho(H) - \rho(0)]/\rho(0)$ used conventionally for multilayers, where the MR can never exceed 100%. Although, with the conventional definition, the MR values in oxomanganates were enor-mous compared with those in metallic multilayers (GMR materials), such high MR effects were obtained only with very large applied magnetic fields [several teslas (T); $T = 10000$ gauss], which limited potential tech-nological applications of oxomanganates.

Most of the CMR oxomanganates are based on the parent perovskite, $LaMnO_3$, an antiferromagnetic (AFM) insulator. When $LaMnO_3$ is doped with alkali metal cations forming $La_{1-x}A_xMnO_3$ ($A = Ca$, Sr, Ba), x amount of the $Mn^{3+}(d^4)$ is oxidised to Mn^{4+} (d^3) (considered as holes containing one less electron than Mn^{3+}) and the material becomes a ferro-magnetic (FM) metal (for example, $La_{1-x}Sr_xMnO_3$ with $x \geq 0.17$).[14] One can also have extensive cation deficiencies of both the A and transition metal sites that can induce dramatic effects in the electrical and mag-netic properties. These FM perovskites have been known since 1950.[15] Their ferromagnetic properties were explained by the so-called double exchange (DE) mechanism of Zener, de Gennes, and Anderson and Hasegawa.[16] It was the discovery in 1994 of the very large MR effect in oxomanganates that created the immediate resurgence of interest in this class of compounds.[12, 17] Although it was realised that, in both multilayers and oxomanganates, the spin-polarised transport is respon-sible for such large MR, the origin of such behaviour in the oxomanga-nates was entirely different. Subsequently, the field of oxomanganates turned into a very active and rich area of research where the funda-mental understanding of the complex interplay between the lattice, phonons, charge and spin was investigated. In addition the potential of these compounds for low field magnetoresistance and room tempera-ture applications is explored. Furthermore, apart from the perovskite oxomanganates, CMR is also exhibited by other manganese containing transition metal oxides. For example, the Ruddlesden-Popper layered perovskites $RE_{1+x}Sr_{2-x}Mn_2O_7$ (where $RE = $ rare earth metal) and the pyrochlore $Tl_2Mn_2O_7$ systems also show large MR.

5.2.4 Tunnelling Magnetoresistance (TMR) and Magnetic Tunnel Junctions (MTJ)

The discovery of tunnelling magnetoresistance (TMR) dates back to the 1970s when spin-dependent tunnelling experiments were performed by Tedrow and Meservey,[18] Jullière,[19] and later in the early 1980s by Maekawa and Gäfvert,[20] but the magnitude of TMR in these experiments was very small. However, large TMR at room temperature was observed more recently by Moodera et al.,[21] and Miyazaki and Tezuka.[22] In its most elementary form a TMR system is obtained when the nonmagnetic barrier layer of a multilayer GMR stack is replaced by an insulating (IS) layer between two FM layers. Such a multilayer stack of FM/IS/FM constitutes a magnetic tunnel junction (MTJ) across which the tunnelling current is measured. The TMR can be defined as a dramatic change in tunnelling current in MJJ when the relative magnetisations of the two FM layers change their alignment.

The TMR of a junction is related to the spin polarisation of the two electrodes. In the simple model of Jullière[19] the TMR ratio, i.e. the ratio of the resistance change between the antiparallel and the parallel state $(R_{\uparrow\downarrow} - R_{\uparrow\uparrow})$ to the resistance in the parallel state $R_{\uparrow\uparrow}$, is directly related to the spin polarisation of the two ferromagnetic electrodes by:

$$\text{TMR} = [R_{\uparrow\downarrow} - R_{\uparrow\uparrow}]/R_{\uparrow\uparrow} = 2P_1P_2/[1 - P_1P_2] \qquad (5.2)$$

where P_1, P_2 are the spin polarisations of the two electrodes. In this scenario, the spin polarisation, P_i, is related to the difference between the density of states at the Fermi level (E_F) of the majority and minority spin electrons as:

$$P_i = [N_\uparrow(E_F) - N_\downarrow(E_F)]/[N_\uparrow(E_F) + N_\downarrow(E_F)] \qquad (5.3)$$

According to Equation 5.3, materials with high spin polarisation will lead to huge TMR ratios.

5.2.5 Powder, Intrinsic and Extrinsic MR

In addition to the above description of MR as AMR, GMR, CMR and TMR, other types of terminologies are also used. An MR is also called powder MR (PMR) when measured on a compact pallet of compressed powders. This is similar to the TMR experiment described above, where the contact regions between the grains act as barrier layers (tunnel barriers). The magnetic orientations of individual grains are random in the absence of an applied magnetic field (H) leading to a high resistance. With application of an external magnetic field the magnetic domains of the grains are oriented in the field direction, leading to a decrease in resistance in accordance with the tunnelling process. This is also known as grain-boundary MR and occurs in many conducting oxides like, CrO_2, $Tl_2Mn_2O_7$ and Fe_3O_4; including the ordered double perovskite, Sr_2FeMoO_6.

Moreover, in case of the CMR of polycrystalline bulk ceramics, the MR originates from intrinsic and extrinsic effects. The intrinsic MR arises due to intragrain effects (related to the intrinsic interactions in the material) and is maximised close to a ferromagnetic-paramagnetic transition temperature (T_C). The extrinsic MR occurs due to transport through barriers (intergrain effect) and higher MR is observed in a wide range of temperatures below T_C and at low fields. Materials that show extrinsic MR are, in general, highly spin polarised, *i.e.* the carriers are either mainly of up-spin, or down-spin character. This extrinsic MR is believed to be due to spin-polarised tunnelling (SPT) or spin-dependent scattering (SDS) processes at grain boundaries.[23] The extrinsic MR appears in polycrystalline materials as well as in artificial structures. According to Equation 5.3, when the spin polarisation, $P = 1$, the material becomes fully spin-polarised and a completely new type of material appears known as a half-metal.

5.3 HALF-METALLICITY

5.3.1 Half-Metallicity in Heusler Alloys

The concept of half-metallicity was first introduced by de Groot and co-workers in the early 1980s based on their band structure calculations on Mn based Heusler alloys NiMnSb.[24] Figure 5.4 shows the band structures of NiMnSb, where the majority-spin electrons are conducting, while the minority-spin electrons are semiconducting. This leads to a situation where the conduction electrons are 100% spin-polarised at the Fermi level (E_F). At the same time, Kübler *et al.* also recognised nearly vanishing minority-spin densities at the E_F and suggested peculiar transport properties in Heusler alloys.[25] However, 100% spin-polarisation was not observed in the initial spin-polarised photoemission studies on NiMnSb.[26]

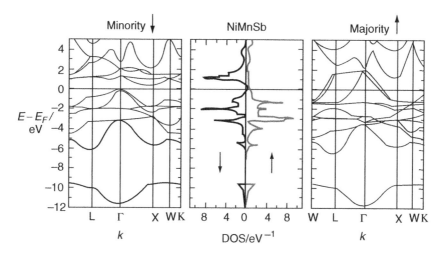

Figure 5.4 Band structure of NiMnSb for the majority-spin direction (right) and the minority-spin direction (left). The middle panel shows the density of states curve (from Felser *et al.*[3]). Reprinted with permission from Fesler *et al.*, *Angew. Chem. Int. Ed.*, **46**, 668 (2007). Copyright (2007) Wiley InterScience

The phenomenon of half-metallicity has gained much interest in order to understand the unusual band structures in various classes of materials and their potential applications in future electronic devices. For example, zinc blend pnictides and chalcogenides (*e.g.* CrAs) are another class of non-oxide materials (apart from Heuslers) in addition to the many oxide classes that are potentially half-metallic materials. Alkali metal doped rare earth oxomanganates, ($RE_{1-x}A_xMnO_3$), rutile-CrO_2, spinel-Fe_3O_4 and Sr_2FeMoO_6 double perovskite oxide are examples of important half-metallic oxides.

5.3.2 Half-Metallic Ferro/Ferrimagnets, Antiferromagnets

Similar to the prediction of half-metallic ferromagnetism (HMF)[24] in Heusler alloys with integer magnetic moments and 100% spin polarisation, another new class of materials known as half-metallic antiferromagnets (HMAFM) were also predicted (in $V_7MnFe_8Sb_7In$)[27] where the half-metallicity is retained with a zero net magnetic moment. These HMAFM should not be confused with the conventional antiferromagnets. The antiferromagnetism in this class of materials is of a truly new nature in the sense that in normal antiferromagnets the electronic structure in the two spin direction are identical by symmetry with no polarisation of conduction electrons, whereas in HMAFM, the half-metallic nature (band gap in one spin channel while the other channel is conducting) prevents such symmetry criteria between the spin-up and spin-down states due to the presence of a band gap in one spin channel.[27] Moreover, due to the gap at E_F for one spin channel, the spin magnetic moment is an integer, which results in zero macroscopic magnetisation when the up and down spin moments cancel exactly. Thus, the HMAFM is a nonmagnetic metal, but its conduction electrons are perfectly spin-polarised. In addition, half-metallic ferrimagnetism was also predicted in a full Heusler compound, Mn_2VAl, with the minority-spin channel being responsible for conduction.[28] The HMF and HMAFM also have been predicted in the double perovskites that are discussed in detail in Section 5.5. The HMAFM is of interest due to its possible application as a probe in spin-polarised scanning tunnelling microscopes (SP-STM) without disturbing the spin character of samples. The HMAFMs are also expected to play a major role in the realisation of spintronic devices that utilise the spin polarisation of the conduction electrons.

5.4   OXIDES EXHIBITING HALF-METALLICITY

5.4.1 CrO_2

Chromium dioxide (CrO_2) was long known and widely used in industry for the manufacture of magnetic recording tapes.[29] It is one among very few oxides that exhibit both metallic conductivity and ferromagnetism (FM). It crystallises in the tetragonal rutile structure[30] (Figure 5.5) and has a high Curie temperature ($T_C = 395$ K). The structure is made of parallel chains of edge-shared CrO_6 octahedra with two short apical bonds and four longer equatorial bonds. The chromium is in 4+ state (Cr^{4+}: $3d^2$) in CrO_2.

Due to its metallic and FM property, similar to double perovskites (see later), CrO_2 is of great interest to material scientists and physicists. A computed band structure of CrO_2 is shown in Figure 5.6. The Fermi level crosses the majority-spin band while it lies in a gap of the minority-spin density of states (DOS). Therefore, CrO_2 is half-metallic on the basis of electronic structure calculations.[32] The same band structure has been reproduced by several groups[33] since it was first demonstrated by

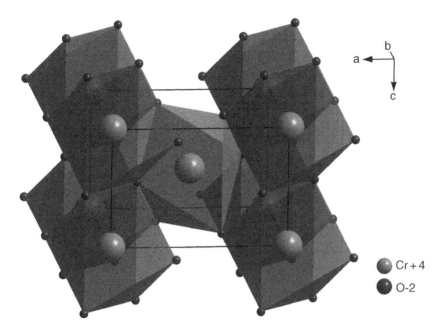

Figure 5.5 The rutile structure of CrO_2

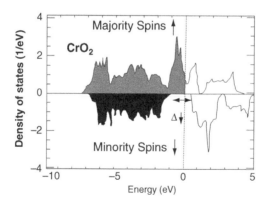

Figure 5.6 Spin-polarised density of states plot of CrO_2. Reprinted with permission from Coey and Venkatesan, 2002 [31]. Copyright (2002) American Institute of Physics

Schwarz.[33] Moreover, this half-metallicity leads to an integral magnetic moment of 2 μ_B/f.u. (f.u. = formula unit), the same as predicted by Hund's rule for the spin moment of Cr^{4+} ($3d^2$) ion. This agrees with the measured saturation moment of 2 μ_B/f.u.[34]

The high T_C of half-metallic CrO_2 prompted further exploration of its properties, especially TMR. Intergrain TMR studies on as grown poly-crystalline CrO_2 films showed significant low-field spin-polarised TMR at low temperatures.[35] This low-field MR can be significantly enhanced by high temperature annealing treatments of the films. The enhanced MR is caused by variation of the effective intergrain tunnelling barrier by surface decomposition of CrO_2 into insulating Cr_2O_3. X-ray and trans-port measurements show that systematic annealing of CrO_2 films lead to an ideal half-metallic system, where the intergrain barrier can be system-atically tuned.[35] Further studies showed CrO_2 to exhibit complex mag-netotransport properties.[36] However, a series of Andreev reflection experiments with CrO_2-superconductor point contacts, demonstrated very high degree of spin-polarisation, $P \approx 80\text{--}97\%$ for CrO_2 in the 1 K temperature range.[37]

5.4.2 Fe_3O_4 and Other Spinel Oxides

Fe_3O_4, also known as the mineral magnetite, is a ferrimagnet with a high T_C of 860 K. Its strong magnetic properties were valued by the ancients and today finds many applications including magnetic record-ing. Fe_3O_4 shows a metal–insulator transition (MIT) at 120 K, which has fascinated the experimentalists and theoreticians for many years. The MIT of Fe_3O_4 is known more famously as the Verwey transition[38] after Verwey who first correlated the transition to be associated with the electron localisation–delocalisation effects. Fe_3O_4 forms in the spinel structure[30] with the general formula AB_2O_4. The spinel structure is formed by cubic close-packed array of oxide anions where 1/8 of the tetrahedral holes are occupied by the A and half of the octahedral holes are filled by the B cations in an ordered fashion. The spinel structure is considered to be a 'normal spinel', after the mineral spinel, $MgAl_2O_4$, where the divalent Mg occupies the tetrahedral site and the trivalent Al occupy the octahedral sites. In contrast, Fe_3O_4 is an 'inverse spinel' where the Fe^{2+} and Fe^{3+} switch positions between the tetrahedral and octahedral sites. In Fe_3O_4, the tetrahedral site (A) is occupied by one Fe^{3+} and the octahedral sites (B) are occupied by equal numbers of Fe^{2+} and Fe^{3+}, which can be represented as $(Fe^{3+})_A[Fe^{2+}Fe^{3+}]_B(O^{2-})_4$. The Fe_3O_4 spinel structure is shown in Figure 5.7.

Electronic structure calculations of Fe_3O_4 in its high temperature cubic form have been carried out with the self-consistent spin-polarised

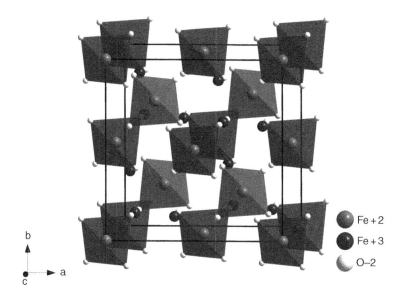

Figure 5.7   The spinel structure of Fe_3O_4

augmented plane wave method with the local spin density approximation (LSDA).[39] The calculations show that Fe_3O_4 is a half-metallic ferrimagnet with the minority spin t_{2g} orbitals forming the conduction band while the majority spin band shows a gap at the Fermi level. Similar band structures are also produced by other methods.[40, 41]

Negative MR with a maximum at coercive field is observed in thin films and compact powders of Fe_3O_4, but not in single crystals.[42] This result is due to PMR, where the magnetisations in the grains are aligned on application of a magnetic field and the MR is related to the intergranular transport of spin-polarised electrons.[42] An MR of 43% at 4.2 K and 13% at room temperature are observed in cobalt-alumina-iron oxide tunnel junctions.[42] The large MR was attributed to the presence of a phase with $Fe_{3-x}O_4$ composition close to that of half-metallic Fe_3O_4 as identified by electron diffraction. The spin-dependent electronic structure of epitaxial thin films of Fe_3O_4 (111) at room temperature was investigated by spin, energy and angle-resolved photoemission spectroscopy.[43] Near the Fermi energy a spin-polarisation of $80 \pm 5\%$ was observed evidencing the half-metallic ferromagnetic state of Fe_3O_4.

The diverse structural, electronic and magnetic properties of Fe_3O_4 have attracted interest to other spinel oxides. $LiMn_2O_4$ is a geometrically frustrated compound with a spinel structure.[44] The magnetisation and

MR studies[45] near the charge ordering temperature ($T = 280$ K) show a large negative MR of ~20% at 7 T. The MR also exhibits field hysteresis, which is not reflected in the magnetisation; a behaviour very different from the CMR oxomanganates.

A half-metallic character has been proposed in another spinel oxide, $LiCr_2O_4$,[46] by spin-polarised first principles density functional calculations (Figure 5.8). Both full potential linearised augmented plane-wave (FP-LAPW) and the LMTO methods were used to calculate the band structures. The lattice and other internal parameters for the cubic spinel structure were optimised with FP-LAPW calculations and subsequently the optimised parameters were used in the band structure calculations. However, the experimental realisation of $Li(Cr^{3+/4+})_2O_4$ is still a challenge to confirm its predicted electronic and magnetic properties.

5.4.3   Perovskite Oxomanganates

Perovskite oxomanganates have already been introduced in the previous section with regard to their CMR properties.[9–13] The electronic band structure calculation of magnetoresistive oxomanganates established that they are half-metallic at low temperature,[47, 48] which led to the idea of close association of half-metallicity (HM) and magnetoresistance (MR). As it was already mentioned for a half-metal, the electronic density of states is completely spin-polarised at the E_F and the conductivity is solely given by single-spin (one type) charge carriers. Despite the theoretical prediction of the HM property of oxomanganates, it was in doubt until recently, when it was demonstrated for the first time by spin-resolved photoemission spectroscopy (SRPES) in the perovskite, $La_{0.7}Sr_{0.3}MnO_3$.[49] Several methods are available to measure the spin polarisations of HM materials. For example, ferromagnet–insulator–ferromagnet (FIF) tunnelling, ferromagnet–insulator–superconductor (FIS) tunnelling and two-dimensional angular correlation of electron–positron radiation (2D-ACAR) are among other methods, in addition to Andreev reflection (AR) at a superconductor–ferromagnet interface and SRPES.[50] An 80% spin-polarisation is obtained by Andreev reflection measurements on a single crystal of $La_{0.67}Sr_{0.33}MnO_3$.[37] Moreover, TMR ratios as high as 1850% are observed, corresponding to a nearly complete spin polarisation of 95%.[51] Many excellent reviews are available on the properties of CMR oxomanganates.[14, 52]

Half-metallic materials, with only one spin direction at the E_F and consequently a spin polarisation of 100%, are potentially attractive for

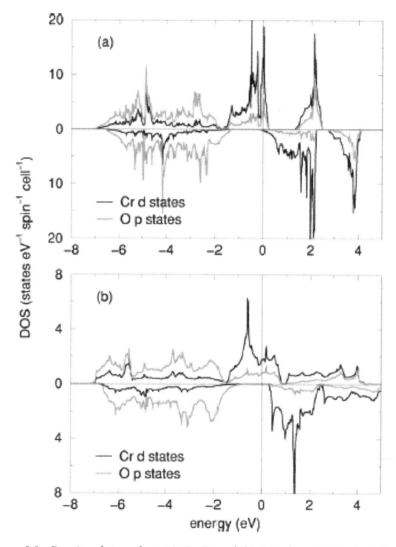

Figure 5.8 Density of states for (a) $LiCr_2O_4$ and (b) CrO_2 by LMTO calculations. Black lines indicate Cr d states and grey lines indicate O p-states. Reprinted with permission from Lauer *et al.*, 2004 [46]. Copyright (2004) American Physical Society

obtaining huge TMR ratios (Section 5.2.4). However, their low T_C ($\sim 360\,K$ for $La_{2/3}Sr_{1/3}MnO_3$) renders their integration into devices for applications at room temperature problematic. This has motivated the search for half-metallic compounds with high T_Cs, well above room temperature.

5.4.4   Double Perovskites

Search for HMF and HMAFM properties in other classes of materials led to the emergence of double perovskite oxides as potential candidates.[53] Computational searches based on density functional theory predicted several new double perovskites[54–56] with HMAFM properties (details on structure and properties of double perovskites are in Section 5.5).

5.4.4.1   Half-Metallic Antiferromagnetic (HMAFM) Double Perovskites

HMAFM in oxides with double perovskite structure was proposed by Warren Pickett for the first time in 1998.[53] Band structure calculations based on local spin-density approximations (LSDA), predicted La_2MnVO_6 double perovskite as a promising candidate for exhibiting HMAFM. The calculations were based on the assumption that both Mn and V ions are trivalent and more importantly, that the Mn^{3+} (d^4) occurs in a low spin configuration ($t_{2g}^4 e_g^0$). The band structure of La_2MnVO_6 show that only the minority t_{2g} bands of both V and Mn contribute to the DOS near the E_F.[53] The antiparallel alignment of low spin Mn^{3+} ($t_{2g}^3\uparrow t_{2g}^1\downarrow$) and V^{3+} ($t_{2g}^2\downarrow$) states would yield a net zero moment. However, low spin Mn^{3+} in oxides is unlikely.

Subsequent to La_2MnVO_6, many other double perovskites containing mixed cations both at the A and B sites have been investigated for possible HMAFMs.[54–56] For example, $LaAVRuO_6$,[54] $LaAVMoO_6$,[55] $LaAVOsO_6$[56] and $LaAMoYO_6$ (A = Ca, Sr, Ba; Y = Re, Tc)[56] were all predicted to show HMAFM properties. It is shown that the HMAFM nature of $LaAVRuO_6$ is very robust regardless of divalent A-cation replacements and cation disorder, in addition to Coulomb correlation.[54] Nevertheless, attempts to realise the predicted HMAFM in the above compounds have not been successful so far. Androulakis *et al.*[57] prepared La_2MnVO_6, which is not an HMAFM due to the absence of complete B-site order and apparent presence of high-spin Mn^{3+} and V^{3+}. Considering the negligible size difference between Mn^{3+} (0.645 Å) and V^{3+} (0.64 Å), their ordering in the double perovskite structure is unexpected.[58] Recently, it was shown that even with a more reduced Mn than Mn^{3+} ($Mn^{(3-\delta)+}$) and oxidised V than V^{3+} ($V^{(3+\delta)+}$), which would enhance the size and charge difference, La_2MnVO_6 does not form in an ordered double perovskite structure.[58] An experimental investigation of the physical properties of $LaAVMoO_6$ (A = Ca, Sr, Ba) was carried out, but a HMAFM state has not been established thus far.[59] Therefore, despite several theoretical predictions, an experimental realisation of HMAFM in double perovskites remains a challenge.

5.4.4.2 Half-Metallic Ferro-/Ferrimagnetic (HMF) Double Perovskites

Double perovskites with ferro/ferrimagnetic and metallic properties have been known from the 1970s. For example, Sr_2FeMoO_6 and Sr_2FeReO_6 are known to show both metallic and ferromagnetic properties.[60] However, it is the half-metallic nature of their band structure and magnetoresistive properties that has drawn extensive attention recently.[61, 62] The details of their half-metallicity and magnetoresistance are discussed in the following section.

5.5 MAGNETORESISTANCE AND HALF-METALLICITY OF DOUBLE PEROVSKITES

5.5.1 Double Perovskite Structure

Before we describe the double perovskite structure, a basic understanding of its parent structure, *i.e.* the perovskite structure is essential. The general composition of perovskite is given by the formula ABO_3, where A is typically an alkali, alkaline earth or rare earth metal and B is a transition metal. The ABO_3 perovskite structure[30, 63] consists of a three-dimensional (3D) network of corner-sharing BO_6 octahedra in which the A cation resides on the dodecahedral sites surrounded by twelve oxide ions (Figure 5.9a). The ideal structure is cubic, but undergoes symmetry lowering distortions when the relative sizes of the two cations are not perfectly matched and the distortion can be quantified in terms of the Goldschmidt's tolerance factor, t.[64] The tolerance factor, t, is defined as $t = (r_A + r_O)/\{\sqrt{2}(r_B + r_O)\}$ where r_A, r_B and r_O are the ionic radii of the respective A, B and oxide ions. An ideal cubic perovskite structure (as in case of $SrTiO_3$) is obtained when the sizes of the A and B cations are such that the $A-O$ and $B-O$ bond distances are perfectly matched *i.e.* $d(A{-}O) = \sqrt{2}\, d(B{-}O)$, and t becomes unity.

With small A cations for which $t < 1$, the cubic structure becomes unstable because the 12-coordinate site is too large for the A cation. The structure adjusts the size of the 12-coordinate site by cooperatively tilting

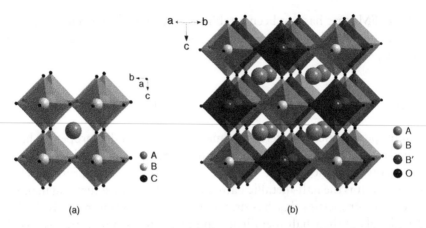

Figure 5.9 (a) A simple perovskite (ABO_3) structure. (b) The rock-salt ordered double perovskite ($A_2BB'O_6$) structure

the BO_6 octahedra and bending the $B-O-B$ bridges.[65] The resulting structure deviates from the ideal cubic structure and adopts lower symmetries, *e.g.* rhombohedral, orthorhombic or monoclinic.[65] For example, LaAlO$_3$ adopts a rhombohedral structure and GdFeO$_3$ forms an orthorhombically distorted perovskite structure, where the tolerance factors are less than unity. Another consequence of octahedral distortion is A-site polyhedron distortion that effectively results in a decrease in the coordination number of the ion.

Besides the octahedral distortions of the perovskite structure due to size effects of A and B cations, other types of distortions solely originating from electronic effects are possible. For example, the first-order Jahn-Teller (FOJT) distortion, which occurs for octahedrally coordinated B-cations with unsymmetrical filling of d-electrons in t$_{2g}$ and e$_g$ orbitals (FOJT is particularly strong for high-spin d^4 and in d^9 configuration), gives rise to elongation, or compression of the octahedra. The orthorhombic distorted structure of LaMnO$_3$ is the manifestation of FOJT distortion of individual MnO$_6$ octahedra that function in a co-operative manner.[66] Moreover, other structural distortions due to the second-order Jahn-Teller (SOJT) can also occur. For example, the out-of-centre distortion of the d^0 cations at the octahedral B site (*e.g.* BaTiO$_3$) is a consequence of electronically driven SOJT distortion. Another electronic off-centring mechanism, often referred to as the inert-pair effect, operates in perovskites with A cations with ns^2 valence electronic configuration (*e.g.* Pb^{2+}, Bi^{3+}).[67] Thus in $BiMnO_3$ or $BiFeO_3$, the energy of the system is lowered by off-centring of the Bi^{3+} with respect to the surrounding oxygens, which leads to a localisation of the lone pair on one side of the Bi ion through hybridisation of the Bi $6s$ and $6p$ states with O $2p$ orbitals.[68] The presence of such stereochemically active lone pairs in $BiFeO_3$ and $BiMnO_3$ are responsible for the ferroelectricity in these multiferroic materials.[68, 69]

The tolerance of the perovskite structure for a wide variety of substitutions both at the A and B sites gives rise to a very large number (several hundreds) of perovskite derivatives with subtle variations in structure. For example, the presence of more than one cation at A and B sites results in compositions such as $AA'B_2O_6$, $A_2BB'O_6$ and $A_3BB'_2O_9$[65, 70] where A, A' or B, B' cations order at the respective sublattices. They are commonly known as double perovskites. Ordering of cations at the A and B sites of double perovskites is an important phenomenon. Various B cation arrangements[71] are known for $A_2BB'O_6$ double perovskites: (i) random distribution resulting in a cubic structure (*e.g.* Ba_2FeNbO_6) or an orthorhombic structure (*e.g.* $SrLaCuRuO_6$), (ii) ordered arrangements of B and B' in a 'rock salt' fashion (Figure 5.9b) in all the three crystallographic directions (*e.g.* Ba_2FeMO_6; $M = Mo$, Re) and (iii) a unique layered ordering of B and B' cations (*e.g.* La_2CuSnO_6).[72]

A double perovskite with a random ordering of B cations have cell parameters similar to that of undistorted cubic perovskites (denoted as a_p) and symmetry lowering structures arise as a result of octahedral tilting distortions as it occurs in simple perovskites.[73] However, the double perovskites originating from undistorted cubic perovskites (space group $Pm\bar{3}m$) with rock salt ordering of B cations have a unit cell twice the size of the perovskite cell and correspondingly the space group changes to $Fm\bar{3}m$. As with the simple perovskite structure, the octahedra in the double perovskite structure also undergo tilting distortions due to small A cation sizes. The consequences of collective effects of octahedral tilting and cation ordering on the double perovskite structure have been extensively descried in the literature.[73] However, extreme care must be taken in assigning space groups for double perovskites. Sometimes the possible space groups cannot be determined with certainty with the data available. For example, the difference between a random orthorhombic and a rock salt monoclinic (with monoclinic angle β close to 90°) double perovskite are very subtle. A common situation occurs in distinguishing orthorhombic $Pnma$ and monoclinic $P2_1/n$ with identical systematic absences. The only reflection condition that justifies a rock salt arrangement with $P2_1/n$ symmetry is the presence of $0kl$: $k = 2n+1$. It is also possible that the monoclinic angle is 90°, but the monoclinic symmetry will be revealed by careful examination of the reciprocal lattice.

Among the $A_2BB'O_6$ ordered double perovskites, the compounds with rock salt ordered B and B' cations are noteworthy. Besides the novelty of their structures, new materials properties ensue depending on the identity of the B and B' atoms as well as the extent of ordering. Thus, rock salt type B cation ordered double perovskite oxides, such as Sr_2FeMoO_6 and Sr_2FeReO_6, give rise to metallic ferrimagnetism and tunnelling magnetoresistance properties.[60–62]

It has been realised that the extent of cation ordering is an important issue in the synthesis of double perovskites where both thermodynamic and kinetic considerations are involved.[74–76] The degree of cation order is crucial for the occurrence of unique magnetic and electrical properties. For example, magnetoresistive properties of Sr_2FeMoO_6 vary depending on the extent of ordering of Fe and Mo.[77]

5.5.2 Ordering and Anti-Site (AS) Disorder in Double Perovskites

As previously mentioned the degree of ordering of the B-site cations of $A_2BB'O_6$ is crucial to their electronic properties. It has been established long ago by Anderson et al. that the difference of size and charge of the cations are the major driving force for their ordering.[71] When a certain fraction of B cations occupy the wrong site, i.e. the B' site, the same amount of B' cation occupies the B site to maintain the 1:1 proportion of the B and B' in a stoichiometric $A_2BB'O_6$. The percentage of misplaced cations is referred to as anti-site (AS) disorder. The AS disorder and the percentage of ordering are related by the relation: $x = (100 - 2y)$, where x is the percentage of ordering and y is the percentage of AS disorder. For example, with 5% AS disorder, the degree of ordering is 90% by this relation. Similarly, a 50% AS disorder will lead to a completely disordered double perovskite. The degree of order is also calculated from the occupation factor as:

$$p(\%) = (q - 0.5) \times 2 \times 100 \qquad (5.4)$$

Here, p is degree of order and q is the occupation factor of B or B' on the B or B' site, respectively, with $q_{(B)}$ (on B) + $q_{(B')}$ (on B) = 1.

AS disorder has a direct influence on the magnetic properties of the double perovskites. It is observed in FM double perovskites that the saturation magnetisation (M_S) decreases with increasing AS disorder. We discuss this point here by considering the electronic and magnetic structure of Sr_2FeMoO_6 with an ordered arrangement of Fe and Mo ions. The localised magnetic moments of Fe and Mo are 5 μ_B ($3d^5$) and 1 μ_B ($4d^1$), respectively. There is one itinerant electron per formula unit and its magnetic moment direction is opposite to the localised Fe moments, which results in the ionic valences of 2.5+ and 5.5+ for the Fe and Mo, respectively and an AFM coupling of the Fe and Mo moments is expected to result in a M_S of 4 μ_B/f.u.[78] However, the reported M_S values are always considerably smaller than 4 μ_B. This is attributed to AS defects[61, 79–81] or anti-phase boundaries[82] present in the compound. Evidence for AS Fe atoms from the Mössbauer spectra[78, 83] in corroboration with Monte Carlo simulations[79] and band structure calculations[84] suggests that the AS disorder effects are responsible for the considerable M_S reduction. For example, systematically lower M_S values of 3.1 μ_B/f.u. and 3.5 μ_B/f.u. are reported for bulk Sr_2FeMoO_6.[61, 85] However, a magnetisation study on Sr_2FeMoO_6 with various degrees of cation ordering reveals M_S to follow an almost linear dependence with respect to the AS defect concentration (Figure 5.10).[81] The linear dependence compares very well with models based on ferrimagnetic arrangements (FIM) and superexchange (SE). In an FIM model, the net magnetisation is given by $M_S = m_B - m_{B'}$, where m_B and $m_{B'}$ are the magnetisations of the B and B′ sublattices, which are antiferromagnetically coupled.[81] With an AS defect concentration of x, the net magnetisation, $M_S = (1 - 2x)m_{Fe} - (1 - 2x) m_{Mo}$, where m_{Fe} and m_{Mo} are the magnetic moments of the Fe and Mo ions, respectively. When considering a spin-only contribution, m_{Fe} (Fe^{3+}) = 5 μ_B and m_{Mo} (Mo^{5+}) = 1 μ_B, the value of net magnetisation leads to:

$$M_S = (4 - 8x)\mu_B \qquad (5.5)$$

Alternatively, in the SE model, the predicted M_S is given by $(4 - 10x)\mu_B$ with local superexchange interactions between $3d^5$-$3d^5$, $4d^1$-$4d^1$ and $3d^5$-$4d^1$. A simple FIM shows a remarkable agreement with the experimental curve (Figure 5.10). The above analysis is also valid with non-integral ionic charges on Fe and Mo, namely, $Fe^{(3-\delta)+}$ and $Mo^{(5+\delta)+}$.[81] Moreover, the observed variation of M_S with AS provides a unique experimental confirmation of the ferrimagnetic ordering in Sr_2FeMoO_6.

On the other hand, to explain the magnetisation and a reduced M_S in n-type metallic double perovskite system, $Sr_{2-x}Ca_xFeMoO_6$ ($0 \leq x \leq 2$), Goodenough and Dass invoked the idea of anti-phase boundaries (APB).

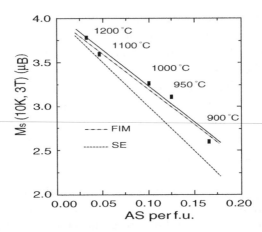

Figure 5.10 Dependence of saturation magnetization (M_S) on anti-site defect concentration, AS of Sr_2FeMoO_6. Solid line is the linear fit to the data. Dotted/dashed lines represent expected M_S dependence according to ferrimagnetic exchange (FIM) and superexchange (SE) models. Reprinted with permission from Balcells *et al.*, 2001 [81]. Copyright (2001) American Institute of Physics

In ordered Sr_2FeMoO_6, different *B*-site ordered domains can exist. In one domain, the Fe can be sitting on the *B* site, while on a *B'* site in another. An APB is created where two such domains meet, and this leads to AFM Fe–O–Fe interactions across the interface (Figure 5.11). In explaining the observed smaller M_S, an APB scenario is contested with the 'AS disorder' picture on the grounds that the latter is unable to explain additional unusual features of magnetisation, namely, the small coercivity and low remanence.[82] At a zero magnetic field, the FM domains on opposite sides of the APB will align antiparallel to each other. Due to the unequal volumes of the domains, there will be a residual magnetisation that gives rise to a small finite remanence. Moreover, a small field is required to align the ferromagnetic domains and upon removal of the field, the exchange forces across the APB realign the domains antiparallel to each other again with a small remnant magnetisation.[82]

It is apparent that correct estimation of the percentage order is important for precisely understanding the magnetic properties. The extent of ordering (or AS disorder) can be determined by the Rietveld refinement of the structure employing powder neutron or X-ray diffraction. In a cubic double perovskite with $Fm\bar{3}m$ space group symmetry, a new set of

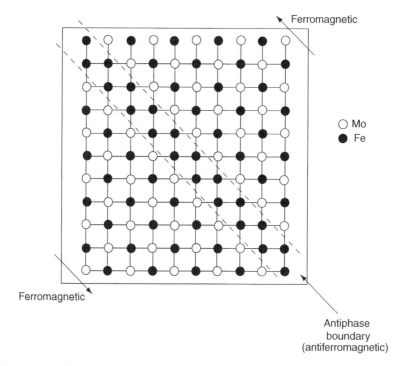

Figure 5.11 Schematic of an anti-phase boundary in an (001) plane of Sr_2FeMoO_6. Reprinted with permission from Goodenough and Dass, 2009 [82]. Copyright (2000) Elsevier Ltd

superstructure reflections (hkl) for all odd h, k and l appear in the X-ray diffraction patterns.[63] For example, (111), (311) and (331) are the first three characteristic superstructure reflections and their intensity gradually decreases with increasing order of the reflection. Moreover, the superstructure peak intensities also decrease with decreasing degree of B-site order, or increase of AS disorder, leading to complete disappearance of superlattice lines with a fully disordered cation arrangement.[63] In addition, an ordered double perovskite with B cations of nearly equal atomic numbers often show weak or almost vanishing superstructure reflections in the powder X-ray diffraction (XRD). Therefore, it is difficult to assess the degree of order from powder XRD in such cases. However, such issues often can be resolved with the help of neutron diffraction studies.

As mentioned in Section 5.5.1, apart from the intrinsic effect of size and charge difference on B-cation ordering in double perovskites, the synthesis condition plays a pivotal role on the degree of ordering.[74–76, 81]

While Balcells et al.[81] reported progressive increase of cation order with rise in sintering temperature, according to Shimada et al.,[76] a kinetic or thermal equilibrium dictates the cation order in Sr_2FeMoO_6. Figure 5.12 depicts the degree of B-cation order (S) as a function of synthesis temperature, T_{syn}, for samples fired at T_{syn} ranging from 900 to 1300 °C. Below 1150 °C, S increases with temperature indicating a kinetic control of ordering.[76] A thermodynamic equilibrium is reached in the vicinity of 1150 °C, where the maximum in S is achieved. Beyond 1150 °C, the ordering is limited by thermodynamics and the S corresponds to thermal equilibrium values at respective temperatures. A highest S of 0.95 was obtained for samples prepared at 1150 °C for 150 h.[76] This suggests that long synthesis periods and moderate temperatures are essential for optimising the degree of order in Sr_2FeMoO_6. Based on these considerations highly ordered samples with record high M_Ss up to 3.9 μ_B were synthesised. The optimal condition of synthesis varies from perovskite to perovskite and generally needs to be determined empirically for every composition.

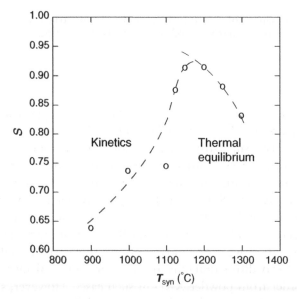

Figure 5.12 B-cation order parameter, S versus synthesis temperature (T_{syn}) for samples of Sr_2FeMoO_6 fired at various T_{syn} for 50 h. The dotted lines are guide for the eye. Reprinted with permission from Shimada et al., 2003 [76]. Copyright (2003) American Chemical Society

5.5.3 Electronic Structure and Magnetic Properties of Double Perovskites

5.5.3.1 General Considerations

The electronic properties of transition metal oxides are dictated by the electronic band structure near the Fermi energy.[86] The position of the band energy, bandwidth and band gaps determine their physical properties. The bandwidth is a crucial parameter that controls the conductivity and the band gap of a material. While, the bandwidth depends on the nature of chemical bonding, *e.g.* type of orbitals; core s, p or valence s, p and the extent of orbital overlap (covalency); it is a general trend that in the $3d$ transition metal oxides, the orbitals are narrower and the overlap is weaker leading to smaller bandwidths as compared with the $4d$ and $5d$ metals.[86] In $4d$ and $5d$ metals the large orbital extension leads to greater overlap and wider bandwidth.[86] These general qualitative trends of band features are applicable for the transition metal perovskites and double perovskites.

In the perovskites (ABO_3) or double perovskites ($A_2BB'O_6$), the electronic structure near the Fermi energy is dominated by the states primarily derived from the transition metal d orbitals and oxygen $2p$ orbitals. It is the corner sharing octahedral framework connected by $B-O-B$ or $B-O-B'$ linkages that form the basis of its electronic structure. The A-site cations do not contribute to the DOS near the Fermi level; the bands associated with the A cations are generally situated deep in the valence bands away from the Fermi energy. However, the nature of the A-site cation (mainly size) has an indirect effect on the band structure; a change in the A-cation size can have significant effect on the tolerance factor leading to octahedral tilting distortions,[73] which affect the bond angles of the $B-O-B$ or $B-O-B'$ linkages. A change in bond angle can alter the effective orbital overlap (covalency) and bandwidth significantly, and thereby, the electronic structure. Other factors, such as, ligand field splitting of d orbitals and exchange splitting (Hund's coupling) also play vital roles for the electronic band structure alterations of double perovskites.

5.5.3.2 Electronic structures of Double Perovskites

Electronic structure calculations for Sr_2FeMoO_6 and Sr_2FeReO_6 with the full-potential augmented plane-wave (FLAPW) method based on the generalised gradient approximation (GGA) produce a half-metallic band structure.[61, 62] In Sr_2FeMoO_6, the up-spin (majority) band has an energy gap of about 1 eV at the Fermi level between the occupied Fe e_g and the unoccupied Mo t_{2g} levels (Figure 5.13). The down-spin (minority) DOS is finite and continuous across the Fermi level with a hybridised Fe (3d) and Mo (4d) character of the charge carriers.[61] Moreover, the important consequence of the band structure is that the electrons are fully spin-polarised, *i.e.* only the down spin electrons are involved in the charge transport of Sr_2FeMoO_6.

The effect of AS disorder on the electronic structure of Sr_2FeMoO_6 has been studied by *ab initio* electronic structure calculations.[84] It is found

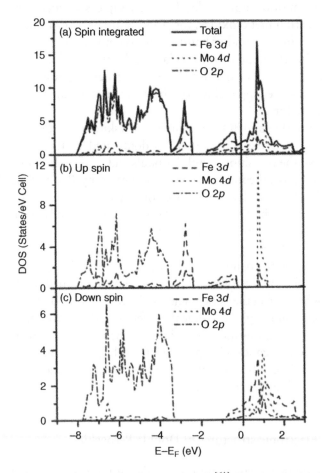

Figure 5.13 Density of states plot for Sr_2FeMoO_6[61]. Reprinted with permission from Sarma, 2001 [98]. Copyright (2001) Elsevier Ltd.

that the presence of AS disorder destroys the HMF state of the ordered compound and consequently reduces the M_S. However, the reduction in M_S is attributed to the strong diminution of individual magnetic moments at each Fe site due to correlation effects, while maintaining the magnetic coupling between various Fe sites ferromagnetic also in the disordered configuration.[84] This is in contrast to the proposition based on dominant antiferromagnetic Fe−O−Fe interactions that grow with AS defects.[79]

As it was already mentioned, the ordered double perovskite Sr_2FeReO_6 is also predicted to have a half-metallic ground state. The observed M_S of 2.7 μ_B/f.u. at 4.2 K is smaller than expected (3 μ_B/f.u.) for an antiferromagnetic coupling between $Fe^{3+}(d^5)$ and $Re^{5+}(d^2)$.[62] Subsequent to the report of the above two compounds with HMF property, an extensive body of research was devoted to double perovskites involving synthesis, structure and electronic, magnetic and magnetoresistive properties, in addition to first principles electronic structure calculations. It should be mentioned that many of these double perovskites have been known since the 1960s.[87] Various cation substitutions at the A site as well as at the B site produced numerous double perovskites with a wide spectrum of structures and properties.[85, 88–92] For example, the A_2FeMoO_6 series with $A = Ca$ and Ba are also reported to show ferromagnetic and metallic behaviours with M_S ranging from 3.5 to 3.9 μ_B/f.u.[85]

The Recontaining double perovskites, A_2FeReO_6 and other A_2MReO_6 ($A = Ca$ or Sr or Ba; $M = Cr$ and Mn) are relatively less studied as compared with the Mo compounds. However, one of the Re containing double perovskites, Sr_2CrReO_6, has gained some interest recently due to its HMF property and a very high T_C (620 K).[93] Electronic structure calculations on Sr_2CrReO_6 in the absence of spin-orbit coupling results in a HM ground state with a total magnetic moment of 1.0 μ_B/f.u.[94] However, a recent experimental study shows that the system is not half-metallic in the true sense, due to a large orbital contribution to the magnetisation.[95] The measured M_S of 1.38 μ_B for Sr_2CrReO_6, obtained only at a very high field of $\sim$20 T, is close to the value previously predicted.[96] Therefore, more critical analysis is required in the Re containing double perovskites before a real half-metallic state can be assigned.

5.5.3.3   Magnetic properties of Double Perovskites

The magnetic properties of half-metallic Sr_2FeMoO_6 and related double perovskites cannot be explained on the basis of conventional DE[16] and

SE[97] mechanisms. In particular, the high magnetic transition temperature ($T_C = 415$ K) of Sr_2FeMoO_6 is unusual on account of the SE
mediated via oxygens between the widely separated magnetic Fe^{3+} ions
in this ordered double perovskite. This suggested a novel origin of magnetism in this compound. Sarma *et al.* invoked a mechanism where the
localised spins are ferromagnetically coupled along with an antiferromagnetic interaction operating between the localised spins and the delocalised electrons.[98] The reduction in M_S with AS disorder can also be
explained by this mechanism.[99]

5.5.4 Magnetoresistance and Half-Metallicity in Double Perovskites

Sr_2FeMoO_6 was first reported by Patterson *et al.*[100] in 1963 as a ferrimagnetic semimetal. After the discovery of TMR[61] in this compound the
double perovskites attracted great experimental and theoretical attention
due to their possible applications in magneto-electronic devices.[60, 62, 101]
Due to its high Curie transition temperature ($T_C = 415$ K), Sr_2FeMoO_6
exhibits a large low-field TMR not only at low temperature but even at
room temperature, and its MR at 4.2 K (7 T) and 300 K (7 T) reach values
as large as 42% and 10%, respectively.[61] As mentioned in the previous
section, the band-structure calculations of Sr_2FeMoO_6 show a mixing of
the spin-down O $2p$, Fe $3d$ and Mo $4d$ bands at the E_F that are responsible
for the conducting channel while the E_F falls in a gap for the up-spin
bands resulting in a half-metallic material. It is the HM nature of
Sr_2FeMoO_6 that gives rise to completely spin-polarised charge carriers,
which by virtue of an intergranular tunnelling process results in low-field
TMR.[61] Subsequently, isostructural Sr_2FeReO_6 was reported to exhibit
intergrain TMR as well, similar to that of Sr_2FeMoO_6.[62]

Magnetoresistive properties of other A_2FeMoO_6 (A = Ca, Ba) double
perovskites also have been reported.[88] Ba_2FeMoO_6 has shown large
intragrain MR above room temperature (near the T_C). Beside intragrain
MR, large values of low-field TMR (37% at 7 T) at 4.2 K are observed in
annealed samples. Moreover, an MR of 4% was obtained at 290 K with a
small field of 0.1 T. Interestingly, Ca_2FeMoO_6 showed an MR of 30% at
room temperature for 9 T. This is significantly larger than reported for
other A_2FeMoO_6 double perovskites (A = Sr, Ba).[88]

In addition to the macroscopic magnetisations, the magnetoresistive property crucially depends on the extent of B-cation ordering.

While an ordered sample ($\sim$91%) of Sr_2FeMoO_6 showed a sharp low-field MR response, it was much lower in a partially disordered ($\sim$31%) sample.[77] Therefore, its absence in the disordered phase indicates the obliteration of HMF property due to large AS disorder.[77] Thus, ordering in double perovskite is crucial for the true nature of the half-metallic ground state and the intrinsic properties of the material.

An investigation of the effect of disorder on the magnetisation of $Sr_2Fe_{1+x}Mo_{1-x}O_6$ ($-1 < x < 0.25$) shows the depolarisation of the conduction band.[102] Moreover, the TMR is strongly reduced by disorder and suppressed in highly disordered samples.[77, 103] Thus, disorder again appears to be detrimental to the half-metallic character giving rise to up-spin states at the E_F and depolarising the conduction band, which eventually reduces the TMR. Apart from these extensively investigated Fe-based double perovskites, a Cr-based double perovskite Sr_2CrWO_6 has also been found to exhibit a large low-field MR and a high T_C of 390 K.[104] Although the MR approached 100% at 5 K in this material, it was drastically diminished to 2% at room temperature in polycrystalline specimens.

5.5.5 High Curie Temperature (T_C) Double Perovskites and Room Temperature MR

It is believed that HM materials, with T_Cs much higher than room temperature (RT) will give high spin-polarisations at RT and will be suitable for magnetoresistive devices. Therefore, double perovskites with high T_Cs are potentially attractive for applications. There is continuous search for new double perovskites with high T_Cs and also various strategies are adopted to enhance the T_Cs among the already known compounds.

An early study investigated the effect of bond distances on T_C in the A_2FeMoO_6 ($A = Ca, Sr, Ba$) system.[105] It showed that maximum T_C was achieved in Sr_2FeMoO_6 and the T_C was found to decrease, when Sr was replaced by either Ca or Ba. The observed variation in T_C has been investigated with the evolution of T_C and crystal structure as a function of average cation radius in A_2FeMoO_6.[106] It was observed that the T_C correlated with the electronic bandwidth (W); the maximum T_C for Sr_2FeMoO_6 corresponded with the largest estimated W.[106]

Similarly, the substitution of Sr by Ca or Ba in A_2CrWO_6 results in a strong reduction of T_C from 458 K (Sr) to 161 K (Ca) and 141 K (Ba). A comparison of the A_2CrWO_6 series to other double perovskites clearly indicates the suppression of T_C with the deviation of tolerance factor, t, from its ideal value of 1. It is found to be a general trend that the compound with t close to 1 has the maximum T_C in a series. However, the suppression of T_C seems to be weak for the A_2FeMoO_6 series [85, 105–108] (T_C varies between 310 and 420 K) as compared with A_2CrWO_6 (A = Ca, Sr, Ba). Moreover, the large size of Ba in Ba_2CrWO_6 causes t to deviate well above 1, which eventually causes a structural transition to a hexagonal structure and strong suppression of the FM interactions.[91]

Among the double perovskites the Re-containing compounds are noteworthy in terms of their high T_Cs [62, 89, 108–111]. However, the suppression of T_C as a function of t is also quite strong in the A_2CrReO_6 series with a T_C of 635 K for Sr_2CrReO_6 and 360 K for Ca_2CrReO_6.[62] Strikingly, the $(Sr_{1-x}Ca_x)_2FeReO_6$[109] is an exception of the trend with Ca_2FeReO_6 having the highest T_C (538 K) although the t decreases continuously from 0.997 for Sr_2FeReO_6 to 0.943 for Ca_2FeReO_6 on substituting Sr by Ca. However, Ca_2FeReO_6 is considered to be a unique material; it is a FM insulator and there may be a different mechanism responsible for the high ordering temperature.[89] Another Cr-based double perovskite, Sr_2CrOsO_6,[87] with a record high T_C of 725 K has been reported, but recently the compound was shown to be an insulator due to the completely filled $5d$ t_{2g} metallic band.[112] It is interesting to note that with increasing number of valence electrons the T_C increases along the series: Sr_2CrWO_6 (450 K), Sr_2CrReO_6 (635 K) and Sr_2CrOsO_6 (725 K). Recently, such an increasing trend in the experimental T_C with increasing number of total valence electrons of the transition metals in selected $A_2BB'O_6$ (A = Ca, Sr, Ba; B = Cr, Fe, Mn; B' = Mo, W, Re, Os) was demonstrated, and has been reproduced by *ab initio* calculations (Figure 5.14).[94]

5.6 SPINTRONICS – THE EMERGING MAGNETO-ELECTRONICS

Present-day electronic devices are based on the manipulation of the charge of the electron. The electronic spin (which can be up or down) is not involved in any electronic input–outputs or signal processing.

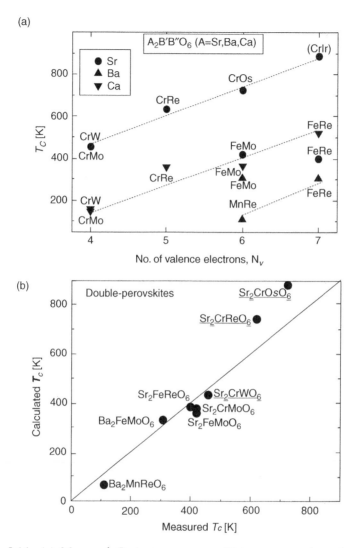

Figure 5.14 (a) Measured Curie temperatures (T_C) *versus* number of valence electrons (N_V). The names of the $B'B''$ elements appear near the data point. CrIr is for a projected value. (b) Plot of calculated *versus* measured T_C. Reprinted with permission from Mandal *et al.*, 2008 [94]. Copyright (2008) American Physical Society

However, with the emergence of new exotic materials and their fascinating physical properties it will be possible to manipulate the spins of the electrons. The manipulation of spin analogous to that of the charge would give rise to a new era of magneto-electronics, which with analogy to the electronics is termed as 'spintronics' (spin-based electronics).[3] The discovery of the GMR effect and large TMR ratios at room temperature laid the foundation stone for development of spin-based electronics.

The challenging task for technical advancements of half-metals is to maintain high spin polarisation at room temperature. Therefore, ferromagnets with higher Curie temperature and wider gap are desirable. It is also interesting to discover highly anisotropic uniaxial half-metals (higher than CrO_2) where, in principle, a large barrier could exist between domain walls facilitating efficient spin-polarised tunnelling. To achieve the full potential of spintronics, exploration of new materials, especially the half-metallic ferromagnets are essential. The HMFs fulfill all the requirements of spintronics due to their characteristic electronic structure.

The HMAFM[27] are still an unexplored area of materials. There is no real HMAFM compound synthesised as yet. Since, these materials also are half-metals, *i.e.* their carriers are 100% spin-polarised without any macroscopic magnetisation and they will be relatively insensitive to applied fields and can even lead to a new sub-area of spintronics.

5.7 SUMMARY

This chapter summarises the fundamental concepts and phenomenon associated with magnetoresistance. The giant and colossal magnetoresistance effects are important for technological applications. In addition, the intrinsic half-metallic property of magnetoresistant materials is characterised by spin-polarised charge carriers. The spin polarisation is crucial for large tunnelling magnetoresistance. Transition metal oxides with different structure types are considered for half-metallicity and magnetoresistance effects. However, in view of current research trends, the double perovskites are emphasised for their excellent half-metallic properties and high Curie temperatures that are required for large room temperature magnetoresistance. In the double perovskites the pivotal role played by the crystal structure, degree of cation order and electronic configuration of the transition metals constitute the major area of research. Other material classes, for example, compounds with the spinel structure are relatively unexplored. Despite significant advances in understanding the electronic and magnetic properties of double perovskites, major challenges remain for synthesising new oxides with half-metallic property and higher Curie temperatures.

ACKNOWLEDGEMENTS

This work was partially supported by an NSF-DMR-0541911 grant. W. H. McCarroll is appreciated for editing the final manuscript.

REFERENCES

[1] C.N.R. Rao and B. Raveau, *Transition Metal Oxides*, Wiley-VCH, Weinheim, 1998.

[2] C.N.R. Rao and J. Gopalakrishnan, *New Directions in Solid State Chemistry*, Cambridge University Press, Cambridge, 1997.

[3] I. Žutić, J. Fabian and S. DasSarma, *Rev. Mod. Phys.*, **76**, 323 (2004); C. Felser, G.H. Fecher and B. Balke, *Angew. Chem. Int. Ed.*, **46**, 668 (2007); A. Fert, *Rev. Mod. Phys.*, **80**, 1517 (2008).

[4] W. Thomson, *Proc. R. Soc. London, Ser. A*, **8**, 546 (1857).

[5] P.A. Grünberg, *Rev. Mod. Phys.*, **80**, 1531 (2008).

[6] M.N. Baibich, J.M. Broto, A. Fert, F. Nguyen Vn Dau, F. Petroff, P. Eitenne, G. Creuzet, P. Friederich and J. Chazelas, *Phys. Rev. Lett.*, **61**, 2472 (1988).

[7] G. Binasch, P. Grünberg, F. Saurenbach and W. Zinn, *Phys. Rev. B*, **39**, 4828 (1989).

[8] S.D. Sarma, J. Fabian, X. Hu and I. Žutić, *IEEE 58th Device Research*, Conference Digest (Cat.o.00TH8526), p. xii+176, 95.

[9] R. von Helmolt, J. Wecker, B. Holzapfel, L. Schultz and K. Samwer, *Phys. Rev. Lett.*, **71**, 2331 (1993).

[10] K. Chahara, T. Ohono, M. Kasai and Y. Kozono, *Appl. Phys. Lett.*, **63**, 1990 (1993).

[11] M. McCormack, S. Jin, T.H. Tiefel, R.M. Fleming, J.M. Phillips and R. Ramesh, *Appl. Phys. Lett.*, **64**, 3045 (1994); S. Jin, T.H. Tiefel, M. McCormack, R.A. Fastnacht, R. Ramesh and L.H. Chen, *Science*, **264**, 413 (1994).

[12] P. Schiffer, A.P. Ramirez, W. Bao and S.-W. Cheong, *Phys. Rev. Lett.*, **75**, 3336 (1995); R. Mahendiran, S.K. Tiwary, A.K. Raychaudhuri, T.V. Ramakrishnan, R. Mahesh, N. Rangavittal and C.N.R. Rao, *Phys. Rev. B*, **53**, 3348 (1996).

[13] Y. Tokura, Y. Urushibara, Y. Moritomo, T. Arima, A. Asamitsu, G. Kido and N. Furukawa, *J. Phys. Soc. Jpn.*, **63**, 3931 (1994); A. Urushibara, Y. Moritomo, T. Arima, A. Asamitsu, G. Kido and Y. Tokura, *Phys. Rev. B*, **51**, 14103 (1995).

[14] C.N.R. Rao, A.K. Cheetham and R. Ramesh, *Chem. Mater.*, **8**, 2421 (1996).

[15] G.H. Jonker and J.H. van Santen, *Physica*, **16**, 337 (1950).

[16] C. Zener, *Phys. Rev.*, **82**, 403 (1951); P.G. de Gennes, *Phys. Rev.*, **118**, 141 (1960); P.W. Anderson and H. Hasegawa, *Phys. Rev.*, **100**, 675 (1955).

[17] P.G. Radaelli, D.E. Cox, M. Marezio, S.-W. Cheong, P.E. Schiffer and A.P. Ramirez, *Phys. Rev. Lett.*, **75**, 4488 (1995).

[18] P.M. Tedrow and R. Meservey, *Phys. Rev. B*, **7**, 318 (1973).

[19] M. Jullière, *Phys. Lett. A*, **54**, 225 (1975).

[20] S. Maekawa and U. Gäfvert, *IEEE Trans. Magn.*, **18**, 707 (1982).

[21] J.S. Moodera, L.R. Kinder, T.M. Wong and R. Meservey, *Phys. Rev. Lett.*, **74**, 3272 (1995).

[22] T. Miyazaki and N. Tezuka, *J. Magn. Magn. Mater.*, **139**, L231 (1995).

[23] H.Y. Hwang, S.-W. Cheong, N.P. Ong and B. Batlogg, *Phys. Rev. Lett.*, 77, 2041 (1996).

[24] R.A. de Groot, F.M. Mueller, P.G. van Engen and K.H.J. Buschow, *Phys. Rev. Lett.*, 50 2024 (1983).

[25] J. Kübler, A.R. William and C.B. Sommers, *Phys. Rev. B*, 28, 1745 (1983).

[26] G. Bona, F. Meier, M. Taborelli, E. Bucher and P.H. Schmidt, *Solid State Commun.*, 56, 391 (1985).

[27] H. van Leuken and R. A. de Groot, *Phys. Rev. Lett.*, 74, 1171 (1995).

[28] R. Weht and W.E. Pickett, *Phys. Rev. B*, 60, 13006 (1999).

[29] B.L. Chamberland, *CRC Crit. Rev. Solid State Sci.*, 7, 1 (1977).

[30] A.F. Wells, *Structural Inorganic Chemistry*, 5th Edn, Clarendon Press, Oxford, 1984.

[31] J.M.D. Coey and M. Venkatesan, *J. Appl. Phys.*, 91, 8345 (2002).

[32] K. Schwarz, *J. Phys. F: Met. Phys.*, 16, L211 (1986).

[33] S.P. Lewis, P.B. Allen and T. Sasaki, *Phys. Rev. B*, 55, 10253 (1997); M.A. Korotin, V.I. Anisimov, D.I. Khomskii and G.A. Sawatzky, *Phys. Rev. Lett.*, 80, 4305 (1998).

[34] T.J. Swoboda, P. Arthur Jr, N.L. Cox, J.N. Ingraham, A.L. Oppegard, and M.S. Sadler, *J. Appl. Phys.*, 32, 374S (1961); H. Brändle, D. Weller, S.S.P. Parkin, J.C. Scott, P. Fumagalli, W. Reim, R.J. Gambino, R. Ruf, and G. Güntherodt, *Phys. Rev. B*, 46, 13 889 (1992).

[35] H.Y. Hwang and S.-W. Cheong, *Science*, 278, 1607 (1997).

[36] S.M. Watts, S. Wirth, S. von Molnar, A. Barry and J.M.D. Coey, *Phys. Rev. B*, 61, 9621 (2000); M.S. Laad, L. Craco and E. Müller-Hartmann, *Phys. Rev. B*, 64, 214421 (2001).

[37] R.J. Soulen Jr, J.M. Byers, M.S. Osofsky, B. Nadgorny, T. Ambrose, S.F. Cheng, P.R. Broussard, C.T. Tanaka, J. Nowak, J.S. Moodera, A. Barry and J.M.D. Coey, *Science*, 282, 85 (1998); W.J. De Sisto, P.R. Broussard, T.F. Ambrose, B.E. Nadgorny, and M.S. Osofsky, *Appl. Phys. Lett.*, 76, 3789 (2000); Y. Ji, G.J. Strijkers, F.Y. Yang, C.L. Chien, J.M. Byers, A. Anguelouch, G. Xiao and A. Gupta, *Phys. Rev. Lett.*, 86, 5585 (2001); J.S. Parker, S.M. Watts, P.G. Ivanov and P. Xiong, *Phys. Rev. Lett.*, 88, 196601 (2002).

[38] E.J.W. Verwey, *Nature*, 144, 327 (1939).

[39] A. Yanase and K. Siratori, *J. Phys. Soc. Jpn.*, 53, 312 (1984).

[40] Z. Zhang and S. Satpathy, *Phys. Rev. B*, 44, 130319 (1991).

[41] M. Penicaud, B. Silberchiot, C.B. Sommers and J. Kubler, *J. Magn. Magn. Mater.*, 103, 212 (1992).

[42] J.M.D. Coey, M. Venkatesan and M.A. Bari, in *Lecture Notes in Physics*, Vol. 595, and C. Berthier, L.P. Levy (Eds) G. Martinez, Springer, Heidelberg, 2002, p. 377.

[43] Y. Dedkov, U. Rüdiger and G. Güntherodt, *Phys. Rev. B*, 65, 064417 (2002).

[44] A.S. Wills, A.P. Raju and J.E. Greedan, *Chem. Mater.*, 11, 1510 (1999).

[45] R. Basu, C. Felser, A. Maignan and R. Seshadri, *J. Mater. Chem.*, 10, 1921 (2000).

[46] M. Lauer, R. Valenti, H.C. Kandpal and R. Seshadri, *Phys. Rev. B*, 69, 075117 (2004).

[47] W.E. Pickett and D.J. Singh, *Physics of Oxomanganates*, Plenum, New York, 1998.

[48] P.K. de Boer, H. van Leuken, R.A. de Groot, T. Rojo and G.E. Barberis, *Solid State Commun.*, 102, 621 (1997); E.A. Livesay, R.N. West, S.B. Dugdale, G. Santi and T. Jarlborg, *J. Phys.: Condens. Matter*, 11, L279 (1999).

[49] J.H. Park, E. Vescovo, H.J. Kim, C. Kwon, R. Ramesh and T. Venkatesan, *Nature*, 392, 794 (1998).

[50] M. Ziese, *Rep. Prog. Phys.*, **65**, 143 (2002).

[51] M. Bowen, M. Bibes, A. Barthélémy, J.-P. Contour, A. Anane, Y. Lemaître and A. Fert, *Appl. Phys. Lett.*, **82**, 233 (2003).

[52] J.M.D. Coey, M. Viret and S. von Molnar, *Adv. Phys.*, **48**, 167 (1999); C.N.R. Rao and A. Arulraj, *Curr. Opin. Solid State Mater. Sci.*, **3**, 23 (1998); Y. Tokura (Ed.) *Colossal-Magnetoresistive Oxides*, Gordon and Breach, Tokyo, 1999.

[53] W.E. Pickett, *Phys. Rev. B*, **57**, 10613 (1998).

[54] J.H. Park, S.K. Kwon and B.I. Min, *Phys. Rev. B*, **65**, 174401 (2002).

[55] M.S. Park and B.I. Min, *Phys. Rev. B*, **71**, 052405 (2005).

[56] Y.K. Wang and G.Y. Guo, *Phys. Rev. B*, **73**, 064424 (2006).

[57] J. Androulakis, N. Katsarakis and J. Giapintzakis, *Solid State Commun.*, **124**, 77 (2002).

[58] T.K. Mandal, M. Croft, J. Hadermann, G. Van Tendeloo, P.W. Stephens and M. Greenblatt, *J. Mater. Chem.*, **19**, 4382 (2009).

[59] M. Uehara, M. Yamada and Y. Kimishima, *Solid State Commun.*, **129**, 385 (2004).

[60] T. Nakagawa, *J. Phys. Soc. Jpn.* **24**, 806 (1968); A.W. Sleight and J.F. Weiher, *J. Phys. Chem. Solids*, **33**, 679 (1972).

[61] K.-I. Kobayashi, T. Kimura, H. Sawada, K. Terakura and Y. Tokura, *Nature*, **395**, 677 (1998).

[62] K.-I. Kobayashi, T. Kimura, H. Sawada, K. Terakura and Y. Tokura, *Phys. Rev. B*, **59**, 11159 (1999); J. Gopalakrishnan, A. Chattopadhyay, S.B. Ogale, T. Venkatesan, R.L. Greene, A.J. Millis, K. Ramesha, B. Hannoyer and G. Marest, *Phys. Rev. B*, **62**, 9538 (2000).

[63] R.H. Mitchell, *Perovskites: Modern and Ancient*, Almaz Press, Thunder Bay, Canada, 2002.

[64] V.M. Goldschmidt, *Geochemische Verteilungs- gesetze der Elementer VII. Skrifter Norske Videnskaps Academi Klasse 1.* Matematisk, Naturvidenskaplig, Klasse, Oslo, 1926.

[65] J.B. Goodenough and J.M. Longo, in *Landolt-Börnstein Numerical Data and Functional Relationships in Science and Technology*, New Series, Group III, Vol. 4a, K.H. Hellwege (Ed.), Springer-Verlag, Berlin, 1970.

[66] E.O. Wollan and W.C. Koehler, *Phys. Rev.*, **100**, 545 (1955); J. Rodriguez-Carvajal, M. Hennion, F. Moussa, A.H. Moudden, L. Pinsard and A. Revcolevschi, *Phys. Rev. B*, **57**, R3189 (1998); M.W. Lufaso and P.M. Woodward, *Acta Crystallogr. B*, **60**, 10 (2004).

[67] N.A. Hill, *Annu. Rev. Mater. Res.*, **32**, 1 (2002).

[68] R. Seshadri and N.A. Hill, *Chem. Mater.*, **13**, 2892 (2001).

[69] J.B. Neaton, C. Ederer, U.V. Waghmare, N.A. Spaldin and K.M. Rabe, *Phys. Rev. B*, **71**, 014113 (2005).

[70] F.S. Galasso, *Perovskites and High T_C Superconductors*, Gordon and Breach, New York, 1990.

[71] M.T. Anderson, K.B. Greenwood, G.A. Taylor and K.R. Poeppelmeier, *Prog. Solid State Chem.*, **22**, 197 (1993).

[72] M.T. Anderson and K.R. Poeppelmeier, *Chem. Mater.*, **3**, 476 (1991).

[73] P.M. Woodward, *Acta Crystallogr. B*, **53**, 32 (1997); P.M. Woodward, *Acta Crystallogr. B*, **53**, 44 (1997); C.J. Howard and H.T. Stokes, *Acta Crystallogr. B*, **54**, 782 (1998); C.J. Howard, B.J. Kennedy and P.M. Woodward, *Acta Crystallogr. B*, **59**, 463 (2003).

[74] P. Woodward, R.D. Hoffman and A.W. Sleight, *J. Mater. Res.*, **9**, 2118 (1994).

[75] P.K. Davies, *Curr. Opin. Solid State Mater. Sci.*, **4**, 467 (1999).

[76] T. Shimada, J. Nakamura, T. Motohashi, H. Yamauchi and M. Karppinen, *Chem. Mater.*, **15**, 4494 (2003).

[77] D.D. Sarma, E.V. Sampathkumaran, S. Ray, R. Nagarajan, S. Majumdar, A. Kumar, G. Nalini and T.N. Guru Row, *Solid State Commun.*, **114**, 465 (2000).

[78] J. Lindén, T. Yamamoto, M. Karppinen, H. Yamauchi and T. Pietari, *Appl. Phys. Lett.*, **76**, 2925 (2000); J. Lindén, M. Karppinen, T. Shimada, Y. Yashukawa and H. Yamauchi, *Phys. Rev. B*, **68**, 174415 (2003).

[79] A.S. Ogale, S.B. Ogale, R. Ramesh and T. Venkatesan, *Appl. Phys. Lett.*, **75**, 537 (1999).

[80] Y. Tomioka, T. Okuda, Y. Okimoto, R. Kumai and K.-I. Kobayashi, *Phys. Rev. B*, **61**, 422 (2000).

[81] L.I. Balcells, J. Navarro, M. Bibes, A. Roig, B. Martinés and J. Fontcuberta, *Appl. Phys. Lett.*, **78**, 781 (2001).

[82] J.B. Goodenough and R.I. Dass, *Int. J. Inorg. Mater.*, **2**, 3 (2000).

[83] J.M. Greneche, M. Venkatesan, R. Suryanarayanan and J.M.D. Coey, *Phys. Rev. B*, **63**, 174403 (2001).

[84] T. Saha-Dasgupta and D.D. Sarma, *Phys. Rev. B*, **64**, 64408 (2001).

[85] B. Garcìa-Landa, C. Ritter, M.R. Ibarra, J. Blasco, P.A. Algarabel, R. Mahendiran and J. García, *Solid State Commun.*, **110**, 435 (1999); R.P. Borges, R.M. Thomas, C. Cullinam, J.M.D. Coey, R. Suryanarayanan, L. Ben-Dor, L. Pinsard-Guadart and A. Revcolevschi, *J. Phys.: Condens. Matter*, **11**, L445 (1999).

[86] P.A. Cox, *The Electronic Structure and Chemistry of Solids*, Oxford University Press, Oxford, 1987.

[87] J.M. Longo and R. Ward, *J. Am. Chem. Soc.*, **83**, 1088 (1961); A.W. Sleight, J.M. Longo and R. Ward, *Inorg. Chem.*, **1**, 245 (1962); F. Patterson, C. Moeller and R. Ward, *Inorg. Chem.*, **2**, 196 (1963).

[88] A. Maignan, B. Raveau, C. Martin and M. Hervieu, *J. Solid State Chem.*, **144**, 224. (1999); J.A. Alonso, M.T. Casais, M.J. Martìnez-Lope, J.L. Martínez, P. Velasco, A. Muñoz and M.T. Fernández-Díaz, *Chem. Mater.*, **12**, 161 (2000).

[89] W. Prellier, V. Smolyaninova, A. Biswas, C. Galley, R.L. Greene, K. Ramesha and J. Gopalakrishnan, *J. Phys.: Condens. Matter*, **12**, 965 (2000); A. Arulraj, K. Ramesha, J. Gopalakarishnan and C.N.R. Rao, *J. Solid State Chem.*, **155**, 233 (2000).

[90] Z. Zeng, I.D. Fawcett, M. Greenblatt and M. Croft, *Mater. Res. Bull.*, **36**, 705 (2001); S. Li and M. Greenblatt, *J. Alloys Compd.*, **338**, 121 (2002); E.N. Caspi, J.D. Jorgensen, M.V. Lobanov and M. Greenblatt, *Phys. Rev. B*, **67**, 134431 (2003); G. Popov, M. Greenblatt and M. Croft, *Phys. Rev. B*, **67**, 024406 (2003); Q. Lin, M. Greenblatt and M. Croft, *J. Solid State Chem.*, **178**, 1356 (2005); R.O. Bune, M.V. Lobanov, G. Popov, M. Greenblatt, C.E. Botez, P.W. Stephens, M. Croft, J. Hadermann and G. Van Tendeloo, *Chem. Mater.*, **18**, 2611 (2006).

[91] J.B. Philipp, P. Majeswki, L. Alff, A. Erb, R. Gross, T. Graf, M.S. Brandt, J. Simon, T. Walther, W. Madder, D. Topwal and D.D. Sarma, *Phys. Rev. B*, **68**, 144431 (2003).

[92] H. Kato, T. Okuda, Y. Okimoto, Y. Tomioka, K. Oikawa, T. Kamiyama and Y. Tokura, *Phys. Rev. B*, **69**, 184412 (2004).

[93] D. Serrate, J.M. DeTeresa and M.R. Ibarra, *J. Phys.: Condens. Matter*, **19**, 023201 (2007).

[94] T.K. Mandal, C. Felser, M. Greenblatt and J. Kübler, *Phys. Rev. B*, **78**, 134431 (2008).

[95] J.M. Michalik, J.M. DeTeresa, C. Ritter, J. Blasco, D. Serrate, M.R. Ibarra, C. Kapusta, J. Freudenberger and N. Kozlova, *Europhys. Lett.*, **78**, 17006 (2007); J.M. DeTeresa, J.M. Michalik, J.Blasco, P.A. Algarabel, M.R. Ibarra, C. Kapusta and U. Zeitler, *Appl. Phys. Lett.*, **90**, 252514 (2007).

[96] G. Vaitheeswaran, V. Kanchana and A. Delin, *Appl. Phys. Lett.*, **86**, 032513 (2005).

[97] P.W. Anderson, *Phys. Rev.*, **79**, 350 (1950).

[98] D.D. Sarma, P. Mahadevan, T. Saha-Dasgupta, S. Ray and A. Kumar, *Phys. Rev. Lett.*, **85**, 2549 (2000); D.D. Sarma, *Curr. Opin. Solid State Mater. Sci.*, **5**, 261 (2001).

[99] S. Ray, A. Kumar, D.D. Sarma, R. Cimino, S. Turchini, S. Zennaro, and N. Zema, *Phys. Rev. Lett.*, **87**, 097204 (2001).

[100] F.K. Patterson, C.W. Moeller and R. Ward, *Inorg. Chem.*, **1**, 196 (1962).

[101] T.H. Kim, M. Uehara, S.W. Cheong and S. Lee, *Appl. Phys. Lett.*, **74**, 1737 (1999); H. Guo, J. Burgess, S. Street and A. Gupta, *Appl. Phys. Lett.*, **89**, 022509 (2006); C. Sakai, Y. Doi and Y. Hinatsu, *J. Alloys Compd.*, **408–12**, 608 (2006); A.K. Azad, S.G. Eriksson, S.A. Ivanov, R. Mathieu and P. Svedlindh, *J. Alloys Compd.*, **364**, 77 (2004); H.J. Xiang and M.H. Whangbo, *Phys. Rev. B*, **75**, 052407 (2007); H.Z. Guo and A. Gupta, *Appl. Phys. Lett.*, **89**, 262503 (2006); A. Masuno, M. Haruta, M. Azuma, H. Kurata and S. Isoda, *Appl. Phys. Lett.*, **89**, 211913 (2006).

[102] D. Topwal, D.D. Sarma, H. Kato, Y. Tokura and M. Avignon, *Phys. Rev. B*, **73**, 094419 (2006).

[103] M. Garcia-Hernandez, J.L. Martìnez, M.J. Martìnez-Lope, M.T. Casais and J.A. Alonso, *Phys. Rev. Lett.*, **86**, 2443 (2001).

[104] J.B. Philipp, D. Reisinger, M. Schonecke, A. Marx, A. Erb, L. Alff, R. Gross and J. Klein, *Appl. Phys. Lett.*, **79**, 3654 (2001).

[105] F. Galasso, F.C. Douglas and R. Kasper, *J. Chem. Phys.*, **44**, 1672 (1966).

[106] C. Ritter, M.R. Ibarra, L. Morellón, J. Blasco, J. Garcia and J.M. De Teresa, *J. Phys.: Condens. Matter*, **12**, 8295 (2000).

[107] W.H. Song, J.M. Dai, S.L. Ye, K.Y. Wang, J.J. Du and Y.P. Sun, *J. Appl. Phys.*, **89**, 7678 (2001).

[108] B.-G. Kim, Y.-S. Hor and S.-W. Cheong, *Appl. Phys. Lett.*, **79**, 388 (2001).

[109] H. Kato, T. Okuda, Y. Okimoto, Y. Tomioka, Y. Takenoya, A. Ohkubo, M. Kawasaki and Y. Tokura, *Appl. Phys. Lett.*, **81**, 328 (2002); H. Kato, T. Okuda, Y. Okimoto, Y. Tomioka, K. Oikawa, T. Kamiyama and Y. Tokura, *Phys. Rev. B*, **65**, 144404 (2002).

[110] T. Alamelu, U.V. Varadaraju, M. Venkatesan, A.P. Douvalis and J.M.D. Coey, *J. Appl. Phys.*, **91**, 8909 (2002).

[111] W. Westerburg, O. Lang, C. Ritter, C. Felser, W. Tremel and G. Jakob, *Solid State Commun.*, **122**, 201 (2002).

[112] Y. Krockenberger, K. Mogare, M. Reehuis, M. Tovar, M. Jansen, G. Vaitheeswaran, V. Kanchana, F. Bultmark, A. Delin, F. Wilhelm, A. Rogalev, A. Winkler and L. Alff, *Phys. Rev. B*, **75**, 020404 (2007).

Index

Functional Oxides Edited by Duncan W. Bruce, Dermot O'Hare and Richard I. Walton
© 2010 John Wiley & Sons, Ltd.

Printed and bound in the UK by
CPI Antony Rowe, Eastbourne

Printed and bound by CPI Group (UK) Ltd, Croydon, CR0 4YY

16/04/2025

14658545-0002